Thermal Processing of Packaged Foods

Second Edition

FOOD ENGINEERING SERIES

Series Editor

Gustavo V. Barbosa-Cánovas, Washington State University

Advisory Board

Titles

Donald Holdsworth and Ricardo Simpson

Thermal Processing of Packaged Foods

Second Edition

 Springer

Donald Holdsworth
Withens
Stretton-Fosse, Glos GL56 9SG
UK
sdholdsworth@ukonline.co.uk

Ricardo Simpson
Depto. Procesos Químicos, Biotecnológicos
  y Ambientales
Universidad Técnica Federico Santa María
Vaparaíso
CHILE
ricardo.simpson@usm.cl

Series Editor:
Gustavo V. Barbosa-Cánovas
Center for Nonthermal Processing of Foods
Washington State University
Pullman, WA 99164-6120
US
Barbosa@wsu.edu

ISBN 978-0-387-72249-8   e-ISBN 978-0-387-72250-4

Library of Congress Control Number: 2007933217

Printed on acid-free paper

9 8 7 6 5 4 3 2 1

springer.com

This book is dedicated to our wives, Margaret and Anita, and family, Christopher, Martin, Giles, Sarah and José Ignacio, María Jesús, and Enrique.

S.D. Holdsworth and Ricardo Simpson

Contents

Preface (First Edition)

My credentials for writing this book are three decades of experience in the canning industry, the research that has supported it, and the establishment of a specialized training course on the thermal processing of packaged foods. My first encounter with the industry was to accompany Tom Gillespy around the various factories of the members of Campden Research Association. He took his annual leave for many years visiting the industry and was dedicated to ensuring that the requirements of good manufacturing practice were observed. The occasion on which I accompanied him, was his last trip before retirement, and I shall always be grateful to him for the kindly advice he gave me on all aspects of canning and food processing. Nobody could have had a better introduction to the industry. In a small way, this book is an appreciation and a memorial to some of his work. He was greatly respected in academic and industrial circles.

This book is concerned with the physical and engineering aspects of the thermal processing of packaged foods—i.e., the heating and cooling of food products hermetically sealed in containers. The two commonest types of container used for this process are glass bottles and cans, although more recently a variety of plastic containers has been added to the list. The main aim of the book is to examine the methods that have been used to establish the time and temperature of processes suitable to achieve adequate sterilization or pasteurization of the packaged food.

It is written from the point of view of the food process engineer, whose principal role is to design, construct, and operate food processing equipment to produce food of acceptable quality and free from public health hazards. The engineering approach requires a knowledge of the microbiological and physico-chemical factors required to solve the necessary equations to establish the safety of the process. In some ways, the canning process is unique, in as much as it requires a mathematical model of the sterilization value to determine the adequacy of the process. Over the last 70 years, a considerable amount of time and energy has been spent around the world on developing suitable mathematical methods to calculate the effectiveness of various processing regimes in order to ensure the safe production of foods. In this book, the various methods and theoretical models on which they are based, for determining adequate times and temperatures for achieving sterility, are discussed and examined.

Most books on canning tend to deal with this subject either by means of a generalized technological description of the process, containers, and products, or from a bacteriological point of view. This book, however, attempts to deal with the more fundamental engineering aspects of the heating and cooling process and the mathematical modeling of the sterilization operation—aspects that are dealt with more briefly elsewhere. Many hundreds of papers have been published on this subject and an untold amount of thermal processing experimental work carried out. Each canning company usually has a person specializing in thermal processing, as well as microbiological laboratory and pilot plant facilities. Much of the academic research work reported is essentially an extension of basic principles, and the development of new, and alternative methods of calculation rather than the discovery of new principles. Some of the work makes a critical comparison of various authors' work and assesses the improvements or otherwise that accrue from using a particular method. Some of it uses new mathematical techniques to perform already established methods, while other work analyzes the errors resulting from the use of different methods of heat penetration. The research and development work is important in training people in the principles of one of the best and well-established methods of making shelf-stable food products.

This book will be of interest to technical managers, process engineers, and research workers as a guide to the literature and the principles underlying thermal processing. It will be of use to those in the industry who are concerned with achieving adequate processes, as well as to those who are concerned with the development of equipment. It will also act as a guide to those who are concerned with the development of legislation, and help them to assess the realities of whatever they wish to impose on the manufacturing industry. Finally, it is hoped that this book will inspire and enthuse research workers to even greater endeavors in this area.

I am most grateful for advice and help from former colleagues, and also to many friends throughout the world.

S.D. HOLDSWORTH

Preface (Second Edition)

In this new edition, the historical perspective of the development of thermal processing has been retained and much new additional material has been added. The development of the subject, as indicated by the amount of research that has been done during the last ten years, has been remarkable, and shows that the technology is very viable and expanding world-wide.

The main developments that have been included are: a) the increased use of new packaging materials, including retortable pouches and the use of containers made from other plastic composite materials, b) the application of newer processing methods which use heat transfer media such as hot water, air/steam, and steam/water, which are necessary for the newer forms of packaging material, c) new methods of theoretically calculating the heat transfer characteristics during processing, including three-dimensional modeling and application of computerized fluid dynamics (CFD) techniques, d) implications of newer models for microbial destruction, e) revised techniques for process evaluation using computer models, including CD software, f) development of process schedules for quality optimization in newer packaging materials, and g) important new aspects of methods of retort control.

Unlike other texts on thermal processing, which very adequately cover the technology of the subject, the unique emphasis of this text is on processing engineering and its relationship to the safety of the processed products.

The authors hope that they have produced an adequate text for encouraging research workers and professional engineers to advance the operation of the manufacturing processes to ensure the production of high quality products with assured safety.

S. D. HOLDSWORTH
R. SIMPSON

1
Introduction

1.1. Thermal Processing Principles

1.1.1. Thermal Processing

A generation ago the title of this book would have contained such terms as *canning*, *bottling*, *sterilization* and heat *preservation*; however, with the passage of time it has become necessary to use a more general title. The term *thermal processing* is used here in a general sense and relates to the determination of heating conditions required to produce microbiologically safe products of acceptable eating quality. It conveys the essential point that this book is concerned with the heating and cooling of packaged food products. The only attempt to produce a generic title has been due to Bitting (1937), who used the term *appertizing*, after the process developed and commercialized by Nicholas Appert (1810) to describe the canning and bottling process. Despite the need for a generic term, rather surprisingly, this has never been used to any great extent in the technical press.

The phrase *packaged foods* is also used in a general sense, and we shall be concerned with a variety of packaging materials, not just tin-plate, aluminum and glass, but also rigid and semi-rigid plastic materials formed into the shape of cans, pouches and bottles. The products known originally as canned or bottled products are now referred to as heat-preserved foods or thermally processed foods.

Thermal Processing is part of a much wider field–that of industrial sterilization–which includes medical and pharmaceutical applications. Those concerned with these subjects will find much of the information in this book will apply directly to their technologies.

1.1.2. The Process

It is necessary to define the word *process*. Generally in engineering, a process is defined as the sequence of events and equipment required to produce a product. Here, however, *process* is a time–temperature schedule, referring to the *temperature* of the heating medium (condensing steam) and the *time* for which it is sustained. Tables of processing schedules are available: In the United States, the National Food Processors' Association produces guides (e.g. NFPA 1982).

In French such schedules are referred to as *barèmes de sterilization* (e.g. Institut Appert 1979).

1.2. Canning Operations

1.2.1. General

Figure 1.1 Illustrates the canning process which consists of five main stages:

Stage 1 Selecting suitable foods, taking them in prime condition at optimum maturity, if appropriate, followed by preparation of the foods as cleanly, rapidly and perfectly as possible with the least damage and loss with regard to the economy of the operation.

Stage 2 Packing the product in hermetically sealable containers–together with appropriate technological aids–followed by removing the air and sealing the containers.

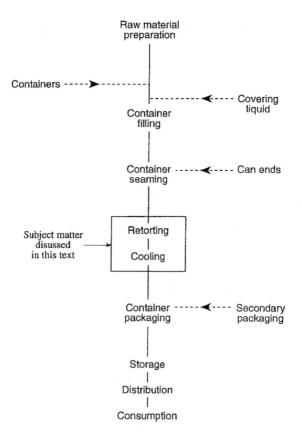

FIGURE 1.1. General Simplified flow diagram for a canning line.

Stage 3 Stabilizing the food by heat, while at the same time achieving the correct degree of sterilization, followed by cooling to below 38°C.

Stage 4 Storing at a suitable temperature (below) 35°C to prevent the growth of food spoilage organisms

Stage 5 Labelling, secondary packaging, distribution, marketing and consumption

The instability of foods at the time they are sealed in containers is due to the presence of living organisms that, if not destroyed, will multiply and produce enzymes that will decompose the food and in some cases produce food-poisoning toxins. Stability, i.e. the production of shelf-stable products, is attained by the application of heat, which will kill all the necessary organisms (For further details, see Section 3.1.2). Of the above listed operations only the stabilization operation, Stage 3, commonly known as processing, will be covered in this text. The technological aspects of the subject are well covered by many texts, among them Jackson and Shinn (1979), Hersom and Hulland (1980), Lopez (1987), Rees and Bettison (1991), and Footit and Lewis (1995). Most of these texts do not elaborate on the subject of this book, which is dealt with only in the monumental works of Ball and Olson (1957) and Stumbo (1973), as well as in the individual specialized texts of Pflug (1982) and of the various food processing centers. Here an attempt is made to review some of the developments in the subject over the last four decades.

1.2.2. Methods of Processing

The most widely used systems are vertical batch retorts, with a lid at the top, which are cylindrical pressure vessels operating at temperatures usually between 100 and 140°C. The sequence of operations consists of putting the cans in baskets, placing them in the retort and closing the lid. Steam is then introduced, leaving the vent valve open, so that the air in the retort can be suitably expelled, thereby leaving an atmosphere of almost pure steam. When the processing temperature has been reached the vent is closed and the temperature maintained for the appropriate time dictated by the given *process*. After the time and temperature requirements have been achieved, cooling water is introduced while maintaining the pressure in the retort using air. Pressurized cooling of this type is required for larger-sized cans so that the pressure differential on the cans is reduced slowly in order not to cause irreversible can deformation. When the pressure has been reduced to atmospheric and the cans sufficiently cooled, the retort is opened and the cans removed. The subsequent operations involve drying, labeling and packaging the cans in the required manner for marketing.

Modifications to the above processing are the use of hot water made by steam injection, either in the retort or externally, and the use of air–steam mixtures for processing retortable pouches of food.

Batch retorts also come in a horizontal format with either square or circular cross-sections, with trolleys on wheels for handling the baskets. Some retorts also have facilities for internal rotation of the cans, or external rotation by end-over-end motion of the retort.

High-speed continuous retorts are now widely used in modern production. There are two main types. With *rotary sterilizers*, the cans pass through mechanical valves into a horizontal, cylindrical steam chamber and rotate around the periphery of the shell. Special pressure valves allow the passage of the heated cans into the cooling shell prior to discharge. *Hydrostatic cookers* are valveless sterilizers in which the pressure in the vertical steam chamber is balanced by water legs of appropriate height to match the temperature of the processing steam. The cans are conveyed through the system on horizontal carrier bars, which pass vertically upwards through the pre-heating leg and vertically downwards through the pre-cooling section. Various different types are available, including facilities for rotating cans in the carrier bar system. Details of the heat transfer in these cookers, and the achievement of the correct processes, are given in Chapter 8.

1.3. Packaging Materials

1.3.1. Introduction

The packaging material and its ability to prevent recontamination (*integrity*) are of paramount importance to the canning industry. A large number of spoilage incidents have been attributed to leaker spoilage, subsequent to processing, due to incorrect sealing or the use of unchlorinated water for cooling the cans. The use of the double-seaming technique and can-lid-lining compounds has been effective in reducing leaker spoilage.

1.3.2. Metal Containers

Cylindrical cans made of metal are the most widely used and in the highest production world-wide. Containers made of tin-plated steel are widely used, although lacquered tin-free steels are gradually replacing them. Aluminum cans, and also thin steel cans with easily opened ends, are widely used for beer and beverage packing. The standard hermetically sealable can, also known as a sanitary can in some countries, has various geometries and consists of a flanged body with one or two seamable ends. In the three-piece version one of the ends is usually—but not always—seamed to the body, and the other is seamed after filling. In the two-piece version, which has steadily increased in use, the body is punched out or drawn in such a way that only one flange and lid are necessary. Cans are usually internally lacquered to prevent corrosion of the body and metal pick-up in the products.

Full details of the fabrication of containers are given in Rees and Bettison (1991) and Footitt and Lewis (1995). Some typical container sizes are given in Tables 1.1, 1.2 and 1.3.

Recent developments have reduced the amount of material used in can manufacture, including the necked-in can, which has the advantage of preventing

TABLE 1.1. A guide to UK & US can sizes (1995 revised 2005).

Imperial size[a] (in)	Metric size[b] (mm)	Gross liquid volume (ml)	Common name
Cylindrical cans			
202 × 108	52 × 38	70	70 g tomato paste
202 × 213	52 × 72	140	Baby food
202 × 308	52 × 90	180	6Z (US) or Jitney
202 × 314	52 × 98	192	6 oz juice
202 × 504	52 × 134	250	25 cl juice
211 × 202	65 × 53	155	5 oz
211 × 205	65 × 58	175	6 oz milk
211 × 300	65 × 100	234	8Z Short (US)
211 × 301	65 × 77	235	Buffet or 8 oz picnic
211 × 304	65 × 81	256	8Z Tall (US)
211 × 400	65 × 100	323	No. 1 Picnic (US)
211 × 400	65 × 101	315	Al–10 oz
211 × 414	65 × 124	400	Al tall – 14 oz No. 211 Cylinder (US)
300 × 108	73 × 38	125	
300 × 201	73 × 515	185	
300 × 204.5	73 × 57.5	213	Nominal $\frac{1}{4}$ kg
300 × 207	73 × 61	230	8T – U8
300 × 213	73 × 71	260	250 g margarine
300 × 303$\frac{1}{2}$	73 × 82	310	400 g (14 oz) SCM
300 × 401	73 × 103	405	14Z (E1)
300 × 405	73 × 110	425	Nominal $\frac{1}{2}$ kg
300 × 407	73 × 113	449	No. 300 (US)
300 × 408$\frac{3}{4}$	73 × 115	445	UT
300 × 410	73 × 118	454	16 oz
300 × 509	73 × 146	572	No. 300 Cylinder (US)
300 × 604	73 × 158	630	
301 × 407[c]	74 × 113	440	
301 × 409	74 × 116	459	No. 1 Tall (UK)
301 × 411	74 × 118	493	No. 1 Tall (US)
303 × 406	74 × 113	498	No. 303 (US)
303 × 509	74 × 141	645	No. 303 Cylinder (US)
307 × 113	83 × 46	215	7 oz
307 × 201	83 × 52	235	
307 × 306	83 × 82	434	No. 2 Vacuum (US)
307 × 403	83 × 106	540	
307 × 408	83 × 114	580	A2
307 × 409	83 × 115	606	No. 2 (US)
307 × 510	63 × 142	761	Jumbo (US)
307 × 512	63 × 144	780	No. 2 Cylinder (US)
401 × 200	99 × 51	325	No. 1.25 (US)
401 × 206	99 × 60	190	
401 × 210	99 × 66	445	
401 × 212	99 × 69	475	1 lb flat
401 × 407	99 × 113	815	
401 × 411	99 × 119	880	A 2$\frac{1}{2}$/nominal kilo No. 2.5. (US)
401 × 509	99 × 141	1025	Litre

(*cont.*)

TABLE 1.1. (Continued)

Imperial size[a] (in)	Metric size[b] (mm)	Gross liquid volume (ml)	Common name
401 × 609	99 × 166	1215	Quart (US)
401 × 711	99 × 195	1430	
404 × 307	104 × 88	571	No. 3 Vacuum (US)
404 × 700	104 × 177	1455	A3 (UK)
404 × 700	104 × 177	1525	No. 3 Cylinder (US)
502 × 510	127 × 140	1996	No. 5 (US)
502 × 612	127 × 172	2040	Milk powder
602 × 700	151 × 178	3709	No. 10 (US)
603 × 304	153 × 83	1335	3 lb
603 × 402	153 × 105	1755	
603 × 600	153 × 152	2630	A6
603 × 700	153 × 178	3110	A10
603 × 910	153 × 245	4500	Nominal 5 kg
606 × 509[c]	159 × 141	2570	6 lb tongue
Rectangular cans			
312 × 115 × 309	93 × 47 × 91	345	12 oz rect. (PLM)
301 × 205 × 311	74 × 56 × 93	345	12 oz corned beef
Beverage cans/beer cans (necked in)			
200/202 × 308	50/52 × 88	150	15 cl
200/202 × 504	50/52 × 134	250	25 cl
209/211 × 315	63/65 × 100	275	10 fl oz
209/211 × 409	63/65 × 115	330	12 fl oz
209/211 × 514	63/65 × 149	440	16 fl oz
209/211 × 610	63/65 × 168	500	$\frac{1}{2}$ litre

[a] External diameter × height. Imperial sizes are quoted with three digits and a possible following fraction: the first refers to whole inches and the rest to sixteenths of an inch. For example, 211 means $2\frac{11}{16}$ in, while $408\frac{3}{4}$ means $4 + \frac{8.75}{16} = 4\frac{35}{64}$ in.
[b] Internal diameter × height.
[c] Non-ISO standard.
Sources: A.I.D. Packaging Services (UK) Ltd, Worcester, Carnaud MB, Wantage, & Can Manufacturers Institute U.S.A. (US).

seam-to-seam contact during storage and handling and has cost-saving benefits. New can seam designs—for example the Euroseam and the Kramer seam, which reduce the seam dimensions, especially the length—have been been reported (Anon 1994). There is also interest in the design of easy-open ends, especially made of less rigid material such as foil seals (Montanari 1995). Two examples, are the Impress Easy Peel® lid, (Isensee 2004) and the Abre-Facil produced by Rojek of Brazil. The latter is a vacuum seal like a closure for a glass jar (May 2004).

1.3.3. Glass Containers

Glass jars are also widely used for packing foods and beverages. They have the advantages of very low interaction with the contents and visibility of the product. However, they require more careful processing, usually in pressurized hot water,

TABLE 1.2. A guide to some European can sizes.

Metric size[a] (mm)	Gross liquid volume (ml)	Common name
55 × 67.8	142	1/6 haute
86 × 35.5	170	1/5
65 × 71.8	212	1/4
83 × 57	283	1/3
65 × 100.1	314	3/8
71.5 × 115.5	425	1/2 haute
73 × 109.5	425	1/2 haute dia. 73
99 × 118.2	850	1/1 dia. 99
100 × 118.5	850	1/1
100 × 225	1700	2/1
153 × 151	2550	3/1
153 × 246	4250	5/1

[a] Internal diameter × height.
Source: Institute Appert, Paris.

and handling. Various types of seals are available, including venting and non-venting types, in sizes from 30 to 110 mm in diameter, and made of either tin or tin-free steel. It is essential to use the correct overpressure during retorting to prevent the lid being distorted. It is also essential to preheat the jars prior to processing to prevent shock breakage.

1.3.4. Rigid Plastic Containers

The main requirement for a plastic material is that it will withstand the rigors of the heating and cooling process. Again it is necessary to control the overpressure correctly to maintain a balance between the internal pressure developed during processing and the pressure of the heating system. The main plastic materials used for heat-processed foods are polypropylene and polyethylene tetraphthalate. These are usually fabricated with an oxygen barrier layer such as ethylvinylalcohol, polyvinylidene chloride, and polyamide. These multilayer materials are used

TABLE 1.3. A guide to some European large rectangular can sizes for meat products.

Size (mm)			Description
105	169	323	12 lb oblong
103	164	305	12 lb oblong LANGEN
95	105	318	Ham mold
105	82	400	Ham mold (long)
115	115	545	16 lb Pullman
115	115	385	11 lb Pullman
100	100	400	8 lb Pullman
100	100	303	6 lb Pullman
100	100	207	4 lb Pullman

Source: Eszes and Rajkó (2004).

to manufacture flexible pouches and semi-rigid containers. The current interest is mainly in the latter, which are used to pack microwavable products. This will be an area of rapid expansion during the next few decades, and thermally processed products, especially ready meals, will have to compete with their chilled and frozen counterparts.

More recent developments have been (i) a cylindrical container which has a polypropylene (PP)/aluminum laminate body with molded ends that are welded together, *Letpak*–Akerlund & Rausing; (ii) ethylene vinyl alcohol (EVOH) oxygen-barrier laminate with double-seamed ends, *Omni Can*—Nacanco; (iii) a bowl shaped plastic container with a double-seamed metal easy-open lid, *Lunch bowl*—Heinz; (iv) a clear plastic can with double-seamed end, *Stepcan*—Metal Box; (v) laminated polypropylene (PP)/ethylene vinyl alcohol (EVOH) bottles with foil laminated caps. and polyvinylidene chloride (PVC)/polypropylene (PP) containers, both with a shelf-life of approximately 12 months; and (vi) polyethylene terephthalate (PTFE) bottles, which can be hot-filled up to 92°C or pasteurized up to 75°C (May 2004).

1.3.5. Retortable Pouches

The retortable pouch is a flexible laminated pouch that can withstand thermal processing temperatures and combines the advantages of metal cans and plastic packages. These consist of laminated materials that provide an oxygen barrier as well as a moisture barrier. Flexible retortable pouches are a unique alternative packaging method for sterile shelf-stable products. Recently, important US companies have commercially succeeded with several products. Pouches may be either pre-made or formed from rolls-stock—the more attractive price alternative. Alternately, the pre-made process permits an increased line speed over that of roll-stock, and mechanical issues of converting roll-stock to pouches at the food plant disappear (Blakiestone 2003).

A typical four ply pouch would have an outer layer of polyethylene terephthalate (PTFE) for heat resistance, aluminum foil for oxygen/light barrier, biaxial orientated nylon for resilience, and an inner-cast poly-propylene for pack sealing. Each layer has an adhesive in between it and the next layer. Clear pouches are also made by using a silicate SiOx layer instead of aluminum foil, and these may be reheated using microwaves. Some typical thicknesses for high-barrier pouch-laminate films are PTFE 12–23 μm, aluminum 9–45 μm, SiOx (Ceramis® - Alcan) 0.1 μm, and o-polyamide 15–25 μm, with either polyethylene or polypropylene sealants 50–150 μm. The possible use of liquid crystal polymers, which have superior oxygen and water vapour barrier properties compared with other polymer films, has drawn considerable interest recently (Taylor 2004).

Various types of pouch geometry are available, such as the *pillow pouch,* which consists of a rectangular-shaped container with one side left open for filling and subsequent sealing. Pillow pouches, which have been manufactured and successfully marketed in Japan, e.g. Toyo Seikan, Yokohama, for many years, are usually distributed in cardboard boxes for outer covers. Apart from products for military purposes, the development and acceptance of pillow pouches has been slow. Another pouch geometry is the *gusset pouch*, which is similar to the above but has a bottom on which the container can stand.

The most important feature of these packages is to produce a contamination-free seal, which will maintain the shelf life of the product. Filling and sealing are, therefore, slow processes if an effective seal is to be achieved. Various tests are used to assess the integrity of the seal: (i) a bursting test by injecting gas under pressure, (ii) seal-thickness measurements and (iii) seal-strength tests. Pouches are usually sterilized in over-pressure retorts.

A retortable plastic laminated box *Tetra-Recart* has been developed and marketed by Tetra Pack (Bergman 2004). This is a more heat resistant carton compared with the company's aseptic packs, and the filled and sealed cartons are processed at temperatures up to 130°C for up to three hours, in over-pressure retorts. A number of commercial products have been presented in this pack, including in-pack sterilized vegetables, hot-filled tomato products and a range of sauces.

Retorts used in processing pouches can be batch or continuous, agitating or non-agitating, and they require air or steam overpressure to control pouch integrity (Blakiestone 2003).

Retortable pouches have several advantages over traditional cans. Slender pouches are more easily disposed of than comparatively bulky cans. Shipping them is easier. In addition, the "fresher" retortable pouch product obviously required significantly less heat to achieve commercial sterility. Furthermore, cooking time is about half that required for traditional cans, resulting in tremendous energy savings. Now that retort pouches of low-acid solid foods appear to have attained some commercial acceptance and recognition of their superior quality and more convenient packaging, the expectation is that other heat-sterilized foods will appear in pouches, creating a new segment within the canned foods category (Brody 2003).

1.4. Some Historical Details

The process of glass packing foods was invented and developed on a small commercial scale by the Frenchman Nicholas Appert in 1810, for which he received a financial award from the French government. Subsequently other members of his family continued the business and received further awards and honors. The original work (Appert 1810) describes the process in excellent detail; however, the reason the process achieved stability of the food and its indefinite shelf-life was not known at that time. It was not until 1860 that Pasteur explained that the heating process killed (nowadays we would say "inactivated") the micro-organisms that limited the shelf-life of food. Very shortly after Appert's publication, an English merchant, Peter Durand, took out a patent—subsequently purchased by Donkin, Hall and Gamble—for the use of metal canisters, which inaugurated the canning industry. The industry developed on a large scale in the United States when an

English immigrant, William Underwood, opened a cannery for fruit products in Boston in 1819. The next major development was the production of continuously seamed cans with the use of the double seam. The American industry developed rapidly after this innovation. It was not until the 1920s that the newly developed technology was available in the United Kingdom. A can-body maker was installed in 1927 in Williamson's factory in Worcester on the recommendation of the staff at the Bristol University Research Station, at Chipping Campden, whose staff at that time included Alfred Appleyard and Fred Hirst. Much of the subsequent technical development of the industry was due to the untiring efforts of Hirst and his colleagues, Bill Adam and Tom Gillespy, as well as the field service staff of The Metal Box Co., the subsequent owners of Williamson's factory.[1]

During the 1920s W.D. Bigelow, the first director of the National Canners' Association in Washington, and colleagues developed a method of determining thermal processes based on heat-penetration measurements in cans of food (Bigelow *et al.* 1920). One of these colleagues, Charles Olin Ball, subsequently developed theoretical methods for the determination of thermal processes (Ball 1923, 1928) and became the acknowledged expert in this subject (Ball & Olson 1957). Subsequent developments in the subject have been largely based on these workers' early concepts.

Following the lead set by the Americans, research workers in the United States and the United Kingdom applied these methods to their canning industries. In the United Kingdom the pioneering work on process determination was done by T.G. Gillespy at Campden Research Station. The establishment of safe procedures in the British canning industry owes much to his scholarship, devotion and integrity. He devoted much of his time to recommending safe processes for each type of sterilizer as it was introduced into the industry. His guide to processes for canning fruits and vegetables (Gillespy 1956) is a model of his clear approach to the subject. He was also very much concerned with the heat resistance of microorganisms, as well as problems of leader spoilage and sanitation in canneries. In fact, he was one of the first workers to identify spoilage by post-process can seam leakage rather than understerilization. Although by today's standards he published relatively little (in fact he regarded much of the published literature with disdain), the clarity of his thought was illustrated by his paper on the principles of heat sterilization (Gillespy 1962), and his two papers on the mathematics of process calculations (Gillespy 1951, 1953). While the nomenclature is often daunting, these publications are well worth mastering as an introduction to the subject.

The contributions of other workers are documented in various parts of this book. It would be an invidious task to detail all the contributions of the world's experts; consequently selection has been necessary. An indexed bibliographic

---

[1] For further historical information see Bitting (1937), Ball and Olson (1957), Metal Box (1960), Adam (1980), and Thorne (1986).

guide to process calculations, which covers much of the work, formed the basis for this book, including heat inactivation of micro-organisms, heat transfer to and in cans, and thermal process calculations (Overington & Holdsworth 1974; Holdsworth 2006).

Some recent works of importance to this text have been produced by Ramaswamy and Singh (1997), Richardson (2001, 2004), Teixiera (1992), and Peleg (2006).

References

Adam, W. B. (1980). *Campden research station, A history 1919–1965.* Chipping Campden, Glos., UK: Campden & Chorleywood Food Research Association.

Anon. (1994). Microseam goes worldwide. *The Canmaker*, 7 November, 21.

Appert, N. (1810). *L'Art de Conserver pendant Plusiers Années Toutes les Substances Animales et Végétales.* Paris: Patris & Co. [A translation by K. G. Bitting was published in 1920 by the Glass Container Association of America, Chicago, and reprinted in S. A. Goldblith, M. A. Joslyn & J. T. R. Nickerson (Eds.), (1961) *An introduction to thermal processing of foods.* Westport, CT: AVI Publishing Company, Inc.]

Ball, C. O. (1923). Thermal process time for canned food. *Bull. Nat. Res. Council*, 7(37), 9–76.

Ball, C. O. (1928). Mathematical solution of problems on thermal processing of canned food. *University of California Publications in Public Health*, 1(2), 145–245.

Ball, C. O. & Olson, F. C. W. (1957). *Sterilization in food technology—Theory, practice and calculations.* New York: McGraw-Hill.

Bergman, O. (2004) Tetra RECART. In G. S. Tucker (Ed.), *Third International Symposium Thermal Processing—Process and Package innovation for convenience foods.* Session 1:3. Chipping Campden UK: Campden & Chorleywood Food Research Association.

Bigelow, W. D., Bohart, G. S., Richardson, A. C., & Ball, C. O. (1920). Heat penetration in processing canned foods. *Bulletin* No. 16L. Washington, DC: National Canners' Association.

Bitting, A. W. (1937). *Appertizing or the art of canning; Its history and development,* San Francisco, CA: The Trade Pressroom.

Blakiestone, B. (2003). Retortable pouches. In *Encyclopaedia of agricultural, food, and biological engineering.* Marcel Dekker. USA.

Brody, A. (2003). Food canning in the 21st Century. *Food Technol.*, 56, 75–79.

Eszes F., & Rajkó. R. (2004). Modelling heat penetration curves in thermal processes. In P. Richardson (Ed.), *Improving the thermal processing of food* (pp. 307–333). Cambridge, UK: Woodhead Publishing Ltd.

Footitt, R. J. & Lewis, A. A. (Eds.), (1995) *The canning of fish and meat.* Glasgow: Blackie Academic and Professional.

Gillespy, T. G. (1951). Estimation of sterilizing values of processes as applied to canned foods. I. Packs heating by conduction. *J. Food Agric.*, 2, 107–125.

Gillespy, T. G. (1953). Estimation of sterilizing values of processes as applied to canned foods. II. Packs heating by conduction: complex processing conditions and value of coming-up time of retort. *J. Food Agric.*, 4, 553–565.

Gillespy, T. G. (1956). The processing of canned fruit and vegetables. *Technical Bulletin* No. 4, Chipping Campden, UK: & Chorleywood Food Research Association.

Gillespy, T. G. (1962). The principles of heat sterilization. In J. Hawthorn & J. M. Leitch (Eds.), *Recent advances in food science Vol. 2 Processing* (pp. 93–105). London: Butterworths.

Hersom, A. C. & Hulland, E. D. (1980). *Canned foods. Thermal processing and microbiology* (7th Edition). Edinburgh: Churchill Livingstone.

Holdsworth, S. D. (1976). A bibliographical guide to thermal process calculations. Supplement No. 1, *Technical Memorandum* No. 155. Chipping Campden, Glos. UK: Campden & Chorleywood Food Research Association.

Holdsworth, S. D. (1979). A bibliographical guide to thermal process calculations. Part 2. *Technical Memorandum* No. 213, Chipping Campden, Glos. UK: Campden & Chorleywood Food Research Association.

Holdsworth, S. D. (1982). A bibliographical guide to thermal process calculations. Part 3. *Technical Memorandum* No. 291, Chipping Campden, Glos. UK: Campden & Chorleywood Food Research Association.

Holdsworth, S. D. (1985). A bibliographical guide to thermal process calculations. Part 4. *Technical Memorandum No. 410*, Chipping Campden, Glos. UK: Campden & Chorleywood Food Research Association.

Holdsworth, S. D. (1988). A bibliographical guide to thermal process calculations. Part 5. *Technical Memorandum No. 498*, Chipping Campden, Glos. UK: Campden & Chorleywood Food Research Association.

Holdsworth, S. D. (1990). A bibliographical guide to thermal process calculations. Part 6. *Technical Bulletin No. 77*, Chipping Campden, Glos. UK: Campden & Chorleywood Food Research Association, Chipping Campden.

Holdsworth, S. D. (2006). Bibliographical guide to thermal processing science. (1996–2006). Personal Publication.

Institut Appert (1979). *Barèmes de Sterilization pour Aliments Appertisés*. Paris: Institut Appert.

Isensee, T. (2004), Peelable ends—market trends. In G. S. Tucker (Ed.), *Third International symposium thermal processing—Process and package innovation for convenience foods* (Session 1:4). Chipping Campden UK: Campden & Chorleywood Food Research Association.

Jackson, J. M. & Shinn, B. M. (1979). *Fundamentals of food canning*. Westport, CT: AVI Publishing.

Lopez, A. (1987). *A complete course in canning* (3 vols.). Baltimore, MD: The Canning Trade Inc.

May, N. (2004). Developments in packaging formats for retort processing. In P. Richardson (Ed.), *Improving the thermal processing of foods* (pp. 138–151). Cambridge: Woodhead Publishing.

Metal Box (1960). *History of canned foods*, London: MB Publications.

Montanari, A., Marmirolig, G., Pezzanni, A., Cassara, A., & LupuI, R. (1995). Easy open ends for food and beverage cans: Definition, manufacture, coating and related problems. *Industria Conserve*, *70*, 410–416.

NFPA (1982). *Thermal processes for low acid canned foods in metal containers*, Bulletin 26–L (12th Edition). Washington, DC: National Food Processors' Association.

Overington, W. J. G. & Holdsworth S. D. (1974). A bibliographical guide to thermal process calculations. *Technical Memorandum No. 139*. Chipping Campden, Glos. UK: Campden & Chorleywood Food Research Association.

Peleg, M. (2006). Advanced quantitative microbiology for foods and biosystems: Models for predicting growth and inactivation. Boca Raton FL. CRC Press.

Pflug, I. J. (1982). *Microbiology and engineering of sterilization processes* (5th Edition.). Minneapolis: University of Minnesota Press.

Ramaswamy, R. P., & Singh, R. P. (1997). Sterilization process engineering. In E. Rotstein, R. P. Singh, & K. Valentas (Eds.), *Handbook of food engineering practice.* (Chapter 2, pp. 1–69). Boca Raton, FL. CRC Press.

Rees, J. A. G. & Bettison, J. (Eds.), (1991). *The processing and packaging of heat preserved foods.* Glasgow, Blackie.

Richardson, P. (Ed.), (2001). *Thermal technologies in food processing.* Cambridge: Woodhead Publishing.

Richardson, P. (Ed.), (2004). *Improving the thermal processing of foods.* Cambridge: Woodhead Publishing.

Stumbo, C. R. (1973). *Thermobacteriology in food processing* (2nd Edition). New York: Academic Press.

Taylor, M. (2004). Innovations in retortable pouches. In G. S. Tucker (Ed.), *Third international symposium Thermal processing—Process and package innovation for convenience foods.* (Session 1:2). Chipping Campden UK: Campden & Chorleywood Food Research Association.

Teixeira, A. (1992). Thermal process calculations. In D. R. Heldman & D. B. Lund (Eds.), *Handbook of food engineering.* (Chapter 11, pp. 563–619.) New York: Marcel Deker.

Thorne, S. (1986). *History of food preservation.* London: Parthenon.

2
Heat Transfer

2.1. Introduction

2.1.1. General Aspects

The main object of this chapter is to give a brief account of the mathematical methods of determining the temperature distribution with time and position in packaged foods while being heated and cooled. This is a prerequisite to establishing a *process* which will ensure the microbiological safety of the product and is also organoleptically acceptable. This requires an examination of the modes of heat transfer in different parts of the processing operation.[1]

2.1.2. Mechanisms of Heat Transfer

There are three modes of heat transfer, which contribute to the overall heat transfer process in differing proportions: conduction, convection and radiation. Conduction is the transfer of heat by molecular motion in solid bodies. Convection is the transfer of heat by fluid flow, created by density differences and buoyancy effects, in fluid products. Radiation is the transfer of electromagnetic energy between two bodies at different temperatures. In Figure 2.1 the main modes for heat transfer in the processing of packaged foods are illustrated.

The first mode is heat transfer to the container or packaging from the heating and cooling medium; the main modes of heat transfer to be considered for the various heating media are given in Table 2.1. Heating with pure steam, or microwaves, is very effective and does not present any appreciable resistance to heat transfer; consequently, it does not need to be taken into account in the overall heat transfer. In the case of all other media it is necessary to take the convective or radiative heat-transfer coefficient into account. Convective-heat transfer rates depend largely on the velocity of flow of the media over the container, and this is an important factor to be controlled in all processing operations. This subject is dealt with in more detail in Chapter 8.

---

[1] The reader is encouraged to consult the very useful text on heat transfer and food products by Hallström *et al.* (1988). This is an excellent guide to the basic principles of heat transfer and its application to food processing.

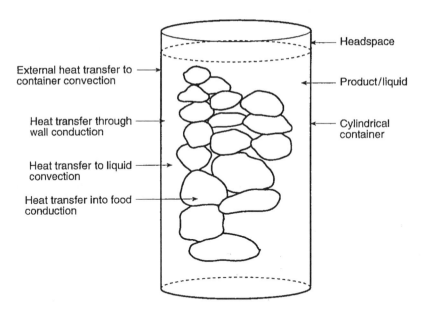

FIGURE 2.1. Heat transfer to food product in a cylindrical container.

The second mode of heat transfer is through the container wall; for metallic containers of normal thickness, the thickness and the thermal conductivity of the material are such that there is no appreciable resistance to heat transfer. However, for glass bottles and plastic containers there is a significant resistance, and this should be considered in determining the overall heat transfer resistance.

The third mode of heat transfer is into the product from the container wall; this depends on the consistency of the food material and is discussed in detail elsewhere (see Chapter 5). Fluid products or solid particulates covered with a fair amount of fluid heat or cool rapidly by convection, while other products of a more solid consistency heat mainly by conduction. In between there are products that heat/cool by a combination of conduction and convection, and some that start with convection heating and finish in conduction mode because of physico-chemical

TABLE 2.1. Heat transfer modes for containers being heated or cooled.

Media	Main mode	Resistance
Steam (air-free)	Condensation	Effectively none
Steam-air mixtures	Convection	Increases with increasing air content
Air	Convection	High
Water, boiling	Convection	Low
Water, hot	Convection	Decreases with increasing water velocity
Water, cold	Convection	Medium
Flame/infrared	Radiation	Low
Fluidized bed	Convection	Medium, depends on degree of agitation
Microwave	Radiation	None

changes. Thus the internal mechanisms of heat transfer are complicated. From a theoretical point of view it is only possible at the present time to deal with simple heat transfer mechanisms; however, empirical methods (see Chapter 5) allow the processor to calculate temperature distributions without being too concerned about the mechanism.

When dealing with heat transfer theory, it should be noted that a distinction is made between (a) steady-state heat transfer, which involves constant temperatures of heat transfer media, and the product, e.g. heating and cooling in continuous-flow heat exchangers; and (b) unsteady-state heat transfer, which implies that the temperatures are continuously changing. It is type (b) with which we are concerned in this book, i.e. the determination of time–temperature profiles at specified points in the container. From a practical point of view, a satisfactory process is determined at the slowest point of heating in the packaged food, and this makes calculation easier, since with conduction heating products, the center point of the food mass is taken as the slowest point of heating, or critical point. It is not sufficient in processing packaged foods just to achieve a given temperature at the slowest point of heating, but to achieve it for a given time, specified either by calculation or experimental investigation.

2.2.   Heat Transfer by Conduction

2.2.1. Introduction

Energy transfer by conduction takes place when different parts of a solid body are at different temperatures. Energy flow in the form of heat takes place from the hotter, more energetic state, to the colder, less energetic state. The quantity of heat transferred under steady-state conditions is given by

$$Q = k \frac{T_1 - T_2}{x} At, \qquad (2.1)$$

where

Q = quantity of heat (J or N m);
T = temperature (K or °C), with subscripts 1, 2 referring to the two parts of the body;
t = time (s);
x = the distance (m) of separation of the two points;
A = the cross-sectional area (m^2) for heat flow;
k = the thermal conductivity (Wm^{-1} K^{-1}).

Differentiating with respect to time gives the rate of heat flow:

$$\frac{dQ}{dt} = k \frac{T_1 - T_2}{x} A, \qquad (2.2)$$

This equation can be written more simply in a differential form

$$\frac{dQ}{dt} = -kA\frac{dT}{dx}.$$ (2.3)

This relates the rate of heat flow dQ/dt to the temperature gradient in the material dT/dx, and is known as the one-dimensional heat conduction equation expressed in Cartesian coordinates. The quantity $(dQ/dt)/A$ is known as the heat flux, and is measured in joules per square meter per second.

2.2.2. Formulation of Problems Involving Conduction Heat Transfer

The main object of this section is to indicate the mathematical basis of the problems encountered in the determination of the temperature distribution in heating canned foods by conduction. The treatment is necessarily brief, and further information can be found in the standard texts, e.g. Ingersoll, Zobel and Ingersoll (1953), Carslaw and Jaeger (1959), Arpaci (1966), Luikov (1968), and Ozisik (1980).

The basis of all unsteady-state conduction heat transfer equations is Fourier's equation, established by the French physicist Jean Baptiste Joseph Fourier (1768–1830) (Fourier 1822) and written as

$$\rho c \frac{\partial T}{\partial t} = \nabla k \nabla T$$ (2.4)

where ρ is the density (kg m^{-3}), c the specific heat or heat capacity (J kg^{-1} K^{-1}) and ∇ the differential operator (del, also known as nabla), where

$$\nabla = \partial/\partial x + \partial/\partial y + \partial/\partial z.$$

Equation (2.4) implies that the thermal conductivity is a function of temperature, an assumption which is not usually made in heat transfer calculations in order to simplify the calculations. Consequently, a simpler equation is generally used,

$$\rho c \frac{\partial T}{\partial t} = k \nabla^2 T$$

or

$$\frac{\partial T}{\partial t} = \alpha \nabla^2 T,$$ (2.5)

where α is the thermal diffusivity, $k/\rho c$ (m^{2}s^{-1}) and ∇^2 is the Laplace operator, given by

$$\nabla^2 = \partial^2/\partial x^2 + \partial^2/\partial y^2 + \partial^2/\partial z^2.$$

The physical significance of this property is associated with the speed of heat propagation into the solid. Materials with high values, such as metals, heat rapidly, whereas food materials and water have comparatively low values and heat much more slowly. Table 2.2 gives some data for food products. More

TABLE 2.2. Some values of thermal diffusivities of various products.

Product	Temperature (°C)	Thermal diffusivity ($\times 10^7 \mathrm{m^2 s^{-1}}$)	Reference
(i) Food products			
Apple sauce	105	1.61	Uno and Hayakawa (1980)
Apple pulp: Golden Delicious	29	1.50–1.62	Bhowmik and Hayakawa (1979)
Cherry tomato pulp	26	1.46–1.50	Bhowmik and Hayakawa (1979)
Tomato ketchup	—	1.20 ± 0.02	Gouvaris and Scholefield (1988)
Tomato: Ace var.	42.9	1.22–1.88	Hayakawa and Succar (1983)
Carrots	138	1.82–1.88	Chang and Toledo (1990)
Pea purée	—	1.59	Bhowmik and Tandon (1987)
Pea purée	—	1.54	Lenz and Lund (1977)
Pea & potato purée	—	1.48	Masilov and Medvedev (1967)
Potato purée	—	1.30 ± 0.04	Gouvaris and Scholefield (1988)
Potato (78% water)	60–100	1.39–1.46	Tung et al. (1989)
Potato	42.9	1.42–1.96	Hayakawa and Succar (1983)
French bean & chicken purée	—	1.62	Patkai et al. (1990)
Mixed vegetables & beef purée	—	1.63	Patkai et al. (1990)
Tuna fish/oil	115	1.64	Banga et.al. (1993)
Mushrooms in brine	—	1.18	Akterian (1995)
Ham, processed	—	0.94	Smith et al. (1967)
Ham salami	—	1.52	Choi and Okos (1986)
Beef purée	—	1.75	Lenz and Lund (1977)
Meat hash	—	1.52	Choi and Okos (1986)
Meat sauce	69–112	1.46 ± 0.05	Olivares et al. (1986)
Meat croquette	59–115	1.98 ± 0.22	Olivares et al. (1986)
Meat, ground	20	1.26–1.82	Tung et al. (1989)
Pork purée	—	1.94	Lenz and Lund (1977)
Meat/tomatoes/potatoes	65–106	1.57 ± 0.20	Olivares et al. (1986)
Meat/potatoes/carrots	58–113	1.77 ± 0.15	Olivares et al. (1986)
Cooked chickpeas/pork sausages	71–114	1.90 ± 0.03	Olivares et al. (1986)
Chicken & rice	65–113	1.93 ± 0.21	Olivares et al. (1986)
Chicken/potatoes/carrots	72–109	1.70 ± 0.03	Olivares et al. (1986)
Lasagne (73.6% water)	60–100	1.32–1.70	Tung (1989)
Water	0–100	1.338–1.713	Evans (1958)
(ii) Simulants			
Acrylic plastic ellipsoids	—	1.19	Smith et al. (1967)
Ammonium chloride	40–100	1.53–1.47	Tung et al. (1989)
Agar-starch/water gels 3–3.5%	40–60	1.38– 1.25	Tung et al. (1989)
Agar-water 5%	54	1.53	Evans (1958)
Bean-bentonite 75% water	115.6	1.72	Evans (1958)
Bentonite 10 Bentonite 10%	120°C	1.77	Uno and Hayakawa (1980)
Bentonite 10%	—	1.90	Bhomik and Tandon (1987)
Bentonite	—	1.86	Peterson and Adams (1983)
Ethylene glycol/water/agar 5%	—	1.11	Evans (1958)
(iii) Container materials			
Polypropylene (PP)		0.071	Shin & Bhowmik (1990, 1993)
Polycarbonate		0.013	Shin & Bhowmik (1990, 1993)
Polyvinylidene chloride (PVDC)		0.062	Shin & Bhowmik (1990. 1993)
Laminate (PP:PVDC:PP)		0.068	Shin & Bhowmik (1990, 1993)

extensive data will be found in the publications of Singh (1982), Okos (1986), Lewis (1987), George (1990), and Eszes and Rajkó (2004). The determination of physical properties from thermometric measurements and a finite element model has been reported by Nahor *et al.* (2001). A computer program, COS-THERM, was developed to predict the thermal properties of food products based on their composition (Beek & Veerkamp 1982; Miles *et al.* 1983). Many foods of high moisture content have values of α ranging from 1.4 to $1.6 \times 10^{-7}\,\mathrm{m^2\,s^{-1}}$. Palazoglu (2006) has reported an interesting study on the effect of convective heat transfer on the heating rate of materials with differing thermal diffusivities including cubic particles of potato and a polymethylpentene polymer. Using the analytical solution for heating a cube with external heat transfer it was shown that the rate of heating depended very much on the combination of heat-transfer coefficient and the thermal conductivity.

Equation (2.5) can be expressed in a variety of forms depending upon the coordinate system being used. Cartesian coordinates –x, y, z – are used for heat transfer in flat plates (equation (2.4)), including slabs where the length is greater than the width, e.g. food in flexible pouches and trays, and for rectangular parallelepipeds or bricks (equation (2.6)), e.g. rectangular-shaped containers both metallic and plastic:

$$\frac{dT}{dt} = \alpha \left[\frac{d^2T}{dx^2} + \frac{d^2T}{dy^2} + \frac{d^2T}{dz^2} \right]. \tag{2.6}$$

Cylindrical coordinates $-x = r\cos b$, $y = r\sin b$, and z – where b is the angle and r the radius for transformation from a Cartesian coordinate system, are used for all containers with a cylindrical geometry, i.e. most canned foods. When transformed the previous equation becomes

$$\frac{dT}{dt} = \alpha \left[\frac{d^2T}{dr^2} + \frac{1}{r}\frac{dT}{dr} + \frac{1}{r^2}\frac{d^2T}{db^2} + \frac{d^2T}{dz^2} \right]. \tag{2.7}$$

For radial flow of heat, i.e. neglecting axial heat flow, the last two terms may be neglected, so that the basic equation to be solved for radial heat transfer into a cylindrical container is (Figure 2.2):

$$\frac{dT}{dt} = \alpha \left[\frac{d^2T}{dr^2} + \frac{1}{r}\frac{dT}{dr} \right]. \tag{2.8}$$

If the temperature is only required at the point of slowest heating, i.e. the center, where at $r = 0$, $dT/dr = 0$ and $dT/r\,dr = d^2T/dr^2$ (see Smith 1974), then equation (2.8) simplifies for the purposes of computation to

$$\frac{dT}{dt} = 2\alpha \left[\frac{d^2T}{dr^2} \right]. \tag{2.9}$$

While there are no containers that approximate to a spherical shape, spherical coordinates $- x = r\cos a\cos b$, $y = r\cos a\sin b$ and $z = r\sin a$ – are useful for predicting the temperature distribution in spherical-shaped food particulates, e.g.

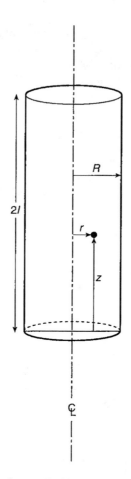

FIGURE 2.2. Coordinate system for a cylindrical can of height $2l$ and diameter $2R$.

canned potatoes in brine. The basic equation (2.5) in spherical coordinates is

$$\frac{1}{\alpha}\frac{dT}{dt} = \frac{1}{r^2}\frac{d}{dr}\left(r^2\frac{dT}{dr}\right) + \frac{1}{r^2\sin b}\frac{d}{db}\left(\sin b\frac{dT}{db}\right) + \frac{1}{r^2\sin^2 b}\frac{d^2 T}{da^2}. \qquad (2.10)$$

If the temperature is only required in the radial direction, the angular terms can be neglected and equation (2.10) may be simplified to give

$$\frac{1}{\alpha}\frac{dT}{dt} = \frac{1}{r^2}\frac{d}{dr}\left(r^2\frac{dT}{dr}\right)$$

or

$$\frac{dT}{dt} = \alpha\left[\frac{d^2 T}{dr^2} + \frac{2}{r}\frac{dT}{dr}\right]. \qquad (2.11)$$

For the central point only, $r = 0$, the equation becomes

$$\frac{dT}{dt} = 3\alpha \left[\frac{d^2 T}{dr^2} \right]. \tag{2.12}$$

In all these cases the problems have been simplified; for more complicated cases the reader is referred to the texts previously mentioned.

A full treatment of the term $\nabla \cdot k\nabla$ is given by Bird *et al.* (1960); for details of the equations in cylindrical and spherical coordinates, see Ruckenstein (1971).

2.2.3. Initial and Boundary Conditions

Temperature representations are often expressed simply as T. However, a more formal method is to give the coordinates of space and time in brackets. Thus a simple one-dimensional temperature distribution would be represented as $T(x, t)$ or $T(r, t)$, two-dimensional distributions as $T(x, y, t)$ and three-dimensional distributions as $T(x, y, z, t)$. For center distributions, where $x = y = z = r = 0$, $T(0, t)$ is used, and for conditions at time zero $T(x, 0)$. Since it is usually obvious what is intended from the context of the equations, this practice is often dispensed with. It will be used in the following discussion where appropriate.

There are two *initial* conditions that may apply to a particular problem:

1. The contents of the container are initially at a uniform temperature T_0 throughout, which is expressed as follows:

$$T = T_0 \quad \text{at} \quad t = 0 \quad \text{or} \quad T = T(x, y, z, 0).$$

   In good canning practice this condition should be achieved, and in calculations it is nearly always assumed.
2. The contents of the container have an initial temperature distribution in space. This is usually expressed as follows:

$$T = f(x) \quad \text{at} \quad t = 0,$$

   or in other suitable ways, e.g.

$$T(x, y, z, 0) = f(x, y, z) \qquad T(r, 0) = f(r),$$

   where $f(x)$ is some function of x. This initial condition is used at the beginning of the cooling period for canned products that have not achieved a uniform temperature distribution at the end of the heating period. It usually applies to large container sizes with conduction-heating products.

Other conditions that have to be taken into account for solving the heat transfer equations are the *boundary* or *end* conditions, the conditions to which the can is exposed during processing. The following boundary conditions are encountered in heat transfer work:

1. The surface temperature is prescribed and does not vary with time, i.e. a surface is exposed to an instantaneous change in temperature. This is referred to as a boundary condition of the first kind by some workers. It applies to steam

heating and is often assumed in heat transfer modelling work. It is the simplest condition to apply and is expressed as

$$dT(x, t)/dx = 0,$$

where x represents the external can diameter, or

$$T(x, t) = \text{constant}.$$

2. The surface temperature is governed by a convective heat coefficient, often referred to as a boundary condition of the third kind. Such a condition applies to cases where the heating medium is not condensing steam, e.g. hot water, a steam–air mixture or a cooling fluid. The surface temperature in these cases depends on the heat-transfer coefficient, which in turn depends on the velocity of the fluid over the surface (see section 2.3). This condition is expressed as follows:

$$-dT(x, t)/dx + h[T_R - T(x, t)] = 0. \tag{2.13}$$

3. The surface temperature is a function of time, i.e. the heating medium heats or cools while the containers are being processed. Three specific cases are used to illustrate this condition:
   (a) Retort temperature change is a *linear function* with time: for example,

$$T_R(t) = T_0 + bt, \tag{2.14}$$

   where T_0 is the initial temperature, T_R is the processing medium temperature, i.e. retort temperature, and b is a constant depending upon the magnitude of the gradient.
   (b) Retort temperature change is an *exponential function* of time:

$$T_R(t) = T_\infty - (T_\infty - T_0)e^{-kt}, \tag{2.15}$$

   where T_∞ is the maximum temperature reached and k a constant. This applies to the initial heating period of cans when placed in a static retort.
   (c) Retort temperature a *harmonic function* of time:

$$T_R(t) = T_\infty \cos(2\pi nt), \tag{2.16}$$

   where n is the frequency of oscillation.

2.2.4. Mean or Volume Average Temperatures

It is necessary to know the exact temperature distribution inside packaged foods in order to calculate the sterilization value; however, there are circumstances in which mass-average temperatures are appropriate – in particular, the determination of a heat-vulnerable component, e.g. a vitamin; for determining some cooling processes; and for determining energy changes.

The average temperature will be signified by putting a bar over the temperature term, thus the volume-average temperature is given by

$$\bar{T}(t) = \frac{1}{V} \int_0^V T(x,t) dV,$$ (2.17)

where V is the volume and dV is the volume element. So for a slab of dimensions $2x, 2y, 2z$, we have

$$\bar{T}(t) = \frac{1}{2x2y2z} \int_{-x}^{x} \int_{-y}^{y} \int_{z}^{-z} T(x,y,z,t) dx, dy, dz;$$ (2.18)

for a one-dimensional slab thickness $2x$:

$$\bar{T}(t) = \frac{1}{2x} \int_{x}^{-x} T(x,t) dx = \frac{1}{x} \int_0^x T(x,t) dx;$$ (2.19)

for a sphere of radius R at any point r:

$$\bar{T}(t) = \frac{3}{R^3} \int_0^R r^2 T(r,t) dr;$$ (2.20)

and for a cylinder of radius R:

$$\bar{T}(t) = \frac{2}{R^2} \int_0^R r^2 T(r,t) dr.$$ (2.21)

The mean temperature may also be used to determine the energy changes in a processing system:

$$Q = c(\bar{T}(t) - T_0)$$ (2.22)

where Q is the amount of energy supplied during time t, c is the specific heat and T_0 is the initial uniform temperature.

2.2.5. Summary of Basic Requirements

The following points need to be considered when attempting to formulate a model to predict the temperature distribution in a packaged food product which is being heated and cooled:

1. Is the product isotropic, i.e. does it have properties the same in all directions? If not, use k_x, k_y, k_z.
2. Do the physical properties vary with temperature, or any other prevailing condition? If so, then use $k(T)$.
3. Is the product at a uniform initial temperature? If not, use $T = f(x, y, z)$.
4. Is the product heated uniformly on all sides? Is the headspace taken into account? See special equations in section 2.5.
5. Does the container or package change shape during the processing? If so, use appropriate dimensions.
6. Is it necessary to consider the resistance to heat transfer through the container wall? If so, use x_w / k_w.

7. Does the heating or cooling medium impose a low heat-transfer coefficient? If so, use heat-transfer boundary condition equation (2.12).
8. Is the surface temperature variable? If so, use $T_R(t)$ as in equations (2.13) and (2.14).

A general rule for proceeding is to apply a simple model first, usually in one dimension, and then a more complex model if the predictions are not in agreement with the experimental results. For many practical factory applications simple models suffice.

2.2.6. Some Analytical Methods for Solving the Equations

There are many methods for solving partial differential equations, and it will suffice here to mention some of those that have been used by researchers in this subject without going into any detail. The first group are the analytical methods and the functions that they use.

2.2.6.1. Method of Separation of Variables

This method assumes that the solution to the partial differential equation, e.g. the simplest one-dimensional unsteady-state equation for the temperature distribution in a slab equation,

$$\frac{\partial T(x,t)}{\partial t} = \alpha \frac{\partial^2 T(x,t)}{\partial^2 x}, \tag{2.23}$$

can be represented as the product of a spatial function $X(x)$ and a time function $T(t)$, viz.

$$T(x,t) = X(x) \cdot T(t). \tag{2.24}$$

Substituting the differentiated forms of (2.24) in (2.23) and separating the variables on either side of the equation results in:

$$\frac{1}{\alpha} \frac{\partial T}{T \partial t} = \frac{1}{X} \frac{\partial^2 X}{\partial x^2}. \tag{2.25}$$

Putting each side equal to a constant, e.g. $-b^2$, it is possible to obtain solutions for $T(t)$ and $X(t)$, viz.

$$T(t) = A e^{-b^2 t}$$

$$X(x) = B \cos(bx) + C \sin(bx).$$

Using equation (2.23), the general solution becomes

$$T(x,t) = [D \cos(bx) + E \sin(bx)]e^{-b^2 t}. \tag{2.26}$$

Using the initial and boundary conditions, a particular solution can then be found for the problem. This involves the use of Fourier sine series, which is discussed in 2.2.6.3.

2.2.6.2. Operational Methods: Integral Transforms and the Laplace Transform

In this method each term of the differential equation is multiplied by $e^{-\beta t}$ and integrated with respect to time from $t = 0$ to $t = \infty$. For the one-dimensional slab this results in

$$\int e^{-\beta t} \cdot \frac{\partial T}{\partial t} \cdot dt = \alpha \int e^{-\beta t} \frac{\partial^2 T}{\partial x^2} \cdot dt. \tag{2.27}$$

The Laplace transform $L[f(t)]$ is defined by

$$f(s) = L[f(t)] = \int f(t)e^{-st} dt, \tag{2.28}$$

and equation (2.27) may be written as

$$L\left[\frac{\partial T}{\partial t}(x, t)\right] = \alpha L\left[\frac{\partial^2 T}{\partial x^2}(x, t)\right]. \tag{2.29}$$

By integrating the terms in equation (2.27) from 0 to ∞ or from standard tables, the partial differential equation is transformed to a second-order differential equation, viz.

$$\alpha \frac{d^2 T}{dx^2}(x, s) - sT(x, s) = 0, \tag{2.30}$$

which has solution

$$T(x, s) = A \cosh\left(\frac{s}{a}\right)^{1/2} + B \sinh\left(\frac{s}{a}\right)^{1/2}. \tag{2.31}$$

This can be transformed back to the original system using the reverse transformation $f(t) = L^{-1}[f(s)]$, resulting in the Fourier series solution.

Tables of transforms are found in all standard texts on heat transfer, e.g. Carslaw and Jaeger (1959), Luikov (1968), Mickley et al. (1957), and Ozisik (1980).

Hayakawa (1964) made elegant use of integral transforms for handling complex boundary conditions experienced in standard canning operations. Various other types of transform are available for handling different situations. The Fourier sine transform is useful for dealing with situations with a prescribed boundary condition and the Hankel transform for dealing with cylindrical geometry (Magnus et al. 1966). The main problem with the integral transform method is finding the reverse function to convert the solution of the derived differential equation.

2.2.6.3. Some Special Transcendental Functions Involved in the Solution
          of Differential Equations

Fourier series. Many of the analytical solutions to heat transfer problems involve the use of series, in particular series of trigonometrical functions. For example, a relationship $y = f(x)$ may be represented, between the limits of $x = 0$ and $x = \pi$, as follows:

$$y = f(x) = a_1 \sin(x) + a_2 \sin(2x) + \cdots + a_n \sin(nx) \tag{2.32}$$

or

$$y = f(x) = a_n \sin(nx)$$

and

$$n = \frac{2}{\pi} \int f(x) \sin(mx) dx$$

If $f(x)$ is a constant, e.g. initial temperature T_0, then

$$a_n = \frac{2}{\pi} T_0 \int \sin(mx) dx = \frac{2}{\pi n} T_0 [1 - (-1)^n]$$

If n is even, then $a_n = 0$, and if n is odd, $a_n = 4T_0/\pi n$ and

$$f(x) = 4T_0 \left(\frac{\sin(x)}{1} + \frac{\sin(3x)}{3} + \frac{\sin(5x)}{5} + \cdots \right). \qquad (2.33)$$

Similar Fourier series are available with cosine terms. The standard analytical solutions for simple slab geometry involve combinations of sine and cosine series. For ease of computation it is essential that the series converge rapidly: In many cases a first-term approximation is satisfactory, especially where long times are involved.

Bessel functions. For problems involving cylindrical geometry – in particular, food and drink cans – the linear second-order equation representing the temperature distribution with time and space is known as Bessel's equation. The analytical solution of this equation requires the use of Bessel functions designated $J_v(x)$, where v is the order.

Bessel functions are defined by

$$J_v(x) = \frac{x^v}{2^v v!} \left[1 - \frac{x^2}{2(2v + 2)} + \frac{x^4}{2 \times 4(2v + 2)(2v + 4)} - \cdots \right]. \qquad (2.34)$$

When $v = 0$ the function $J_0(x)$ is known as a Bessel function of the first kind and of order zero, i.e.

$$J_0(x) = 1 - \frac{x^2}{2^2} + \frac{x^4}{2^2 \cdot 4^2} - \frac{x^6}{2^2 \cdot 4^2 \cdot 6^2} + \cdots \qquad (2.35)$$

$J_0(x_n)$ is a continuous function of x and cuts the $y = 0$ axis at various points known as the roots, x_n:

Root number n	1	2	3	4	5
Roots of $J_0(x_n)$	2.405	5.520	8.654	11.792	14.931
Roots of $J_1(x_n)$	3.832	7.015	10.173	13.324	16.470

The first differential of $J_0(x)$ is $-J_1(x)$, known as a Bessel function of the first kind and of the first order.

Bessel functions may be calculated using the following equations:

$$J_0(x) = A \cos(x - 0.25\pi) \tag{2.36a}$$

$$J_1(x) = A \cos(x - 0.75\pi), \tag{2.36b}$$

where $A = (2/\pi x)^{1/2}$.

2.2.6.4. Duhamel's Theorem

A useful method of dealing with time-dependent boundary conditions, e.g. ramp and exponential functions, is to use Duhamel's theorem to convert the step-response solution to the required solution. This method has been widely used in the solution of canning problems (Riedel 1947; Gillespy 1953; Hayakawa 1964; Hayakawa & Giannoni-Succar 1996).

If the step-function temperature distribution is given by $\Phi(t)$ and the required solution for a time-dependent boundary condition $\Theta(t)$, then Duhamel's theorem states that the relationship between the two is given by

$$\Theta(t) = \int_0^\tau f(t) \cdot \frac{\partial \Phi}{\partial t}(x, t - \tau) d\tau, \tag{2.37}$$

where $f(t)$ is the temperature distribution equation for the time-dependent boundary condition and τ is the time limit for integration.

Duhamel's theorem has also been used in a direct approach to decoupling temperature data from specific boundary conditions, in order to predict data for different experimental conditions. A theoretical inverse superposition solution for the calculation of internal product temperatures in containers in retorts subjected to varying retort temperature profiles (Stoforos *et al.* 1997).

2.2.7. Some Numerical Techniques of Solution

2.2.7.1. Introduction

Many of the mathematical models for heat transfer into cylindrical containers have complex boundary conditions which do not permit simple analytical solutions to be obtained in a form which can easily be manipulated. Consequently, numerical methods have been developed and are now extensively used because of their suitability for modern computing. They require neither the solution of complex transcendental equations nor the functions outlined above. These methods are based on iterative estimations of temperatures using approximate methods. It is not possible to obtain directly solutions that show the interrelationship of the variables, and the solutions are essentially in the form of time-temperature data. In view of the fact that the temperature distributions obtained are used to determine the achieved lethality, it is necessary to check that the method used is of sufficient accuracy to prevent a sub-lethal process being recommended. There is always a possibility of cumulative error.

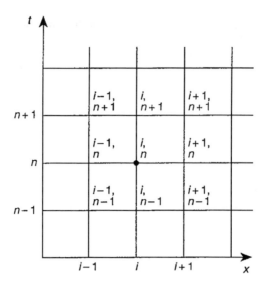

FIGURE 2.3. Temperature nodal points on space–time grid.

2.2.7.2. Finite-Difference Approximation Method

In the finite-difference methods the derivative functions are replaced by approximate values expressed by values of a function at certain discrete points known as "nodal points." The result of this operation is to produce an equivalent finite-difference equation which may be solved by simple algebraic or arithmetic manipulation.

For unsteady-state heat transfer it is necessary to construct a space–time grid (see Figure 2.3) in which the temperatures at the nodal points are defined in terms of time (t) and space (x) coordinates.

Considering the basic one-dimensional heat transfer equation:

$$\frac{1}{\alpha}\frac{\partial T}{\partial t} = \frac{\partial^2 T}{\partial x^2}. \tag{2.38}$$

Using a Taylor series expansion the value of $T(x, t)$ may be expressed as

$$T_{(x,t+\Delta t)} = T_{(x,t)} + \Delta T \cdot \frac{\partial T}{\partial t} + \frac{\Delta T^2}{2} \cdot \frac{\partial^2 T}{\partial t^2} + \frac{\Delta T^3}{3} \cdot \frac{\partial^3 T}{\partial t^3}. \tag{2.39}$$

If the increment is sufficiently small then the terms higher than T may be neglected, thus

$$\frac{\partial T}{\partial t} = \frac{T_{(x,t+\Delta t)} - T_{(x,t)}}{\Delta t} = \frac{T_{i,n+1} - T_{i,n}}{\Delta t}. \tag{2.40}$$

Two series are required for $\partial^2 T/\partial x^2$: these are the expansion of $T_{(x+\Delta x,t)}$ and $T_{(x-\Delta x,t)}$, from which

$$\frac{\partial^2 T}{\partial x^2} = \frac{T_{(x+\Delta x,t)} - 2T_{(x,t)} + T_{(x-\Delta x,t)}}{\Delta x^2} = \frac{T_{i+1,n} - T_{i,n} + T_{i-1n}}{\Delta x^2}. \qquad (2.41)$$

Equation (2.39) then becomes

$$T_{i,n+1} = T_{i,n} + \frac{\alpha \Delta t}{\Delta x^2}[T_{i+1,n} - 2T_{i,n} + T_{i-1,n}]. \qquad (2.42)$$

For a better approximation the term $T_{(x,t+2\Delta T)}$ can be expanded in the same way.

Two important aspects of this method of solution are the convergence and the stability criteria, without which full confidence in the solution cannot be guaranteed. Convergence implies that the finite-difference solution will reduce to the exact solution when the size increments, e.g. Δx and ΔT, are infinitesimally small. Stability implies that the errors associated with the use of increments of finite size, round-off errors and numerical mistakes will not increase as the calculations proceed. Tests for these are given in the standard books on numerical analysis.

This method is usually referred to as the *explicit* method. The time step t has to be kept very small and the method is only valid for $0 < t/\Delta x^2 \leq \frac{1}{2}$, i.e. $\Delta t \leq \frac{1}{2}\Delta x^2$, and Δx has to be kept small in order to obtain reasonable accuracy.

A more suitable, but computationally more demanding, method is the *implicit* method in which the nodal points are replaced by the values taken at $n+1$ rather than n, although this results in three unknown temperatures, and N equations, where N is the number of spatial nodes representing the conductive body. The nodal temperatures are obtained by simultaneous solution of the equations. The basic equation for the relationship between the nodal points is given by

$$T_{i,n+1} = T_{i,n} + \frac{\alpha \Delta t}{\Delta x^2}[T_{i+1,n+1} - 2T_{i,n+1} + T_{i-1,n+1}]. \qquad (2.43)$$

Several other methods, which are intermediate between the explicit and implicit methods, have been developed, e.g. the Crank–Nicholson method, Jacobi method, Gauss–Seidel method and "over-relaxation" methods. Discussions of the application finite-difference methods are available in a large number of text-books: See Smith (1974), Adams and Rogers (1973), Croft and Lilley (1977), Carnahan *et al.* (1969), and Minkowycz *et al.* (1988).

The finite-difference technique has been applied to a wide range of canning problems (see Tables 2.3, 2.4 and 2.5). Tucker and Badley (1990) have developed a commercial center temperature prediction program for heat sterilization processes, known as CTemp.

Welt *et al.* (1997) have developed a no-capacitance surface node NCSN procedure for heat transfer simulation an, which can be used for process design. This method when used with simulation steps of 10 s was found to provide a better fit to the experimental data compared with capacitance surface node technique (CNS).

TABLE 2.3. Some conduction-heating models for predicting temperatures in cylindrical cans of food.

Form of solution	Product/container	Process conditions	Comments	Reference
Analytical equations	Geometrical objects	Linear surface heating & step-change	Any position & center	Williamson and Adams (1919)
Analytical equations	Metal cans & glass jars, fruit & vegetables	Linear surface heating & step-change	Determination of effective thermal diffusivity	Thompson (1919)
Analytical equations	Cans: fruit & vegetables	Variable surface temperature	Duhamel's theorem	Thompson (1922)
Analytical equations	Cans: various fish	Step-change	Thermal properties	Langstroth (1931)
Analytical equations	Cans: fish	Step-change	Based on Langstroth (1931) first-term approximation	Cooper (1937)
Analytical equations	Cans: fish	Step-change		Tani (1938a)
Analytical equations	Cans: fish	Step-change	Cooling	Okada (1940a)
Analytical equations	Cans: food	Step-change	Based on Williamson and Adams (1919)	Taggert and Farrow (1941, 1942)
Tables of numerical values	Geometrical objects	Step-change	Very useful method	Olson and Schultz (1942)
Analytical equations	Cans: food	Step-change & initial temperature distribution	Classical equation for cylindrical can	Olson and Jackson (1942)
Analytical equations	Cans: food	Various heating profiles	Duhamel's theorem	Riedel (1947)
Analytical equations	Cans: food	Step-change & initial temperature distribution	Heating & cooling Duhamel's theorem	Hicks (1951)
Analytical equations with numerical tables	Cans: food	Step-change & variable surface temperature	Duhamel's theorem	Gillespy (1951, 1953)
Analytical equations	Cans: meat	Step-change & initial temperature distribution	Heating & cooling derivation given	Hurwicz and Tischer (1952)
Analytical equations	Cans: food	Step-change & finite-surface heat-transfer coefficient	Effect of headspace on temperature distribution	Evans and Board (1954)
Analytical equations	Cans: food	Step-change	Hyperbolic secant	Jakobsen (1954)

Method	Product	Condition	Comment	Reference
Analytical equations	Cans: food	Variable surface temperature profiles	A major contribution to the theoretical analysis	Hayakawa (1964, 1969, 1970)
Analytical equations	Cans: food	Step-change	f_h/j analysis	Hayakawa and Ball (1968)
Analytical equations	Cans: food	Step-change	Heating & cooling	Hayakawa & Ball (1969a)
Analytical equations	Cans: food	Step-change	Cooling curve	Hayakawa & Ball (1969b)
Numerical solution	Cans: food	Step-change	2D finite-difference equation	Teixeira et al. (1969)
Analytical equations	Cans: food	Step-change	Jakobsen's equation (1954)	Shiga (1970)
Analytical equations	Cans: food	Variable surface temperature profiles	Based on Hayakawa (1964)	Hayakawa and Ball (1971)
Analytical equations	Cans: food	Variable surface temperature profiles	Duhamel's theorem	Hayakawa (1971)
Analytical equations	Cans: food	Step-change	Heating & cooling	Flambert & Deltour (1971, 1973a, b)
Analytical equations	Cans: food	Multiple step-changes	Duhamel's theorem	Wang et al. (1972)
Response charts	Cans: simulant & food	Variable surface temperature	Comparison with Gillespy (1953); mass average temps	Hayakawa (1974)
Analytical equations & response charts	Cans: food	Step-change	Central & average temperatures	Leonhardt (1976a, b)
Analytical & numerical solutions	Cans: food	Step-change	Lethality–Fourier number method	Lenz and Lund (1977)
Analytical equations	Cans: food	Variable surface temperature	f_h/j concept	Ikegami (1978)
Finite-element solution	Cans: model	Surface heat-transfer coefficient	Galerkin residual method of transform used	De Baerdemaeker et al. (1977)
Analytical equations	Cans: food	Step-change, center temperature	Second-order linear system	Skinner (1979)
Analytical equations	Cans: food	Surface heat-transfer coefficient	Simplified equations	Ramaswamy et al. (1982)
Analytical equations	Cans: food	Surface heat-transfer coefficients	Thermocouple errors	Larkin and Steffe (1982)
Finite-element solution	Glass jar: apple sauce, can: salmon	Surface heat-transfer coefficient	Cooling effects	Naveh et al. (1983b)
Analytical equations	Cans: food	Heating & cooling	Lethality–Fourier number method	Lund and Norback (1983)

(cont.)

TABLE 2.3. (continued)

Form of solution	Product/container	Process conditions	Comments	Reference
Analytical equations, computer programs	Geometrical objects	Heating & cooling	Based on Olson and Schultz (1942)	Newman and Holdsworth (1989)
Analytical equations	Cans: food	Surface heat-transfer coefficient	Cooling effects	Datta et al. (1984)
Analytical equations	Cans: food	Non-homogeneous food	Effective thermal diffusivity	Olivares et al. (1986)
Analytical solution	Cans: food	Surface heat-transfer coefficient	Effect of air on the can base coefficient	Tan and Ling (1988)
Finite difference	Cans: food	Surface heat-transfer coefficient	Applicable for process deviations	Mohamed (2003)
Numerical solution	Cans: food	Surface heat-transfer coefficient	Finite-difference model	Richardson and Holdsworth (1989)
Analytical equations, computer programs	Geometrical objects, cans: potato purée	Surface heat-transfer coefficient	Effect of l/d ratio	Thorne (1989)
Finite difference	Cans: simulant	Surface heat-transfer coefficient	Cooling	Tucker & Clark (1989, 1990)
Finite element	Cans: fish/oil	Step-change	Anisotropic model, solid & liquid layer	Perez-Martin et al. (1990)
Finite difference	Plastic: 8% bentonite	Wall resistance	ADI technique	Shin and Bhowmik (1990)
Finite difference	Cans: model	Surface heat-transfer coefficient	Step process	Silva et al. (1992)
Finite element	Cans: fish/oil	Step-change, solid & liquid layer	Anisotropic model	Banga et al. (1993)
Analytical equations	Cans: sea food	Surface heat-transfer coefficient	z-transfer function	Salvadori et al. (1994a, b)
Analytical	Model	Finite surface resistance	3D model	Cuesta and Lamu (1995)
Analytical equations	Cans: mashed potato	Step change	Linear recursive model	Lanoiselle et al. (1995); Chiheb et al. (1994)
Numerical	Cans: fruit/syrup	Variable boundary conditions	z-transfer function	Marquez et al. (1998)
Zone Modeling	Cans: baked beans	Heating & cooling	Uncertain data	Johns (1992a, b)
Zone Modeling	Cans	Heating & cooling	Simple model	Tucker et al. (1992)
Finite element	Cans: tomato concentrate	Surface heat-transfer coefficient	Stochastic boundary conditions	Nicolaï & De Baerdemaeker (1992)
Finite element/Monte Carlo	A1 can: tomato concentrate	Surface heat-transfer coefficient	Parameter fluctuations	Nicolaï & De Baerdemaeker (1997); Nicolaï et al. (1998)
Various	Cans: model	Various	Comparison of techniques	Noronha et al. (1995)

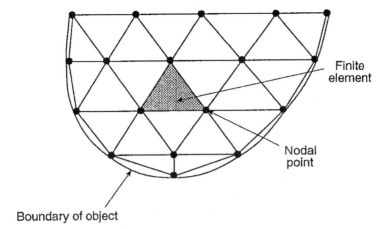

FIGURE 2.4. Discretization of an object into finite elements.

Chen and Ramaswamy (2002a, b, c) have also developed a method of modelling and optimization based neural networks and genetic algorithms.

2.2.7.3. The Finite-Element Method

With this method the body under investigation is divided up into an assembly of subdivisions called *elements* which are interconnected at *nodes* (see Figure 2.4). This stage is called *discretization*. Each element then has an equation governing the transfer of heat, and system equations are developed for the whole assembly. These take the form

$$q = kT \tag{2.44}$$

where k is a square matrix, known as the stiffness or conductance matrix, q is the vector of applied nodal forces, i.e. heat flows, and T is the vector of (unknown) nodal temperatures.

In the case of one-dimensional heat conduction, the heat in is given by

$$q_i = -\frac{kA}{L}(T_j - T_i)$$

and the heat out by

$$q_j = -\frac{kA}{L}(T_i - T_j)$$

which in matrix form is

$$\begin{bmatrix} q_i \\ q_j \end{bmatrix} = \frac{kA}{L} \begin{bmatrix} 1 & -1 \\ -1 & 1 \end{bmatrix} \begin{bmatrix} T_i \\ T_j \end{bmatrix}. \tag{2.45}$$

The stages in developing a finite-element model are as follows:

1. *Discretization*. For two-dimensional solid bodies that are axisymmetrical, triangular or rectangular elements may be used; whereas for three-dimensional
   objects, cubes or prisms may be used. In problems involving curved areas,
   shells of appropriate curved geometry are chosen.
2. *Size and Number of Elements*. These are inversely related: As the number
   of elements increases, the accuracy increases. It is generally more useful to
   have a higher density of mesh elements, especially where temperatures are
   changing rapidly. This requires careful planning at the outset of the analysis.
   It is important that all the nodes are connected at the end of the mesh.
3. *Location of Nodes*. It is essential that where there is a discontinuity in the
   material, the nodes should join the two areas.
4. *Node and Element Numbering*. Two different methods are used, either horizontal or vertical numbering for the nodes; the elements are given numbers in
   brackets.
5. *Method of Solution*. Most finite-element computer programs use wavefront
   analysis; however, the Gaussian elimination technique may also be used. The
   reader is recommended to consult the texts by Segerlind (1984) and Fagan
   (1992) for further information in relation to the application of the method.
   The ANSYS (1988) computer software package is very useful for problem-
   solving.

Relatively few applications of the technique to food processing problems have
been reported. General discussions and overviews have been given by Singh and
Segerlind (1974), De Baerdemaeker *et al.* (1977), Naveh (1982), Naveh *et al.*
(1983), Puri and Anantheswaran (1993), and Nicolai *et al.* (2001).

The method has been applied to the heating of baby foods in glass jars (Naveh
1982) and the heating of irregular-shaped objects, e.g. canned mushrooms in
brine (Sastry *et al.* 1985). The temperature distribution during the cooling of
canned foods has been analyzed by Naveh *et al.* (1983b), and Nicolai and De
Baerdemaeker (1992) have modelled the heat transfer into foods, with stochastic
initial and final boundary conditions. Nicolai *et al.* (1995) have determined the
temperature distribution in lasagne during heating.

2.2.7.4. Some Other Methods

Hendrickx (1988) applied transmission line matrix (TLM) modelling to food
engineering problems. This method, like the finite-difference model, operates on
a mesh structure, but the computation is not directly in terms of approximate field
quantities at the nodes. The method operates on numbers, called pulses, which are
incident upon and reflected from the nodes. The approximate temperatures at the
nodes are expressed in terms of pulses. A pulse is injected into the network and
the response of the system determined at the nodes.

For systems with complex boundary conditions, such as those found in canning
operations, the equations have been solved using the response of the linear system
to a disturbance in the system, in this case a triangular or double ramp pulse.
The solution uses both Laplace transforms and z-transfer functions. The formal

solution of the heat-transfer equation for a finite cylinder with complex boundary conditions was derived by Salvadori *et al.* (1994). Márquez *et al.* (1998) have applied the technique to the study of a particulate/liquid system, viz., pasteurizing fruit in syrup.

2.2.8. Some Analytical Solutions of the Heat Transfer Equation

2.2.8.1. Simple Geometrical Shapes

The simplest cases are the temperature distributions in one dimension for an infinite slab, an infinite cylinder and a sphere. From the first, two more complex solutions can be obtained (see Section 2.2.8.2).

2.2.8.1.1. The Infinite Slab

The one-dimensional flow of heat in a slab is given by

$$\frac{\partial T}{\partial t} = \alpha \frac{\partial^2 T}{\partial x^2} \tag{2.46}$$

(see equation (2.38)). The general form of the solution $F(t)$ is given by

$$T = T_R - (T_R - T_0)F(t), \tag{2.47}$$

where T is the temperature distribution at space coordinate x and time t, often written as $T(x, t)$; T_R is the retort temperature and T_0 is the initial temperature of the solid body, at time $t = 0$.

For a slab of thickness $2X$, the solution at any point is

$$F(t) = \frac{4}{\pi} \sum_{n=1}^{\infty} \frac{(-1)^{n+1}}{2n-1} \cos\left((2n-1)\frac{\pi x}{2X}\right) e^{[-(2n-1)^2\pi^2\alpha t/4X^2]}, \tag{2.48}$$

which, at the center, becomes

$$F(t) = \frac{4}{\pi} \sum_{n=1}^{\infty} \frac{(-1)^{n+1}}{2n-1} e^{[-(2n-1)^2\pi^2\alpha t/4x^2]}. \tag{2.49}$$

It is often conveniently designated by $S(\theta)$, where θ is the dimensionless Fourier number $\alpha t/l^2$ where l is the thickness of the body, i.e. $2X$:

$$S(\theta) = \frac{4}{\pi} \sum_{n=1}^{\infty} \frac{(-1)^{n+1}}{2n-1} e^{[-(2n-1)^2\pi^2\theta]} \tag{2.50}$$

$$= \frac{4}{\pi}\left[e^{-\pi^2\theta} - \frac{1}{3}e^{-9\pi^2\theta} + \frac{1}{5}e^{-25\pi^2\theta} - \cdots\right]. \tag{2.51}$$

This subject is discussed in detail by Ingersoll *et al.* (1953), Olson and Schultz (1942), and Newman and Holdsworth (1989). The latter includes a range of useful computer programs in BASIC.

If the body has an initial temperature distribution $f(\lambda)$, then the temperature distribution is given by

$$T_R - T = \frac{2}{l} \sum_{n=1}^{\infty} (n\pi x/l) \int_0^\lambda f(\lambda) \sin(n\pi\lambda/l) d\lambda \cdot e^{-n^2\pi^2\theta}. \tag{2.52}$$

This equation ultimately reverts to (2.50), when $f(\lambda)$ is a uniform temperature $T_0 - T_R$.

If the body has a surface heat-transfer coefficient, h, then the temperature distribution is given by

$$F(t) = 2 \sum_{n=1}^{\infty} \frac{\sin M_n \cos(M_n(x/X))}{M_n + \sin M_n \cos M_n} e^{-M_n^2\theta} \tag{2.53}$$

and for the center temperature, this becomes

$$F(t) = 2 \sum_{n=1}^{\infty} \frac{\sin M_n}{M_n + \sin M_n \cos M_n} e^{-M_n^2\theta} \tag{2.54}$$

where M_n is obtained from the solution of $M_n = B_i \cot M_n$ and B_i is the Biot number hX/k. This is also known as the Nusselt number in heat-transfer correlations.

2.2.8.1.2. Infinite Cylinder

The basic one-dimensional heat transfer equation is given by

$$\frac{dT}{dt} = \alpha \left[\frac{d^2 T}{dr^2} + \frac{1}{r} \frac{dT}{dr} \right]. \tag{2.55}$$

The solution for a rod of radius $2a$, at any radial point r, for a constant retort temperature T_R and a uniform initial temperature distribution T_0 is given by

$$F(t) = \sum_{i=1}^{n} A_i J_0(R_i r/a) e^{-R_i^2\theta}, \tag{2.56}$$

where $A_n = 2/[R_n J_1(R_n)]$, $J_0(x)$ is a Bessel function of zero order, $J_1(x)$ a Bessel function of the first order (see Section 2.2.6.3), R_n is the nth root of the characteristic equation $J_0(x) = 0$, and $\theta = \alpha t/a^2$ is the dimensionless Fourier number.

For the center point it is possible to define $C(\theta)$, similarly to $S(\theta)$ in equation (2.50):

$$C(\theta) = 2 \sum_{i=1}^{n} A_i e^{-R_i^2\theta}, \tag{2.57}$$

i.e.

$$C(\theta) = 2[A_1 e^{-R_1^2\theta} + A_2 e^{-R_2^2\theta} + A_3 e^{-R_3^3\theta} + \cdots].$$

If the initial temperature distribution at time $t = 0$ is $f(r)$, then the temperature distribution after time t is given by

$$T_R - T = \sum_{i=1}^{n} A_i' J_0(R_i r/a) e^{-R_i^2 \theta} \cdot \int_0^r r f(r) J_0(R_n r/a) dr, \qquad (2.58)$$

where $A_n' = 2/[a^2 J_1(R_n)]$.

If there is a finite surface heat coefficient on the outside, then the temperature distribution is given by

$$F(t) = A_n'' J_0(R_n \cdot r/a) e^{-[R_n^2 \theta]}, \qquad (2.59)$$

where

$$A_n'' = [2 J_1(R_n)/R_n[J_0^2(R_n) + J_1^2(R_n)],$$

$$R_n J_1(R_n) = J_0(R_n) Bi,$$

$$Bi = hr/k.$$

2.2.8.1.3. A Spherical Object

The basic heat transfer equation for determining the temperature distribution in a spherical object of radius a is given by

$$\frac{dT}{dt} = \alpha \left[\frac{d^2 T}{dr^2} + \frac{2}{r} \frac{dT}{dr} \right]. \qquad (2.60)$$

The solution for the simplest case of a step function is given by

$$F(t) = (2a/\pi r) \sum_{n=1}^{\infty} (-1)^{n+1} \sin(n\pi r/a) e^{-n^2 \pi^2 \theta} \cdot \qquad (2.61)$$

where $\theta = \pi t/a^2$.

For the central temperature only the equation reduces to

$$F(t) = 2 \sum_{n=1}^{\infty} (-1)^{n+1} e^{-n^2 \pi^2 \theta}. \qquad (2.62)$$

A function $B(x)$ is defined in terms of $F(t)$ for later use as follows:

$$F(t) = B(x), \qquad (2.63)$$

where $x = \pi^2 \theta$.

For the case of external heat transfer the following equation is applicable:

$$F(t) = \frac{2r}{a} \sum_{n=1}^{\infty} \frac{\sin M_n - M_n \cos M_n}{M_n - \sin M_n - \cos M_n} \cdot \frac{\sin(M_n(r/a))}{M_n} e^{-[M_n^2 \theta]}, \qquad (2.64)$$

where $\tan M_n = M_n/(1 - Bi)$.

2.2.8.2. More Complex Geometries

2.2.8.2.1. Rectangular Parallelepiped or Brick

The simple analytical solution for this case is obtained using the $S(\theta)$ function given in equation (2.50). The temperature distribution is given by

$$F(t) = S(\theta_x)S(\theta_y)S(\theta_z), \tag{2.65}$$

where $S(\theta_n) = \alpha t/n^2$ and n is the overall side dimension. For a cube of side a, the temperature is given by

$$F(t) = [S(\theta)]^3. \tag{2.66}$$

Applications of these formulae to various conditions for heating canned foods in rectangular metallic or plastic containers are presented in Table 2.4.

2.2.8.2.2. Finite Cylinder

The simple analytical solution for a cylinder of radius r and length l is given by

$$F(t) = S(\alpha t/l^2)C(\alpha t/r^2). \tag{2.67}$$

Tables for $S(\theta)$, $C(\theta)$ and $B(x)$ are given in Ingersoll *et al.* (1953), Olson and Schultz (1942), and Newman and Holdsworth (1989).

The complex analytical equations that are required to interpret heat penetration data are as follows. For a can with infinite surface heat-transfer coefficient and uniform initial temperature, radius a and length $2l$ (Cowell & Evans 1961)

$$F(t) = \frac{8}{\pi} \sum_{n=0}^{\infty} \sum_{m=1}^{\infty} \frac{(-1)^n}{2n+1} \cos(2n+1) \frac{\pi x}{2l} \frac{J_0(R_m r/a)}{R_m J_1(R_m)} e^{-\psi \alpha t}, \tag{2.68}$$

where

$$\psi = \frac{(2n+1)^2 \pi^2}{4l^2} + \frac{R_m^2}{a^2}.$$

At the center point

$$F(t) = \frac{8}{\pi} \sum_{n=0}^{\infty} \sum_{m=1}^{\infty} \frac{(-1)^n}{2n+1} \frac{1}{R_m J_1(R_m)} e^{-\psi \alpha t}. \tag{2.69}$$

A slightly different form of this equation is given by Hurwicz and Tischer (1952).

For a can with infinite surface heat-transfer coefficient and initial temperature distribution $f(r, \theta, z)$, radius a and height $2b$, where z is any point on the height axis and r is any point on the radial axis, the equation takes the form given by Ball and Olson (1957), based on Olson and Jackson (1942) (see also Carslaw & Jaeger 1959):

$$F(t) = \sum_{j=1}^{\infty} \sum_{m=1}^{\infty} \sum_{n=1}^{\infty} [A_{jmn} \cos(n\theta) + B_{jmn} \sin(n\theta)]$$

$$\times J_n(R_j r/a) \sin\left(\frac{m\pi}{2b}(z+b)\right) e^{-\psi'\pi t}, \tag{2.70}$$

where

$$\psi' = \frac{m^2\pi^2}{b^2} + \frac{R_j^2}{a^2}$$

and A_{jmn} and B_{jmn} are factors depending upon the initial temperature distribution. For the center point $r = 0$

$$F(t) = A_{110} \sin\left(\frac{\pi(z+b)}{2b}\right) J_0(R_1 r/a) e^{-\psi''\alpha t}, \tag{2.71}$$

where

$$\psi'' = \frac{\pi^2}{4b^2} + \frac{R_1^2}{a^2}$$

and

$$A_{110} = \frac{2}{\pi^2 a^2 l [J_0'(R_n)]^2} \int_0^a R J_0(R_1 r/a) dr \int_{-b}^{+b} \sin\left(\frac{\pi(z+b)}{2b}\right) dz$$

$$\times \int_{-\pi}^{+\pi} (\cos\theta) f(r,\theta,z) d\theta. \tag{2.72}$$

For a can with finite surface heat-transfer coefficient and an initial constant temperature,

$$F(t) = \sum_{n=1}^{\infty} \sum_{m=1}^{\infty} A_{n,1} A_{m,2} J_0(R_{n,1} r/a) \cos(R_{m,2} z/l) e^{-\psi'''\alpha t}, \tag{2.73}$$

where

$$A_{n,1} = \frac{2Bi_1}{J_0(R_{n,1})(R_{n,1}^2 + Bi_1^2)}$$

(for tables, see Luikov 1968: 273),

$$A_{m,2} = (-1)^{n+1} \frac{2Bi_2 \left(Bi_2^2 + R_{m,2}^2\right)^{1/2}}{R_{m,2} \left(Bi_2^2 + Bi_2 + R_{m,2}^2\right)}$$

(for tables, see Luikov 1968: 224), $R_{n,1}$ and $R_{m,2}$ are the roots of the corresponding characteristic equations (for extended tables, see Hayakawa 1975; Peralta Rodriguez 1987b), and

$$\psi''' = \frac{R_{n,1}^2}{a^2} + \frac{R_{m,2}^2}{l^2}.$$

For cans with other boundary conditions, Table 2.4 lists a number of models which have been used to determine the temperature distribution in canned foods with a wide range of boundary conditions. The effect of variable surface temperature profiles, including various combinations of exponential and linear heating profiles

TABLE 2.4. Some conduction-heating models for predicting temperatures in slab- or brick-shaped containers of food.

Form of solution	Product/container	Process conditions	Comments	Reference
Analytical equations	Cans: fish	Finite surface resistance	3D model	Okada (1940c)
Analytical equations	Cans: fish	Thermal conductivity varies with temperature	1D nonlinear model	Fujita (1952)
Analytical equations	Pouches: food	Finite surface resistance	3D model	Chapman and McKernan (1963)
Analytical/numerical	Model	Time-variable boundary conditions	1D model	Mirespassi (1965)
Finite difference	Model	Step-change	3D model	Manson et al. (1970)
Analytical equations	Model	Step-change	3D model	Alles and Cowell (1971)
Analytical equations	Model	Centre & average temperatures	3D Model	Leonhardt (1976a, b)
Analytical equations	Model	Finite surface resistance	1D non-symmetric model	Uno and Hayakawa (1979)
Analytical equations	Pouches: simulant	Step-change	3D model	Castillo et al. (1980)
Analytical equations	Metal brick	Retort profile	3D model	Ramaswamy et al. (1983)
Finite differencer	Pouches: bentonite	Finite surface resistance	3D model	McGinnis (1986)
Finite difference	Pouches: bentonite	Finite surface resistance	3D model	Bhowmik and Tandon (1987)
Finite difference	Cans: food	Finite surface resistance & wall thickness	3D model	Tucker & Holdsworth (1991a, b)
Finite difference	Model	Ramp/hold/step-cool	1D model	Hendrickx et al. (1993)
Analytical equations	Model	Finite surface resistance	z-transfer function	Salvadori et al. (1994)
Finite difference	Pouches: model	Step-change	2- & 3D model	Silva et al. (1994)
Analytical	Model	Finite surface resistance	3D model	Cuesta and Lamua (1995)
Finite element	Lasagne: brick	Finite surface resistance	3D model	Nicolai et al. (1995);
				Nicolai & De Baerdemaeker (1996)

for heating and cooling phases, has been dealt with analytically by Hayakawa (1964). The effect of air in the headspace has been analyzed by Evans and Board (1954).

In many solutions to heat transfer problems it has often been assumed that the solution for a solid body may be reasonably approximated by an infinite body. For example, Olson and Schultz (1942) considered that a length $4\times$ the diameter of a cylinder or $4\times$ the shortest dimension of a square rod would be sufficient to use the infinite geometry approximation. Erdogdu and coworkers have shown that the infinite assumption ratio depends on the Biot number (hd/k) and the above assumptions do not always hold, Turnham and Erdogdu (2003, 2004) and Erdogdu and Turnham (2006).

2.2.8.2.3. Other Geometrical Shapes of Container

A number of other shapes for containers are used, e.g. pear- or oval-shaped cans for fish products and tapered cans for corned beef and fish products. Table 2.5 lists the solutions to the heat transfer equation which have been obtained.

Smith *et al.* (1967) used a generalized temperature distribution model,

$$F(t) = C \, e^{-M^2\theta}, \tag{2.74}$$

where C is a pre-exponential factor and M a general shape modulus given by

$$M^2 = G\pi^2, \tag{2.75}$$

in which G is the geometry index, which takes the value 1.000 for a sphere, 0.836 for a cylinder and 0.750 for a cube. A general formula for G is

$$G = 0.25 + 3C^{-2}/8 + 3E^{-2}/8, \tag{2.76}$$

where $C = R_s/(\pi a^2)$ and $E = R_1/(\pi a^2)$, in which R_s is the smallest cross-sectional area of the body which includes the line segment a, R_1 is the largest cross-sectional area of the body which is orthogonal to R_s. This approach has been discussed in detail by Hayakawa and Villalobos (1989) and Heldman and Singh (1980).

2.2.8.3. Heating and Cooling

While it is important to know the temperature distribution during heating and to establish an adequate *process* on this basis, it is also important to know what contribution the cooling phase makes in order to prevent overheating and to optimize the process for maximum quality retention. For smaller-sized cans the contents of the can reach the processing temperature during the heating period; however, for larger cans there is a temperature distribution within the can at the onset of cooling. In fact the temperature of the center point continues to rise for some time before the effect of the cooling of the outer layers is felt. This being so, it is necessary to know at what stage to commence cooling to achieve a satisfactory process. This can best be done by studying the equations which govern the heating and cooling process. The technique used is to derive

TABLE 2.5. Some conduction-heating models for predicting temperatures in arbitrary-shaped products & containers.

Geometry	Product/container	Process conditions	Comments	Reference
Any shape	Model	Finite surface resistance	Analytical solution	Smith et al. (1967)
Any shape	Model	Series of step-changes, finite surface resistance	Analytical solution	Wang et al. (1972)
Any shape	Model	Finite surface resistance	Analytical solution	Thijssen et al. (1978), Thijssen and Kochen (1980)
Any shape	Model	Finite surface resistance	Analytical solution	Ramaswamy et al. (1983)
Any shape	Model, particulates in cans	Finite surface resistance	Numerical solution	Lekwauwa and Hayakawa (1986)
Any shape	Model	Finite surface resistance	Analytical solution based on Smith et al. (1968)	Hayakawa and Villalobos (1989)
Any shape	Model	Finite surface resistance	Finite element	Akterian and Fikiin (1994)
Any shape	Model	Finite surface resistance	Analytical solution	Cuesta and Lamua (1995)
Cubes	Polycarbonate/heated in water	Step-change	Finite difference	Kim and Teixeira (1997)
Cylinders	Polycarbonate/heated in water	Step-change	Finite difference	Kim and Teixeira (1997)
Ellipsoidal	Sweet potato	Finite surface resistance	ADI: finite difference	Wadsworth and Spirado (1970)
Ellipsoidal	Plastic model, processed ham	Finite surface resistance	Analytical solution	Smith (1966); Smith et al. (1967, 1968), Clary and Nelson (1970), Clary et al. (1971)
Elliptical cross-section	Shrimp	Finite surface resistance	Heat flow lines	Erdogdu et al. (1998a, b)
Oval-shaped	Cans	Step-change	Analytical solution	Iwata (1940)
Oval-shaped	Cans	Step-change	Finite difference	Simpson et al. (1989)
Oval-shaped	Solid shape	Surface heat transfer	volume element	Erdogdu et al. (2001a)
Oval-shaped	Polycarbonate/heated in water	Step-change	Finite difference	Kim and Teixeira (1997)
Pear-shaped	Cans	Step-change	Based on Smith et al. (1966)	Manson et al. (1974)
Mushroom-shaped	Model	Finite surface resistance	Finite difference	Sastry et al. (1985)
Mushroom-shaped	Model	Finite surface resistance	Finite difference	Akterian (1995)
Shrimp-shaped	Shrimp	Finite surface resistance	Finite difference	Chau and Snyder (1988)
Elliptical cross-section	Shrimp	Finite surface resistance	Heat flow lines	Erdogdu et al. (1998a, b)
Elliptical cylinders	Solid shape	Infinite surface resistance	Volume element	Erdogdu et al. (2001a)
Conical shape	Acrylic cones	Finite surface resistance	Volume element	Pornchaloempong et al. (2001, 2003)
Cone frustrums	Fish/pouch	Finite surface resistance	Finite difference	Simpson et al. (2004)
Bowl-shaped, plastic	Apple sauce, bentonite	Finite surface resistance	ADI: finite difference	Sheen et al. (1993)
Irregular shaped particles/liquid	Model	Finite surface resistance	Numerical	Califarno and Zaritzky (1993)

an equation for the cooling period with an initial temperature distribution, and substitute the temperature distribution at the end of heating in this equation. A rigorous derivation of the equation for a cylindrical container has been given by Hurwicz and Tischer (1952).

The heating stage is represented by the equation

$$T = T_R - (T_R - T_0)F(t_h), \tag{2.76a}$$

where t_h is the time for heating, T the temperature at time t at the center (or spatially distributed), T_R the retort temperature, i.e. process temperature, and T_0 the initial temperature of can contents. Heating followed by cooling is given by

$$T = T_R - (T_R - T_0)F(t) + (T_C - T_R)(1 - F(t - t_h)), \tag{2.77}$$

where t is the total heating and cooling time, i.e. $t_h + t_c$, and T_C the temperature of the cooling water, assumed to be constant at the surface of the container. This equation has been used to study the location of the point of slowest heating by various workers: see Hicks (1951), Hayakawa and Ball (1969a), and Flambert and Deltour (1971, 1973a, b). It should be noted that if first-term approximation of the heat-transfer equation for the heating effect is used, it is less reliable to use the same for the cooling period. The first-term approximation is only applicable for determining the temperature after the long heating times and is not applicable to estimating the temperatures during the early stages of the cooling period. The latter require many terms in the summation series of the solution in order to obtain convergence.

In general, and especially for large sized containers, the temperature distribution at the end of heating is not uniform and therefore the condition of a uniform initial temperature does not apply.

2.2.8.4. Computer Programs for Analytical Heat Transfer Calculations

Newman and Holdsworth (1989) presented a number of computer programs in BASIC for determining the temperature distributions in objects of various geometric shapes. These were based on the analytical solutions for the case of infinite surface heat-transfer coefficients and applied essentially to the case of a finite cylinder and a parallelepiped. The programs also calculate the lethality of the process (see Chapter 6). Thorne (1989) extended the range of available computer programs to the case of external heat-transfer coefficients. These programs operated under MS-DOS. A number of computer simulation programs used for engineering applications are given in 6.9.

2.2.9. Heat Transfer in Packaged Foods by Microwave Heating

The applications of microwave heating in the food industry are numerous, and several processes, including tempering and thawing, have been developed commercially (Metaxas & Meredith 1983; Decareau 1985; Decareau & Peterson

1986). However, applications to the pasteurization and sterilization of food products are at present in the early stages of development.

The interaction of microwave energy and food products causes internal heat generation. The rapidly alternating electromagnetic field produces intraparticle collisions in the material, and the translational kinetic energy is converted into heat. For many food products the heating is uneven; the outer layers heat most rapidly, depending on the depth of penetration of the energy, and the heat is subsequently conducted into the body of the food. Current research is concerned with achieving uniform heating, especially in relation to pasteurization and sterilization of foods, where non-uniform heating could result in a failure to achieve a safe process. For materials that are electrical conductors – e.g. metals, which have a very low resistivity – microwave energy is not absorbed but reflected, and heating does not occur. Short-circuiting may result unless the container is suitably designed and positioned. Metallic containers and trays can effectively improve the uniformity of heating (George 1993; George & Campbell 1994). Currently most packages are made of plastic materials which are transparent to microwave energy.

The amount of heat generated in microwave heating depends upon the dielectric properties of the food and the loss factor (see below), which are affected by the food composition, the temperature and the frequency of the microwave energy. For tables of electrical properties of food, and discussion of their application, see Bengtsson and Risman (1971), Ohlsson and Bengtsson (1975), Mudgett (1986a, b), Kent (1987), Ryynänen (1995), and Calay et al. (1995).

Many mathematical models have been developed for microwave heating. The most basic are Maxwell's electromagnetic wave propagation equations. These are difficult to solve for many applications, and a simpler volumetric heating model involving the exponential decrease of the rate of heat generation in the product is used. The basic equation for this model is

$$q = q_0 e^{-x/\delta}, \tag{2.78}$$

where q is volumetric heat generation (W m^{-3}), q_0 heat generated in the surface (W m^{-3}), x the position coordinate in the product, and δ the penetration depth based on the decay of the heating rate.

The general equation for the temperature distribution based on microwave heating and conduction into the product is

$$\partial T / \partial t = \nabla^2 T + q_0 e^{-x/\delta}. \tag{2.79}$$

The rate of heat generation is given by

$$q_0 = \pi f \varepsilon_0 \varepsilon'' |E^2|, \tag{2.80}$$

where f is frequency (Hz), ε_0 the permittivity of free space (F m^{-1}), ε'' the dielectric loss, and $|E^2|$ the root-mean-square value of the electric field (V m^{-1}).

The penetration depth x in the product is obtained from the equation

$$x = \lambda_0 / [2\pi (\varepsilon' \tan \delta)^{1/2}]. \tag{2.81}$$

The term tan δ is the ratio of the dielectric loss ε'' to the dielectric constant ε', and is known variously as the loss tangent, loss factor or dissipation constant. It is listed in tables of physical property data, e.g. Kent (1987).

The main frequency bands used are 2450 and 896 MHz in Europe and 915 MHz in the USA. Greater penetration and more uniform heating are obtained at the longer wavelengths for food products with low loss factors.

Datta and Liu (1992) have compared microwave and conventional heating of foods and concluded that microwave heating is not always the most effective method, especially for nutrient preservation. The effect depends on a variety of properties of the system.

Burfoot *et al.* (1988) examined the microwave pasteurization of prepared meals using a continuous tunnel device. The product was heated to 80–85°C for a few minutes, sufficient to inactivate vegetative pathogenic bacteria, e.g. Salmonella and Campylobacter, but not bacterial spores. The latter are controlled by storing the product below 10°C. This type of product is not shelf-stable at room temperature and a full sterilization process would be necessary with low-acid products of this type to obtain a stable product. Microwave tunnels for this purpose would have to be pressurized to maintain the integrity of the package when sterilizing temperatures (121°C) had been achieved. A general-purpose plant known as Multitherm has been developed by AlfaStar Ab, Tumba, Sweden (Hallström *et al.* 1988). Burfoot *et al.* (1996) have modeled the pasteurization of simulated prepared meals in plastic trays with microwaves. Large differences between actual and predicted temperatures were found at some points.

For measuring temperatures in microwave systems an invasive fibre-optic probe has been developed, which uses the change in color with temperature of a crystal situated at the end of a glass fibre. Fluoroptic probes are manufactured by Luxtron Corp., CA, USA.

2.2.10. Dielectric Heating

Dielectric heating is performed at radio frequencies of 13.56. 27.12. or 40.68 MHz. This distinguishes from microwave heating, which uses $\cong$ 900 or 2450 MHz. (Rowley 2001). Wang *et al.* (2004) have emphasized the advantage of the volumetric heating characteristics of dielectric heating, especially for heating sealed polymeric and pouches. In a study on the heating of 6 lb. capacity polymeric trays filled with whey protein gels, the production of M-1 (see §6) was used as a chemical marker to determine the degree of lethality and cooking value achieve by RF dielectric heating and steam.

2.3. Heat Transfer by Convection

2.3.1. Introduction

Convective heat transfer inside containers results either from the natural effects of changes in density in the liquid induced by changes in temperature at the container walls (free or natural convection) or by creating motion in the container contents by axial or end-over-end rotation (forced convection).

The process of natural convection initially involves heat transferred by conduction into the outer layers of fluid adjacent to the heated wall; this results in a decrease in the density, and the heated fluid layer rises. When it reaches the top of the liquid at the headspace, the induced fluid motion causes it to fall in the central core, the displaced hot fluid at the wall being replaced by colder fluid in the core. As the temperature of the can contents becomes more uniform and the driving force smaller, the fluid velocity tends to decrease and eventually, when the fluid becomes uniformly heated, the motion ceases.

The mechanism can also be represented by an initially thin boundary layer of liquid in which the fluid is rising, the thickness of which increases as the process continues. According to experimental investigations by Datta and Teixeira (1988), after 30 seconds heating the thickness of this layer was 2 mm and the temperature drop from the can wall to the boundary layer edge was from 121 to 45°C.

With more viscous fluids the process is somewhat slower, although the viscosity and the density are, in general, reduced by a rise in temperature. There will also be an increase in the conductive component with thicker fluids. Ball and Olson (1957) devised a qualitative method of determining the intensity of convection currents in containers based on the ratio of the conductive to convective heat transfer components. This situation should not be confused with the following phenomenon.

For both pure convection and pure conduction heating in canned foods the "heating curve" plot of the logarithm of the reduced temperature against time is more or less linear, after an initial lag period (see Chapter 5). However, for some products the plot shows two lines with differing slope, known as a "broken heating curve". Products showing this type of effect usually start as convective heating packs, but due to physico-chemical changes, e.g. starch gelation, they finally heat by a highly conductive mechanism.

Convection currents inside a container can be visualized, in an idealized form, as shown in Figure 2.5. In the upper part of the container the hot liquid is being pumped by the heated fluid rising in the boundary layer, and being placed on the cold liquid in the core. Simultaneously, in the case of a vertical container, there is also heat rising from the bottom end of the can, which produces mixing eddies in the bulk of the fluid. Datta (1985) has shown that as result of instabilities in the temperature distribution, regular bursts of hot liquid occur on the base of the container. This phenomenon is known as Bernard convection.

The mechanism of unsteady-state convection is very complex and varies with time of heating and/or cooling; consequently it is very difficult to model precisely. Hiddink (1975) used a flow visualizing technique with metallic powders in liquids in containers with light-transmitting walls, to highlight the streamlines in the bulk of the fluid. Other workers, e.g. Blaisdell (1963), have used thermocouples at several points in the containers to plot the temperature profiles.

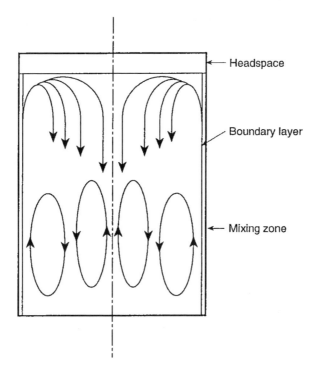

FIGURE 2.5. Idealized representation of convection currents in a can of liquid.

Guerrei (1993) has discussed the internal flow patterns in convection-heating bottles of cylindrical and square shape, using the concept of a double-layer system of rising hot liquid on the wall of the container which discharge into a central volume. The point-of-slowest-heating was shown to be 0.006 m from the wall of an 0.06 m internal diameter container.

The most important point, from a practical safety point of view, is where the slowest heating point is situated. This subject is dealt with in Chapter 5.

While the discussion has been concerned with convection heating/cooling inside containers, there is also the problem of convection heating on the outside, from the heating or cooling medium to the container wall. In the case of pure steam heating, condensation of steam on the container wall surface raises the temperature of the surface almost immediately to that of the steam and, consequently, no problem arises. If, however, the steam contains air, either as an adulterant or intentionally, then the velocity of the mixture over the surface will affect the temperature of the surface and a more complex situation will arise. Similarly, if water is used for heating, and also for cooling, then the wall temperature will be affected by the velocity of the heat transfer medium. In all cases, except for condensing pure steam, it is necessary to consider external convective heat transfer to the outer surface of the container.

2.3.2. Basic Concepts in Convection Heat Transfer

Mathematical models for the prediction of temperatures in the heating of canned foods by convection are necessary in order to determine the process requirements. However, they are much less easy to obtain than in the case of heating by conduction. There are three approaches to convective heat transfer: the film theory; the use of dimensionless numbers; and the more rigorous mathematical treatment of the basic fluid dynamic and heat transfer models. Reviews of this subject have been presented by Ball and Olson (1957), Holdsworth and Overington (1975), Rao and Anantheswaran (1988), as well as in various theses, e.g. Blaisdell (1963), Stevens (1972), Hiddink (1975), and Zechman (1983).

2.3.2.1. Film Theory

The basic heat transfer equation for convection is

$$Q = h_s A(T_b - T_s), \tag{2.82}$$

where Q is the quantity of heat flowing (J s^{-1}), h_s is the surface heat-transfer coefficient (W m^{-2} K^{-1}), A is the surface area (m^2), T_b is the bulk fluid temperature (K) and T_s is the surface temperature.

The idealized temperature profile for a material being heated by a fluid and separated by a container wall is shown in Figure 2.6. On either side of the wall boundary layers or films may be visualized, through which the bulk of the temperature drop takes place. Thus if T_1 is the temperature of the heating medium and T_4 is the temperature in the bulk of the fluid being heated, and the

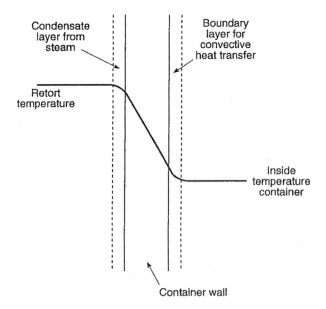

FIGURE 2.6. Temperature profile across container wall.

average film temperatures are T_2 and T_3, then an overall heat-transfer coefficient U (alternatively denoted by H) can be defined as follows

$$Q = UA(T_1 - T_4), \tag{2.83}$$

where

$$\frac{1}{U} = \frac{1}{h_s} + \frac{x_w}{k_w} + \frac{1}{h_b}, \tag{2.84}$$

and h_s and h_b are the heat-transfer coefficients for the surface being heated and the bulk of the fluid, x_w is the thickness of the wall (m), and k_w is the thermal conductivity of the wall material (W m^{-1}K^{-1}).

Some typical values for h_s (W m^{-2} K^{-1}) are: steam, 12 000; steam + 3% air, 3500; steam + 6% air, 1200; cold water, 500; air moving at 3 m s^{-1}, 30; still air, 6. From these values it can be seen that the resistance to heat transfer, the reciprocal of the heat transfer coefficent, is extremely small for condensing steam; consequently this can be ignored in all heat transfer calculations.

Some values for the thermal conductivities of container materials (W m^{-1} K^{-1}) are: ferrous metal containers, 40–400; aluminum, 220; glass, 1–2; poly-vinylchloride, 0.29; polyethylene, 0.55; water, 0.57. From these values it can be seen that it may be necessary, when using container materials other than metals, to consider the x_w/k_w term in the overall heat transfer equation.

2.3.2.2. Correlations for Predicting Heat-Transfer Coefficients

Engineering practice makes use of dimensionless numbers for correlating heat-transfer coefficients with the physical circumstances of heat transfer and the physical properties and flow conditions of the fluids involved. The four dimensionless numbers used in heat transfer studies are:

(a) Reynolds number, used for flow characterization,

$$Re = vd\rho/\mu;$$

(b) Nusselt number, ratio of the heat transferred by convection to that transferred by conduction,

$$Nu = hd/k;$$

(c) Prandtl number, influence of physical properties

$$Pr = c\rho/k;$$

(d) Grashof number, influence of buoyancy forces in natural convection

$$Gr = d^3\rho^2\beta g\Delta T/\mu^2;$$

(e) Rayleigh number, product of Grashof and Prandtl numbers

$$Ra = Gr \cdot Pr$$

where

c = specific heat at constant pressure (J kg^{-1} K^{-1});

d = characteristic linear dimension – thickness, length or diameter, as appropriate (m);

g = gravitation constant (m s^{-2});

h = heat-transfer coefficient (W m^{-2} K^{-1});

k = thermal conductivity (W m^{-1} K^{-1});

T = temperature difference (K);

v = velocity (m s^{-1});

β = coefficient of thermal expansion (K^{-1});

μ = dynamic viscosity (Pa s);

ρ = density (kg m^{-3}).

For example, for forced convection we have the correlation

$$Nu = A(Re)^a (Pr)^b.$$

For gas flow across cylinders, with Re in the range 10^3–10^5, values of $A = 0.26$, $a = 0.60$ and $b = 0.30$ have been obtained. For liquids, with Re between 4×10^3 and 4×10^4, we have $A = 0.193$, $a = 0.618$ and $b = 0.333$.

For natural convection, the relationship is

$$Nu = A(Pr)^c (Gr)^d.$$

For horizontal or vertical cylinders with streamline flow and $Re < 2100$, values of $A = 0.47$ and $c = d = 0.25$ have been obtained. With turbulent flow and $Re > 10000$, we have $A = 0.10$, and $c = d = 0.33$. All the fluid properties are measured at the mean film temperature $\frac{1}{2}(T_w + T_b)$, where the subscripts b and w to the temperature refer to the bulk fluid and the wall respectively.

Several other dimensionless numbers are encountered in heat transfer studies and these will be defined where appropriate.

Since the numerical values for the indices and coefficients are determined from experimental results, it is unwise to use dimensionless correlations outside the ranges for which the experimental results have been obtained. Extrapolation of the data requires careful consideration.

For methods of deriving dimensionless correlations and for further information on different applications, the standard engineering texts should be consulted, e.g. Burmeister (1983), Coulson and Richardson (1984), Hallström et al. (1988), Jaluria (1980), and Perry and Green (1984).

2.3.3. Models for Convection Heat Transfer

Some of the models that have been used to predict temperature distributions and velocity profiles in heated and cooled can liquid products are shown in Tables 2.6–2.9.

The models may be classified as follows:

2.3.3.1. Energy Balance Model

This is the simplest of the models; it was first proposed by Jones (1931) and subsequently used by many other workers. It is often known as the Schultz–Olson model in recognition of the contribution of the American Can Company workers (Schultz & Olson 1938).

By equating the overall rate of heat transfer into the can with the rate of accumulation of heat inside the can, an energy balance equation can be established. For the heat transfer into the can,

$$Q = UA(T_s - T_m), \tag{2.85}$$

and the rate of accumulation of heat

$$Q = mc\frac{dT}{dt}, \tag{2.86}$$

where

Q = rate of heat transfer (W);
U = overall heat-transfer coefficient (W m^{-2} K^{-1});
A = surface area (m^2);
T_s = temperature of the heating medium (K);
T_m = temperature of the product (mass average) (K);
c = specific heat (J kg^{-1} K^{-1});
t = time (s);
m = mass (kg).

Equating the left-hand sides of (2.85) and rearranging we obtain:

$$\frac{mc}{A}\frac{dT}{T_s - T_m} = U\,dt. \tag{2.87}$$

Integrating,

$$\int_{T_0}^{T_m} \frac{mc}{A}\frac{dT}{T_s - T_m} = \int_0^t U\,dt. \tag{2.88}$$

leads to

$$\log_e \frac{(T_s - T_m)}{(T_s - T_0)} = \frac{UA}{mc}t \tag{2.89}$$

or

$$T_m = T_s - (T_s - T_0)\,e^{-[(UA/mc)]t} \tag{2.90}$$

This equation shows that the rate of heating is an exponential function, which depends on the overall heat-transfer coefficient U (or the internal heat-transfer coefficient h_{int} when steam is used as the heating medium with metallic cans), the surface area A, the mass of the contents and their specific heat, as well as the temperature of the heating medium and the initial temperature of the contents T_0.

This approach has been modified by Videv (1972) and applied by Akterian (1995) to a range of canned and bottled foods heating by convection. Their approach uses the concept of thermal inertia E for the term mc_p/UA. This characterizes the temperature lag between the liquid temperature and the heating medium. The theoretical treatment has been extended to the case of an external linear heating profile (Videv 1972). The variation in heat transfer rate into canned products has been discussed theoretically by Dincer et al. (1993).

2.3.3.1.1. Slow Come-Up Time with Perfect Mixing

Equation (2.91) has been derived from equation (2.87), solving an ordinary differential equation and assuming a linear retort temperature profile (i.e. simulating temperature profile during come-up time) (Simpson et al. 2003).

$$T_m = a\left(1 - \exp\left(-\frac{UA}{mc}t\right)\right) + b\left(t - \frac{mc}{UA}\right) + b\frac{mc}{UA}\exp\left(-\frac{UA}{mc}t\right)$$

$$+ T_0 \exp\left(-\frac{UA}{mc}t\right) \tag{2.91}$$

Where retort temperature is time dependent and expressed as: $T_S(t) = a + bt$ and equation (2.91) is valid for: $0 < t \le CUT$. For $t > CUT$, temperature T (or T_m) can be expressed by equation (2.89) using an appropriate initial temperature (constant T_S).

Provided that f_h is defined as $\ln 10 \times [MCp/UA]$ (Merson et al. 1978), equation (2.91) can be re-arranged and expressed as:

$$T_m = (a + bt) - b\frac{f_h}{\ln 10} + \left(-a + b\frac{f_h}{\ln 10} + T_0\right)\exp\frac{-t \ln 10}{f_h} \tag{2.92}$$

Further working on equation (2.92) renders:

$$\frac{T_m - (a + bt) + b\frac{f_h}{\ln 10}}{\left(-a + b\frac{f_h}{\ln 10} + T_0\right)} = \exp\frac{-t \ln 10}{f_h} \tag{2.93}$$

From equation (2.93), the dimensionless temperature ratio can be expressed as:

$$\frac{T_m - (a + bt) + b\left(\frac{f_h}{\ln 10}\right)}{T_0 - a + b\left(\frac{f_h}{\ln 10}\right)} = \frac{T'_m - (a' + b't) + b'\left(\frac{f_h}{\ln 10}\right)}{T'_0 - a' + b'\left(\frac{f_h}{\ln 10}\right)} = \text{Constant} \tag{2.94}$$

2.3.3.2. Effective Thermal Diffusivity Model

This model, first discussed by Thompson (1919), makes use of the unsteady-state conduction model solutions and an apparent or effective thermal diffusivity. Thompson found that this depended on the ratio of solids to liquid in the container. Tani (1938b) also used this approach to studying the temperature profiles in heated 1 lb tall cans filled with water. More recently, Teixeira et al. (1992) have used a similar method with the j-value concept (see chapter 5). Early Japanese work was discussed by Olson (1947).

2.3.3.3. Transport Equation Model

This is the most rigorous approach to determining the temperature distributions and the velocity profiles in containers filled with liquids. The early work in this area was reviewed by Ede (1967) and particularly useful work was reported by Evans and Stefany (1966) and Evans *et al.* (1968). The form of the equations depends on the geometry of the container; care should be taken to use the most appropriate coordinate system (Bird *et al.* 1960; Ruckenstein 1971).

The equations which have to be solved in relation to the container boundaries are the equation of continuity:

$$\frac{1}{r}\frac{d}{dr}(r\rho v) + \frac{d}{dz}(\rho u) = 0; \tag{2.95}$$

the equation of momentum in the radial direction:

$$\rho\left(\frac{du}{dt} + v\frac{du}{dr} + u\frac{du}{dz}\right) = -\frac{dP}{dz} + \mu\left\{\frac{1}{r}\frac{d}{dr}\left(r\frac{d\mu}{dr}\right) + \frac{d^2\mu}{dz^2}\right\} + \rho g; \tag{2.96}$$

the equation of momentum in the radial direction:

$$\rho\left(\frac{dv}{dt} + v\frac{dv}{dr} + u\frac{dv}{dz}\right) = -\frac{dP}{dr} + \mu\left\{\frac{d}{dr}\left(\frac{1}{r}\frac{d}{dr}(rv)\right) + \frac{d^2v}{dz^2}\right\}; \tag{2.97}$$

and the energy equation to determine the temperature distribution:

$$\frac{dT}{dt} + v\frac{dT}{dr} + u\frac{dT}{dz} = \frac{k}{\rho c_p}\frac{1}{r}\frac{d}{dr}\left(r\frac{dT}{dr}\right) + \frac{d^2T}{dz^2}. \tag{2.98}$$

The boundary conditions required for the solution are, for the side wall,

$$r = R, \quad 0 < z < H, \quad T = T_1, \quad u = 0, \quad v = 0;$$

for the center line,

$$r = 0, \quad 0 < z < H, \quad dT/dr = 0, \quad du/dr = 0;$$

for the bottom wall,

$$0 < r < R, \quad z = 0, \quad T = T_1, \quad u = 0, \quad v = 0;$$

and for the top wall,

$$0 < r < R, \quad z = H, \quad T = T_1, \quad u = 0, \quad v = 0.$$

The initial conditions are that the fluid is at rest and at a uniform temperature:

$$0 < r < R, \quad 0 < z < H, \quad T = T_0, \quad u = 0, \quad v = 0.$$

Numerical solutions for these equations have been attempted (see Table 2.6). The work of Datta (1985) and Datta and Teixeira (1987, 1988) represents the most advanced approach at the present time.

2.3.4. Some Experimental Work and Correlations

Here some of the experimental work to determine internal heat-transfer coefficients will be considered. Both homogeneous and heterogenous products heated in static and rotary retorts will be mentioned. Tables 2.6 to 2.9 give further references to work in this area.

2.3.4.1. Homogeneous Products Heated in Batch Systems

Much of the early work on heat penetration into cans of liquid products was aimed at determining the time – temperature profiles. Jones (1931) was one of the first workers to apply a simple model to determine temperature profiles in canned liquid products. Blaisdell (1963) made an extensive survey of the literature and carried out experimental work on the heating of a 50% sucrose solution in glass jars of varying sizes. The results were correlated using the heat penetration parameters f_h and j (see Chapter 5). Evans and Stefany (1966), using a range of liquids, obtained a simple correlation for predicting internal heat-transfer coefficients,

$$Nu = 0.55(Gr \cdot Pr)^{0.25},$$

using the length of the container as the characteristic length. Hiddink (1975), in a similar study, obtained a correlation

$$Nu = 0.31(Gr \cdot Pr)^{0.268}.$$

Verboven et al. (2004) have described various applications of CFD to optimizing thermal processes and have given details of the techniques used in various applications.

Further work is listed in Table 2.6.

2.3.4.2. Homogenous Systems Heated with Agitation

Many products are now processed in retorts (see Chapter 8) that impart either axial or end-over-end motion to the container contents. By choosing a suitable headspace it is possible to increase the rate of heat transfer and thereby reduce the time required for processing. The success of this type of process depends on the suitability of the product and careful control of the headspace during the filling operation. Table 2.7 gives some examples of the types of correlation which have been applied to the estimation of heat-transfer coefficients. Some important studies have been carried out, from which quantitative correlations for heat-transfer coefficients have been obtained. Some of these will now be discussed briefly; however, other matters relating to external heat-transfer coefficients from the heating medium will be dealt with in Chapter 8.

TABLE 2.6. Some studies on convection heat transfer for non-agitated homogeneous canned foods.

Product	Process	Model	Comments	Reference
Water	100°C: No. 2 & No. 3 cans	Energy balance	Film model	Jones (1931)
Water	Cooling: No. 2 & No. 3 cans	Energy balance	Newton's law	Jones (1932)
Water	100°C: 1 lb tall glass	Conduction (modified)	Apparent thermal diffusivity	Tani (1938b)
Water	100°C: 1 lb tall glass	Navier–Stokes equation	Dimensionless; data from Tani (1938)	Okada (1940b)
Water/aluminum powder	100°C: 1 lb glass, two sizes	Energy balance	Flow visualization	Ban and Kaziwara (1941)
Ideal fluid ($Pr = 1$)	Heating & cooling	Navier–Stokes equation	Analytical solution	Hammitt and Chu (1962)
Water/glycerol, butanol, methanol	100°C: 2″ & 2.5″ d cylinders	Energy balance	Dimensionless groups, heat transfer coefficients	Evans and Stefany (1966)
Water	Heating: closed container	Semi-empirical model	Turbulent free convection	Tatom and Carlson (1966)
Ideal fluid	Heating: closed container	Navier–Stokes equation	Laminar natural convection	Barakat and Clark (1966)
Water & glycerol	Heating: cylinder	Energy balance	Mixing/main core study	Evans et al. (1968)
Water/sucrose solution	< 83°C: 8–22 oz glass jars	Navier–Stokes equation	Dimensionless groups, flow visualization	Blaisdell (1963)
Ethylene glycol	100°C: 401 × 411 can	Navier–Stokes equation	Velocity & temperature profiles	Stevens (1972)
Water & sucrose solution	100°C: 300 × 410 can	Energy balance	Rayleigh nos v. temperature	Jowitt and Mynott (1974)
Water & siliconc fluid	115°C: half cylinder	Energy & stratification	Flow visualization	Hiddink (1975); Hiddink et al. (1976)
Water	100°C: 3 kg can 156 × 150 mm	Energy balance	Temperature distribution	Miglioli et al. (1983)
Beer	100°C: bottle	Finite-element model (CFD)	Temperature/velocity profiles	Engleman and Sani (1983)
Water	121°C: 303 × 406 can	Navier–Stokes equation	Temperature/velocity profiles	Datta & Teixeira (1987, 1988)
Water, xanthan gum, oils	120°C: 300 × 401 can	Conduction (modified)	Apparent thermal diffusivity	Bera (1988)
Water/CMC	121°C: 81 × 111 mm can	Finite volume model (CFD)	Temperature distributions	Abdul Ghani et al. (1999a, b)
Liquid	121°C: 4 × 12 × 22 cm pouch	Finite volume model (CFD)	Temperature distributions	Abdul Ghani et al. (2002a)
Starch/water	121°C: 303 × 406 can	Fluid dynamics analysis	Viscosity study of gelation	Yang and Rao (1998)
3.5% corn starch/water	121°C: 303 × 406 can	2D-finite element	Energy balance	Tattiyakul et al. (2001, 2002a)
3.5% starch/water	110–121°C: 162 × 111 mm can	2D-finite element	Energy balance	Tattiyakul et al. (2002b)

(cont.)

TABLE 2.6. (Continued)

Product	Process	Model	Comments	Reference
5% waxy maize starch/water	110–121°C: 162 × 111 mm can	2D-finite element	Energy balance	Tattiyakul et al. (2002b)
5.6° Brix tomato juice	110–121°C: 162 × 111 mm can	2D-finite element	Energy balance	Tattiyakul et al. (2002b)
5% Tomato paste solids	121°C: cans	Distributed parameter model	Energy balance	Bichier et al. (1995)
Carrot/orange soup	121°C: can	Finite volume model (CFD)	Velocity/temperature profiles	Abdul Ghani et al. (2002a)
Carrot/orange soup	121°C: 4 × 12 × 22 cm pouch	Finite volume model (CFD)	Temperature distributions	Abdul Ghani et al. (2001)
Carrot/orange soup	121°C: 72 × 106 mm can	CFD PHOENICS	Temperature/flow profile	Ghani et al. (2001); Abdul Ghani et al. (2002b)
Tomato dip product	Holding 90°C & cooling 70, 50, 20°C: 690 gm. jars	Finite difference	Temperature distribution	Plazl et al. (2006)
Non-Newtonian simulant	121°C: 303 × 406 can	Finite-element model (CFD)	Temperature/velocity profiles	Kumar et al. (1990) Kumar and Battacharya (1991)
Non-newtonian, CMC/water	125°C: 75 × 115 mm	CFD; CFX-4	Transient natural convection	Quarini and Scott (1997)
CMC. 0.85% w/w	121°C: 307 × 408 & conical can	Finite element model (CFD)	Temperature/velocity profiles	Varma and Kannan (2006)

TABLE 2.7. Some studies on convection heat transfer for agitated homogeneous canned foods.

Product	Process	Model	Comments	Reference
Sucrose solution, juices & fruit purées	Spin-cooker/cooler 98°C: axial rotation 300 × 406; 401 × 411; 604 × 614 cans	Energy balance	Heat transfer coefficients	Quast and Siozawa (1974)
Silicone oil	Laboratory device 120°C: axial & end-over-end	Energy balance	Heat-transfer coefficients – duration of a transfer unit	Bimbenet and Michiels (1974)
Water & sucrose solution	FMC Steritort 121°C: axial rotation 303 × 406; 608 × 700	Energy balance	Temperature profiles, heat-transfer coefficients	Lenz and Lund (1978)
Aqueous & organic liquids	Hydrolock pilot sterilizer 130°C: end-over-end 54 × 90; 73 × 52; 73 × 107; 99.4 × 111.2 mm cans	Energy balance	Heat-transfer coefficients, surface tension consideration	Duquenoy (1980); Duquenoy (1984)
Glucose syrup 84° brix	Stork simulator; various modes 125°C: 105 × 112 mm can	Energy balance modified for process deviations	Heat-transfer coefficients	Naveh and Kopelman (1980)
Water, glycerol & sucrose solutions 30, 50 & 60%	Agitating sterilizer 100°C: 303 × 406 can; end-over-end	Energy balance	Heat-transfer coefficients, dimensionless analysis	Anantheswaran & Rao (1985a)
3.5% corn starch/water	FMC Steritort. Intermittent agitation 121°C: 303 × 406 can	Energy balance	2D-finite element	Tattiyakul et al. (2001, 2002a)
3.5% starch/water	FMC Steritort: 110–121°C: 162 × 111 mm can	Energy balance	2D-finite element	Tattiyakul et al. (2002b)
5% waxy maize starch/water	FMC Steritort: 110–121°C: 162 × 111 mm can	Energy balance	2D-finite element	Tattiyakul et al. (2002b)
5.6° Brix tomato juice	FMC Steritort: 110–121°C: 162 × 111 mm can	Energy balance	2D-finite element	Tattiyakul et al. (2002b)
5% Tomato paste solids	FMC Steritort: 5 rpm: 121°C: cans	Energy balance	Distributed parameter model	Bichier et al. (1995)
Aqueous solution guar gum, various conc.	Agitating sterilizer 100°C: 303 × 406 can; end-over-end	Energy balance	Heat-transfer coefficients, dimensionless analysis	Anantheswaran & Rao (1985b)
Water, silicone oil	Rotary simulator 100°C: axial rotation 303 × 406; 401 × 411; 404 × 700 cans	Energy balance	Heat-transfer coefficients	Soule and Merson (1985)
Aqueous solution guar gum	Rotary device/batch retort 121°C; end-over-end; 303 × 406 can	Energy balance	Heat-transfer coefficients	Price and Bhowmik (1994)

Naveh and Kopelman (1980) studied a variety of rotational modes, based on end-over-end and axial rotation modes, some in off-set positions, using 105×112 mm containers of glucose syrup $84°$ brix. A modified energy balance model was used to determine the overall heat-transfer coefficient, using a modified energy balance model, as follows:

$$U = \frac{1}{n} \sum_{i=1}^{n} \frac{mc_p}{At_i} \log \left(\frac{T_R - T}{T_R - T_0} \right)_i \qquad (2.99)$$

where U is the average momentary overall heat-transfer coefficient, taken at intervals of time t_i, m is the mass of can and contents, c_p is the specific heat of the product, A is the surface area for heat transfer, T is the temperature of the food, T_R is the retort temperature and T_0 is the initial temperature of the product.

Duquenoy (1980, 1984) determined heat-transfer coefficients for cans containing a range of aqueous and organic liquids agitated in end-over-end mode. The correlation obtained included a dependence on surface tension of the liquid as well as filling ratio.

Anantheswaran and Rao (1985a) determined heat-transfer coefficients for end-over-end can rotation for a range of Newtonian fluids in 303×406 cans. They obtained the correlation

$$Nu = 2.9 Re^{0.436} Pr^{0.287}, \qquad (2.100)$$

where $Nu = h_i d_r / k$, $Re = d_r^2 N \rho / \mu$ (rotational), and $Pr = \mu / k$, in which d_r is the diameter of rotation, and N the speed of rotation (s^{-1}).

The correlation was not improved by trying to incorporate natural convection effects, length to diameter ratio, or headspace volume. The arithmetic average heat-transfer coefficient was determined from instantaneous heat-transfer coefficients by integration of the basic energy equation.

Subsequently these workers (Anantheswaran & Rao 1985b) reported experimental work using non-Newtonian fluids (guar gum of various concentrations) and obtained a general correlation:

$$Nu = 1.41 GRe^{0.482} GPr^{0.355}, \qquad (2.101)$$

where the generalized Reynolds GRe and Prandtl GPr numbers were obtained using the non-Newtonian apparent viscosity term $(8/N)^{n-1} K[(3n + 1)4n]^n$ in place of the viscosity term. N is the rotational speed (s^{-1}). It was found that other factors, e.g. can length to diameter ratio and the flow index, had no statistically significant influence on the heat-transfer coefficient.

Soule and Merson (1985) obtained an overall heat transfer correlation for axial rotation of cans filled with water and silicone oils of varying viscosity. They found a dependency on both the can dimension ratio L/D and the viscosity ratio μ_b/μ_w, where the subscripts b and w refer to the viscosities evaluated at the average bulk temperature and can wall temperature, respectively, taken over the whole process. Their correlation is:

$$Nu = 0.434 Re^{0.571} Pr^{0.278} (L/D)^{0.3565} (\mu_b/\mu_w)^{0.154}. \qquad (2.102)$$

The average internal heat-transfer coefficient, h_i, was estimated from the overall coefficient, U, from the following equation:

$$h_i = (1.07 \pm 0.04)U. \tag{2.103}$$

Datta and Teixeira (1987, 1988) solved the basic fluid flow and heat transfer equations numerically. The solution compared well, both qualitatively and quantitatively, with the available experimental data. The fluid flow patterns revealed liquid rising at the wall because of the buoyancy of the heated boundary layer, radial flow and mixing at the top of the liquid and uniform core flow near the axis.

Kumar *et al.* (1990) used a finite-element technique and computational fluid dynamics to study the flow of a thick viscous liquid heated in cans. The liquid had non-Newtonian flow characteristics and obeyed a simple power law model. A flow model with the viscosity dependent on temperature was used.

Further examples are referred to in Table 2.7

2.3.4.3. Heterogeneous Products Heated in Batch Retorts

To estimate the ratio of conduction/convection heat transfer within the container, the following equation has been proposed (Ball & Olson 1957):

$$K_{va} = \frac{K_v f_h - f_{h'}}{f_h K_v - f_h K_a} \tag{2.104}$$

Where:

K_{va}: Conduction-convection heat transfer index (dimensionless)
K_v: Can conduction index (dimensionless)
K_a: Can convection index (dimensionless)
f_h: slope index of the first can (min)
$f_{h'}$: slope index of the second can (min)

According to equation (2.104), when K_{va} index tends to 1 the heat transfer is mainly carried out by conduction. On the other hand, if K_{va} index tends to 0 the main heat transfer mechanism is convection.

2.3.4.4. Heterogenous Products Heated in Batch Retorts

Most workers, dealing with convection packs consisting of particulates in a covering liquid, carry out experimental trials to determine process safety with a thermocouple placed at a convenient point, often considered to be the point of slowest heating in the liquid. While this is adequate for particulates of small size, larger-sized constituents may not receive an adequate process at their centers. For some vegetable products this could result in a sub-botulinum process (see Chapter 4). Potatoes have been shown to receive inadequate processes in their centers (Thorpe & Atherton 1969; Atherton *et al.* 1970). While, in practice, it is necessary to make a judgement on the efficacy of a given process, it may well be, as in the case of canned carrots, that many years of industrial practice dictate that this criterion is not always essential. However, with many other products, e.g. artichoke hearts, microbiological challenge tests are required to establish the safety of the process.

Models for predicting the temperature distributions in both the particulates and the covering liquid have been developed by relatively few workers. However, some of the work will be briefly mentioned here; other work is listed in Table 2.8.

Ikegami (1977) developed an equation based on heat transfer parameters f_h and j for both solid and liquid phases, which could be used to predict the temperature profile:

$$\frac{T_s - T_R}{T_0 - T_R} = \frac{jj_2 f_2}{f_2 - f} 10^{-t/f_2}, \tag{2.105}$$

where j and f are the heat transfer parameters for the solid, and j_2 and f_2 for the liquid, and T_s, T_R and T_0 are, respectively, the solid food, retort and initial food temperatures.

Rumsey (1984) used a numerical approach to solve an energy balance model, which had been modified by the addition of a term representing the temperature gradient in the particulate:

$$m_1 c_1 \frac{dT_1}{dt} + n_p m_p c_p \frac{dT_p}{dt} = U_c A_c (T_R - T_1), \tag{2.106}$$

where m is the mass, c is the specific heat, t the time, T the temperature, n the number of particulates; the subscripts referring to 1 the liquid, p the particulate, R the retort and c the container. The temperature distribution in the particulate, T_p, is obtained using the solution of the unsteady-state heat conduction equation for the particular geometry.

One of the more important studies was due to Lekwauwa and Hayakawa (1986), who developed a model assuming an initially uniform temperature in the particulates, which was not necessarily the same as the covering liquid; the solid and liquid components had homogeneous, isotropic and temperature-independent properties; the particulate sizes were statistically distributed (approximately gamma); the heat-transfer coefficient between the solid surface and the bulk liquid was constant; and the heat-transfer coefficients for cooling conditions were different from those for heating. The technique used for determining the particle temperature distributions was to apply Duhamel's theorem (Section 2.2.6.4) to the conduction step response (modeled in terms of f_h and j) with the change in liquid temperature. The energy balance on the whole system was established considering the heat transfer between the external heating medium and the liquid in the container, and also between the liquid and the particulates. A computer algorithm was presented for this model. Experimental results for the heating of aspheroidally shaped white potatoes in distilled water were in good agreement with those from the theoretical models, for both analytical and numerical solutions. Hayakawa et al. (1997) have used Duhamel's theorem to determine the temperature distribution in irregular shaped objects, subjected to an external heat transfer coefficent. This involved using the experimentally determined values of f_h and j (see Chapter 3) The method was also applicable to deviant processes.

TABLE 2.8. Some studies on convection heat transfer for non-agitated heterogeneous canned foods.

Product	Process	Model	Comments	Reference
Glass spheres/water/silicone fluid	115°C: half cylinder	Fluid dynamic model	Temperature distribution	Hiddink (1975)
Bamboo shoots, sausages, baby clams, mushrooms/liquid	JCS Nos 1, 2, 4 & 7 cans	f_h & j model	Heat penetration data	Ikegami (1977)
Mushroom-shaped aluminum, mushrooms	115.6–126.7°C: 211 × 212 can	Energy balance	Heat transfer coefficients, effect of temperature & particle size	Sastry (1984)
Aluminum, PTFE & potato spheres/liquids	100°C: see Hassan (1984)	Energy balance–particles/liquids	Heat transfer coefficients	Rumsey (1984)
Mushroom-shaped aluminum in water	80°C: heating bath	Energy balance	Heat transfer coefficients	Alhamdan et al. (1990)
Mushroom-shaped aluminum in carboxymethylcellulose	80°C: heating bath	Energy balance	Dimensionless correlations	Alhamdan and Sastry (1990)
Mushrooms/brine	121°C: pilot retort Omnia jars 0–8	Energy balance–thermal inertia	Temperature profile	Akterian (1995)
Fruits in syrup	93.3°C: retort; Omnia jars 105 mm diam	Empirical model	f_h & heat-transfer coefficients	Akterian (1996)
Vegetables in brine	121°C: retort; Omnia 105 mm diam	Empirical model	f_h & heat-transfer coefficients	Akterian (1996)
Food particles/CMC	cans	Energy balance	Surface heat-transfer coefficients	Awuh et al. (1993)

The energy balance model was also used by Sastry (1984) to determine heat-transfer coefficients to canned aluminum mushroom-shaped particles packed in water. The heat-transfer coefficients were correlated using the relationship:

$$Nu = 0.01561(Gr \cdot Pr)^{0.529}.$$

The convective heat-transfer coefficient was found to be time-dependent, increasing to a maximum during the early stages of heating and steadily declining. The mean heat-transfer coefficients varied between 396 and 593 W m^{-2} K^{-1}.

Another application of the energy balance model was to determine the heat-transfer coefficients between irregular-shaped particulates and Newtonian as well as non-Newtonian covering liquids (Alhamdan et al. 1990; Alhamdan and Sastry 1990). For water, dimensionless correlations were, for heating,

$$Nu = 5.53(Gr \cdot Pr)^{0.21},$$

and for cooling,

$$Nu = 0.08(Gr \cdot Pr)^{0.27}.$$

The average heat-transfer coefficients ranged between 652 and 850 W m^{-2} K^{-1} for heating and 384–616 W m^{-2} K^{-1}. For a non-Newtonian liquid, carboxymethyl cellulose solution, the heat-transfer coefficients were for heating 75–310 W m^{-2} K^{-1} and for cooling 22–153 W m^{-2} K^{-1}. The heat-transfer coefficients decreased with increasing apparent viscosity.

Stoforos and Merson (1990) developed a technique for estimating heat-transfer coefficients in liquid/particulate canned foods using only liquid temperature data. They solved the energy balance equation, modified for particulates, analytically using Laplace transformations.

2.3.4.5. Heterogeneous Products Heated with Agitation

In this category there are a large number of products containing particulates and covering liquid, which can benefit from being processed in rotary retorts (see Chapter 8). Processes are shorter, energy can be saved, and textural and other quality attributes can be improved. Using rotary processes requires the contents of each container to receive the same rotary motion; this means that the ratio of solids to liquids and the headspace in each container have to be carefully controlled. Table 2.9 gives some examples of experimental work that has been done to investigate the heat transfer into rotating and agitated containers. Relatively little work has been done on the quantitative aspects of the subject: three theses by Hassan (1984), Deniston (1984), and Stoforos (1988) and an overview by Stoforos and Merson (1990) cover what has been done. The technology has received a thorough treatment in the work of Manfried Eisner (1988). Some of the more important contributions will now be discussed.

Lenz and Lund (1978) derived a useful heat-transfer correlation for spherical particulates 9.65–38.1 mm in diameter and void fraction 0.32–0.41 in covering liquids of both water and 60 wt% sucrose solution using 303 × 406 and

TABLE 2.9. Some studies on convection heat transfer for agitated heterogeneous canned foods.

Product	Process	Model	Comments	Reference
Silicone oil, sausage	Laboratory device, 120°C: axial & end-over-end	Energy balance & particle conduction	Heat-transfer coefficients – duration of transfer unit	Bimbenet and Michiels (1974)
Water & sucrose solution, peas & lead shot	FMC Steritort, 121°C: axial rotation 303 × 406; 608 × 700 can	Energy balance & particle conduction	Temperature profiles – heat-transfer coefficients	Lenz and Lund (1978)
Potatoes/water	Stock Pilot Rotor-900, 121°C: end-over-end 307 × 409 cans	Energy balance & particle conduction	Effect of size distribution & surface heat transfer	Lekwauwa and Hayakawa (1986)
Water, silicone fluid, PTFE, aluminum, potato	Rotary single can simulator, 50°C: axial rotation	Energy balance & particle conduction	Analytical solution, heat-transfer coefficients	Deniston et al. (1987), Stoforos (1988), Merson and Stoforos (1990)
Spheres	303 × 406 can			Stoforos & Merson (1990, 1992)
Water, sucrose solution, raw snap beans & aluminum	FMC Steritort, 115.6°C: axial rotation 303 × 406	Energy balance & particle conduction	Heat-transfer coefficients – dimensionless correlation	Fernandez et al. (1988)
Sucrose syrup/particles	100 × 119 mm cans; axial rotation; ambient temperature	Position emission particle tracking	Head space & mixing study	Cox et al. (2003)
Spheres, cylinders, cubes/liquid	Stock Rotomat–PR 900	Flow visualization	Heat-transfer coefficients–mixing studies	Sablani and Ramaswamy (1998)
Nylon spheres/water & /oil	Stock Rotomat–PR 900	Experimental correlations	Heat-transfer coefficients– influence of particle size & concentration	Sablani and Ramaswamy (1997)
Poly-propylene spheres/water	110–130°C: full immersion water sterilizer	Experimetal correlations	Heat-transfer coefficients; effect of head space	Sablani and Ramaswamy (1996)
Spheres, cylinders, cubes/liquid	Stock Rotomat–PR 900	Flow visualization	Heat-transfer coefficients & mixing studies	Sablani & Ramaswamy (1993, 1998)
Potato shape 7 aluminum spheres	Axial rotation	Energy balance & particle conduction	Heat-transfer coefficients & 4th order Runge-Kutta model	Stoforos & Merson (1992, 1995)

608×700 cans. These were processed in an FMC Steritort pilot simulator. The correlations were, for liquids only

$$Nu = 115 + 15Re^{0.3}Pr^{0.08}, \tag{2.107}$$

for liquids and particulates,

$$Nu = -33 + 53Re^{0.28}Pr^{0.14}X, \tag{2.108}$$

where $X = D_p/[S(1-\varepsilon)]$ and D_p is the particle diameter, S is the radius of the reel in the simulator, and ε is the fraction of the container volume occupied by the fluid. Values for the particulate–liquid heat-transfer coefficients ranged from 180 to 970 W m^{-2} K^{-1}.

Fernandez *et al.* (1988) used the simple energy balance model to determine the heat-transfer coefficients between canned snap beans and the covering liquid in 303×406 cans. The best correlation obtained for the experimental data was

$$Nu = 2.7 \times 10^4 Re^{0.294}Pr^{0.33}(S_f)^{6.98}, \tag{2.109}$$

where the Reynolds number was in a rotational form $D^2 N\rho/\mu$, in which D is the can diameter and N the number of rotations per minute, and S_f is a shape factor given by $\pi(6V_p/\pi)^{2/3}A_p^{-1}$, with V_p and A_p being, respectively, the particulate volume and area.

Merson and Stoforos (1990) studied the effect of axial rotation on heat-transfer coefficient and the effect of the relative velocity between the particle and the liquid. They obtained a more complex correlation, in which the first term on the right includes particulate conduction, the second, heat transfer in the laminar boundary layer on the upstream side of the sphere, and the third term is the heat transfer in the turbulent wake.

$$Nu = 2 + 0.4Re^{0.5}Pr^{0.4}(\mu/\mu_s)^{0.25} + 0.06Re^{0.66}Pr^{0.4}(\mu/\mu_s)^{0.25}. \tag{2.110}$$

A semi-analytical solution for determining particle temperature and a semi-numerical solution for determining fluid temperature has been applied to the solution of heat transfer in axially rotating liquid/particle canned foods. The time dependent boundary condition was modeled using Duhamel's theorem and a 4th-order Runge-Kutta technique was used for the solution of the finite difference mode. (Storofos & Merson 1995).

This measurement of particulate temperatures is difficult. The presence of a thinwire thermocouple prevents free motion in the particle, and consequently could seriously affect the actual value of the heat-transfer coefficients. Liquid crystal sensors incorporated in the particulates have been used, which allows unrestricted movement of the monitored particulates (Storfos and Merson 1991, 1992). For a critical overview concerning heat-transfer coefficients to particulates, see Maesmans *et al.* (1992).

Sablani *et al.* (1997a) have used dimensionless correlations to determine the heat-transfer coefficients for a range of different plastic particles in both oil and water undergoing end-over-end agitation. The correlation for multiparticles involves the ratio of the particle diameter (d_e) to that of the can diameter (D_c) see equation (2.111).

$$Nu = 0.71 Re^{0.44} Pr^{0.36} (\varepsilon/100 - \varepsilon)^{-0.37} (d_e/D_c)^{-0.11} \Psi^{0.24} \qquad (2.111)$$

where ε is the particle concentration and Ψ is the particle sphericity.

2.3.5. Conclusions

While a considerable amount of effort has been put into modelling convective heating of liquids and liquids/particulates in containers, there are wide discrepancies in the results obtained by different workers. With the development of more advanced techniques for monitoring temperatures inside containers and with the ability to solve coupled hydrodynamic and heat transfer problems using finite-element techniques and computational fluid dynamics, it should be possible in the future to obtain more reliable models capable of extrapolation to a range of can sizes. At the present time the work on convective heating has led to a better understanding of the heat-transfer regimes inside canned foods. This in turn has given us a better understanding of the factors that must be considered in order to be assured of the microbiological safety of canned foods. This is particularly important in reducing the severity of processes to give better-quality products, where this is appropriate, and to reduce the energy requirements. It is important to consider carefully the meaning of heat-transfer coefficients; whereas it is relatively easy to obtain an overall coefficient, which includes the internal coefficient from the wall to the covering liquid, the average heat transfer from the liquid to the particulate, the container wall resistance and the external heat-transfer coefficient from the heating medium to the external wall, it is more difficult to quantify the instantaneous local distribution of heat transfer depending upon the particle motion relative to the fluid motion. It may be, because of the unpredictability of the relative liquid/particulate motion inside cans, that we have reached the limit of the usefulness of this approach.

2.4. Radiation Heating

Thermal radiation has a wavelength from 0.8 to 400 μ m and requires no medium to transmit its energy. The transfer of energy by radiation involves three processes: first, the conversion of the thermal energy of a hot source into electromagnetic waves; second, the passage of the waves from the hot source to the cold receiver; and third, the absorption and reconversion of the electromagnetic waves into thermal energy.

The quantity of energy radiated from the surface per unit time is the emissive power E of the surface. For a perfect radiator, known as a black body, the emissive power is proportional to the fourth power of the temperature T:

$$E = \sigma T^4, \qquad (2.112)$$

where σ is the Stefan–Boltzmann constant, $5.670 \times 10^{-8}\,\mathrm{W\,m^{-2}\,K^{-4}}$, and the heat-transfer equation for the net exchange of heat between two bodies at temperatures T_1 and T_2, respectively, is given by

$$Q = \varepsilon\sigma(T_1^4 - T_2^4), \tag{2.113}$$

where Q is the net radiation per unit area, and ε is the emissivity, i.e. the ratio of the energy emitted by a grey body to that emitted by a black body. For a highly polished surface ε can be as low as 0.03, and for a black body ε is 1.

The heat-transfer coefficient for radiation can be obtained from equation (2.113) as follows:

$$Q = h_r(T_1 - T_2)$$

$$h_r = \frac{Q}{T_1 - T_2} = \frac{\varepsilon\sigma}{T_1 - T_2}(T_1^4 - T_2^4). \tag{2.114}$$

The main application for radiation theory is in relation to can sterilization using gaseous flames. The cooling of cans to the atmosphere also involves loss of heat by radiation as well as convection.

A more complex heat-transfer model for canned foods heated in a flame sterilizer was developed by Peralta Rodriguez and Merson (1983). The following assumptions were made in developing the model:

1. The flow of combustion gases past the can surface is uniform.
2. A single average heat-transfer coefficient applies to the whole of the can surface.
3. Heat transfer through the end of the can is neglected.
4. The flame is considered to be a regular-shaped gaseous body.
5. Radiation from the hot gases after they have left the flame is neglected.
6. Radiation exchange only occurs between the flame and the can, and from the can to the surroundings.
7. The temperature of the flame is the adiabatic temperature calculated on the dissociation of the combustion species.
8. The dissociated species do not combine on the can surface.

The heat balance equation is

$$Q_T = Q_C + Q_R, \tag{2.115}$$

where Q is the rate of heat transfer and the subscripts T, C, R refer to total, convection and radiation, respectively.

The convective contribution is given by

$$Q_C = h_c A_c(T_g - T_{\mathrm{sur}}), \tag{2.116}$$

where h_c is the average film heat-transfer coefficient for the surface, A_c is the surface area of the can, T_g is the gas temperature, and T_{sur} is the temperature of the surface of the can.

The radiative contribution is given by

$$Q_R = Q_i - Q_e - Q_r, \qquad (2.117)$$

where Q_i is the radial energy reaching the surface, Q_e is the energy re-emitted, and Q_r is the reflected energy. These three terms are given by the following equations:

$$Q_i = \varepsilon_f A F_{\text{fc}} \sigma T_f^4, \qquad (2.118)$$

where the subscript f refers to the flame properties and F_{fc} is the dimensionless view factor from the flame to the can;

$$Q_e = \varepsilon_c A_c \sigma T_{\text{sur}}^4, \qquad (2.119)$$

where the subscript c refers to the can; and

$$Q_r = \rho_c \varepsilon_f A_f F_{\text{fc}} \sigma T_f^4, \qquad (2.120)$$

where ρ_c is the reflectivity of the can. Combining these equations we obtain

$$Q_R = (1 - \rho_c) \varepsilon_f A_f F_{\text{fc}} \sigma T_f^4 - \varepsilon_c A_c \sigma T_{\text{sur}}^4. \qquad (2.121)$$

From Kirchhoff's laws of radiation, the absorptivity α_c is equal to the emissivity ε_c. This, and using the reciprocity relation, $A_f F_{\text{fc}} = A_c F_{\text{cf}}$, give

$$Q_R = \alpha_c A_c \sigma (\varepsilon_f F_{\text{cf}} T_f^4 - T_c^4). \qquad (2.122)$$

Using these equations and other basic heat-transfer equations, the temperature of the liquid T_1 was derived:

$$\frac{T_1 - T_g}{T_0 - T_g} = \exp(-Xt) - \Psi[1 - \exp(-Xt)], \qquad (2.123)$$

where T_0 is the initial temperature of the fluid, T_g is the temperature of the radiating gas, and

$$X = A_c U_c / Mc$$

$$\Psi = [\alpha_c \sigma (\varepsilon_f F_{\text{cf}} T_f^4 - T_{\text{sur}}^4)]/[h_c(T_g - T_0)],$$

in which M is the mass of fluid, c is the specific heat, and U_c is the overall heat-transfer coefficient, based on the log mean cylindrical area.

For a 60% sucrose solution in a 304 × 406 can and using an average flame temperature up to 1000°C, the external heat-transfer coefficient was 26.38 W m^{-2} K^{-1} and the internal coefficient ranged from 220 to 621 W m^{-2} K^{-1}, depending on the rotational speed of the can. The overall heat-transfer coefficient, U_c, varied from 23.86 to 25.12 W m^{-2}K^{-1} and did not increase with increasing rotational speeds above 60 rpm. The flame emissivity was taken as 0.0241 and the absorptivity of the can body 0.055. From this work it was concluded that the external heat-transfer coefficient, unlike steam sterilization, was the rate-controlling step. Table 2.10 gives some information about other work

TABLE 2.10. Some heat transfer studies for radiation heating from flames.

Product	Container size	Process conditions	Comments	Reference
Water Bentonite/water 4 & 10%	$401 \times 411(A2\frac{1}{2})$	Flame sterilization Single can Axial rotation 0–60 rpm	Heating curves External temperature Profile	Paulus and Ojo (1974)
Bentonite 1% Sucrose syrup 60%	$303 \times 406(\frac{1}{2}\,kg)$	Flame sterilization Single can Axial rotation 30 & 50 rpm	Energy balance model Internal heat transfer Coefficients	Merson *et al.* (1981)
Sucrose syrup 60% Silicone fluids	$303 \times 406(\frac{1}{2}\,kg)$	Flame sterilization Single can Axial rotation 10–200 rpm	Radiation & convective Convective model Heat-transfer coefficients	Peralta Rodriguez (1982, 1987a) Peralta Rodriguez & Merson (1982, 1983)
Water Carboxymethylcellulose/water 1.5–2.5%	$303 \times 406(\frac{1}{2}\,kg)$	Flame sterilization Single can Axial rotation 50–200 rpm	Energy balance model Heat-transfer coefficients	Teixeira Neto (1982)
Bentonite/water 1%	209×413 (aluminum)	Flame pasteurization Single can Axial rotation 40 rpm	Energy balance model Temperature profiles Heat-transfer coefficients	Noh *et al.* (1986)

on heat transfer to cans being sterilized by flame heating. Further aspects of flame sterilization are dealt with in chapter 8.

2.5. Some Computer Programs

One of the most important developments in the study of thermal processing has been the application of computer simulation programs. This has enabled the operation of processing equipment to be improved by elucidating the mechanisms of heat transfer and fluid flow in processing systems. The following selected programs have been applied, not only to process evaluation but also to control problems.

2.5.1. Conduction Heat Transfer Analysis Programs

(i) ANSYS Finite Element Software (deSalvo & Gorman 1989).   Used by a number of workers, including, Hendrickx et al. (1992) to determine the center sterilization value in a food can and to optimize the nutrient retention, and Christianini and Massaguer (2002) for evaluating the thermal process for tuna/brine in institutional retort pouches. The latter work is interesting because of a comparison between the analytical solution given in §2.2.8.2. and both a r-D and a 3-D Finite element solution. All three models give good agreement for the heating phase; however, for the cooling phase the finite-element models were better than the analytical solution.

(ii) CHAMPSPACK Uses Matlab and Fortran routines for finite element computations of coupled heat and mass transfer problems with internal heat generation and convective boundary conditions. It was developed for the design and simulation of thermal process applications for foods and agricultural products (Scheerlinck & Nicolaï 1999). The package includes facilities for insertion of kinetic data for microbial destruction and quality factor inactivation. For application to enzyme inactivation in vegetables see Martens et al. (2001).

(iii) SPEEDUP[TM] (Simulating Program for Evaluating and Evolutionary Design of Unsteady State Process) Alwis et al. (1992) used the SPEEDUP[TM] system developed by Sargent et al. (1982) for optimizing quality and processing effects of a conduction-heating meat product. This showed the need for higher temperatures and shorter times for obtaining better meat quality.

(iv) NEURAL Networks   Neural networks have been used to predict optimum sterilization temperatures and corresponding process time and quality factors. The method has the advantage over conventional regression models that all three parameters can be determined simultaneously (Sablini et al. 1995). Neural network modeling has also been used to study heat transfer to liquid particle mixtures in cans undergoing end-over-end processing (Sablani et al. 1997b).

2.5.1.1. Computational Fluid Dynamics (CFD) programs

(i) PHOENICS (Cham Ltd., London, UK)   Simulations using a PHOENICS package were used to determine the relative concentration profiles of the spores and also the temperature profiles. An application to microbial inaction is discussed by Ghani *et al.* (1999a, b; 2002a, c), Ghani *et al.* (2001b).

(ii) CFX/TASCflow (CFX International, AEA Technology, Harwell, UK This has been used to simulate a range of canning applications, e.g., transient natural convection in containers filled with non-Newtonian CMC solution (Quarini & Scott 1997), the axial and end-over-end rotation in canned foods (Emond & Tucker 2001; James 2001; Hughes *et al.* 2003; Tucker 2004) and also natural convection heating of canned foods in conical and cylindrical containers (Varma *et al.* 2006).

(iii) FLOWPACK (ICI, Ltd., UK) – Zone Modeling   Tucker *et al.* (1992) have developed zone modeling as a method of determining temperature profile in simple geometrically shaped canned food. Best *et al.* (1994a, b) have presented a steady-state and a dynamic simulation of the processing of cans in a hydrolock sterilizer using zone modeling and a computer program "Flowpack."

(iv) Others STAR-CD (Computational Dynamics Ltd. London UK.

Fluent/FIDAP (Lebanon, NH, USA).
Fluent v. 6.1.has been used by Jun and Sastry (2005) to study the effect pulsed ohmic heating has on food in flexible pouches.

References

Adams, J. A., & Rogers, D. F. (1973). *Computer-aided heat transfer analysis*. New York: McGraw-Hill.

Akterian, S. G. (1995). Numerical simulation of unsteady heat transfer in canned mushrooms in brine during sterilization processes. *J. Food Eng.*, *25*, 45–53.

Akterian, S. G. (1996). Studying and controlling thermal sterilization of convection-heated canned foods using functions of sensitivity. *J. Food Eng.*, *29*(3/4), 125–143.

Akterian, S. G., & Fikiin, K. A. (1994). Numerical simulation of unsteady heat conduction in canned arbitrary shaped foods during sterilization processes. *J. Food Eng.*, *21*, 343–354.

Alhamdan, A., & Sastry, S. K. (1990). Natural convection heat transfer between non-Newtonian fluids and an irregular-shaped particle. *J. Food Process Eng.*, *13*, 113–124.

Alhamdan, A., Sastry, S. K., & Blaisell, J. L. (1990). Natural convection heat transfer between water and an irregular-shaped particle. *Trans. ASAE*, *33*(2), 620–624.

Alles, L. A. C., & Cowell, N. D. (1971). Heat penetration into rectangular food containers. *Lebensm.-Wiss. u. -Technol.*, *4*(2), 50–53.

Alwis, A. A. P. de, Varley, J., Niranjan, K. (1992) Simulation of thermal food processing operations. In W. Hamm & S. D. Holdsworth (Eds.), "*Food Engineering in a Computer Climate*" (pp. 253–262). Rugby, U.K: Institution of Chemical Engineers.

Anantheswaran, R. C., & Rao, M. A. (1985a) Heat transfer to model Newtonian liquid foods in cans during end-over-end rotation. *J. Food Eng.*, *4*(4), 1–19.

Anantheswaran, R. C., & Rao, M. A. (1985b). Heat transfer to model non-Newtonian liquid foods in cans during end-over-end rotation. *J. Food Eng.*, *4*(4), 21–35.

ANSYS (1968). *ANSYS Engineering Analysis System*. Houston, PA: Swanson Analysis System.

Arpaci, V. S. (1966). *Conduction heat transfer*. New York: Addison-Wesley.

Atherton, D., Overington, W. J. G., & Thorpe, R. H. (1970). Processing studies on canned new potatoes. In *Potatoes for canning*, AMDEC Annual Progress Report 1970. Chipping Campden, Glos., UK: Campden & Chorleywood Food Research Association.

Awuah, G. B., Ramaswamy, H. S., & Simpson B. K. (1993). Surface heat transfer coefficients associated with heating of food particles in CMC solution. *J. Food Process Eng.*, *16*(1), 39–57.

Ball, C. O., & Olson, F. C. W. (1957). *Sterilization in food technology – Theory, practice and calculations*. New York: McGraw-Hill.

Ban, H., & Kaziwara, Y. (1941). Cooling of cans filled with water due to internal convection. *Bull. Jap. Soc. Sci. Fish.*, *10*, 38–42. [in Japanese].

Banga, J. R., Alonso, A. A., Gallardo, J. M., & Perez-Martin, R. I. (1993). Mathematical modelling and simulation of the thermal processing of anisotropic and non-homogeneous conduction-heated canned foods: application to canned tuna. *J. Food Eng.*, *18*(4), 369–387.

Barakat, H. Z., & Clark, J. A. (1966). Analytical and experimental study of the transient laminar natural convection flows in partially filled liquid containers. *Proc. 3rd. Int. Heat Transfer Conf. (ASME)*, Vol. 2, pp. 152–162.

Beek, G. van, & Veerkamp, C. H. (1982). A program for the calculation of the thermal properties of foodstuffs. *Voedingsmiddelentechnologie*, *15*, 19–23.

Bengtsson, N. E., & Risman, P. O. (1971). Dielectric properties of foods at 3 GHz as determined by cavity perturbation technique. 2. Measurement on food materials. *J. Microwave Power*, *6*, 107–123.

Bera, F. (1988). Break in the study of heat transfer by convection in cans. In *Progress in Food Preservation Processes*, Vol. 1, pp. 59–68, Brussels: CERIA.

Best, R. J., Bewaji, E. O., & Johns, W. R. (1994a). Dynamic simulation for waste reduction in hydrostatic canning retort operations. In *Proc. 4th. Bath Conference Food Process Engineering* Vol. 1 (pp. 261–268). Rugby, UK: Institution of Chemical Engineers.

Best, R. J., Bewaji, E. O., Johns, W. R., & Kassim, H. O. (1994b). Steady-state and dynamic simulation of food processes including spatial effects. *Food and Bioproducts Processing, Trans. Inst. Chem. Eng.* 72C2, 79–85.

Bhowmik, S. R., & Hayakawa, K. (1979). A new method for determining the apparent thermal diffusivity of thermally conductive food. *J. Food Sci.*, *44*(2), 469–474.

Bhowmik, S. R., & Tandon, S. (1987). A method of thermal process evaluation of conduction heated foods in retortable pouches. *J. Food Sci.*, *52*(1), 202–209.

Bimbenet, J. J., & Michiels, L. (1974). Convective heat transfer in canning process. *Proc. IVth Int. Congress Food Sci. Technol.*, Vol. IV, pp. 363–379.

Bird, R. B., Stewart, W. E., & Lightfoot, E. N. (1960) *Transport phenomena*. New York: Wiley.

Blaisdell, J. L. (1963) Natural convection heating of liquids undergoing sterilization in unagitated food containers. PhD thesis, Dept of Agric. Eng., Michigan State Univ.

Burfoot, D., Griffin, W. J. and James, S. J. (1988). Microwave pasteurization of prepared meals. *J. Food Eng.*, *8*(3), 145–156.

Burfoot, D., Railton, C. J., Foster, A. M., & Reavell, S. R. (1996). Modelling the pasteurisation of prepared meals with microwaves at 896 MHz. *J. Food Eng.*, *30*(1/2), 117–133.

Burmeister, L. C. (1983). *Convective heat transfer*. New York: Wiley.

Calay, R. K., Newborough, M., Probert, D., & Calay, P. S. (1995). Predictive equations for the dielectric properties of foods. *Int. J. Food Sci. Technol.*, *29*, 699–713.

Califano, A. N., & Zaritzky, N. E. (1993). A numerical method of simulating heat transfer in heterogeneous and irregularly shaped foodstuffs. *J. Food Process Eng.*, *16*(3), 159–171.

Carnahan, B., Luther, H. A., & and Wilkes, J. O. (1969). *Applied numerical methods*. New York: Wiley.

Carslaw, H. R., & Jaeger, J. C. (1959) *Conduction of heat in solids* (2nd Edition). Oxford: Oxford University Press.

Castillo, P. F., Barreiro, J. A., & Salas, G. R. (1980). Prediction of nutrient retention in thermally processed heat conduction food packaged in retortable pouches. *J. Food Sci.*, *45*, 1513–1516, 1528.

Chang, S. Y., & Toledo, R. T. (1990). Simultaneous determination of thermal diffusivity and heat-transfer coefficient during sterilization of carrot dices in a packed bed. *J. Food Sci.*, *55*(1), 199–205.

Chapman, S., & McKernan, B. J. (1963). Heat conduction in plastic food containers. *Food Technol.*, *17*, 1159–1162.

Charm, S. (1961) A method for calculating the temperature distribution and mass average temperature in conduction-heated canned foods during water cooling. *Food Technol.*, *15*, 248–253.

Chau, K. V., & Snyder, G. V. (1988). Mathematical model for temperature distribution of thermally processed shrimp. *Trans. ASAE*, *31*(2), 608–612.

Chen, C. R., & Ramaswamy, H. S. (2002a). Dynamic modelling of thermal processing using artificial neural networks. *J. Food Proc. & Preserv.*, *26*, 91–111.

Chen, C. R., & Ramaswamy, H. S. (2002b). Modeling and optimization of constant retort temperature (CRT) thermal processing using neural networks and genetic algorithms. *J. Food Proc. Eng.*, *25*(4). 351–379.

Chen, C. R., & Ramaswamy, H. S. (2002c). Modeling and optimization of variable retort temperature (VRT) thermal processing using coupled neural networks and genetic algorithms. *J. Food Eng*,. *53*(3), 209–220.

Chiheb, A., Debray, E., Le Jean, G., & Piar, G. (1994). Linear model for predicting transient temperature during sterilization of a food product. *J. Food Sci.*, *59*(2), 441–446.

Choi, Y., & Okos, M. R. (1986). Thermal properties of foods – Review. In M. R. Okos (Ed.), *Physical and chemical properties of foods*, St Joseph, MI: ASAE.

Christianini M., & Massaguer, P. R. (2002). Thermal process evaluation of retortable pouches filled with conduction heated food. *J. Food Proc. Eng.* *25*(5): 395–405.

Clary, B. L., & Nelson, G. L. (1970). Determining convective heat-transfer coefficients from ellipsoidal shapes. *Trans. ASAE*, *13*(3), 309–314.

Clary, B. L., Nelson, G. L., & Smith, R. E. (1971). The application of geometry analysis technique in determining the heat transfer rates from biological materials. *Trans. ASAE*, *14*(3), 586–589.

Cooper, D. LeB. (1937). The application of thermal data to the calculation of sterilizing times for canned fish. *J. Biol. Bd. Can.*, *3*(2), 100–107.

Coulson, J. M., & Richardson, J. F. (1985). *Chemical engineering, Vol. 1. Fluid flow, heat transfer and mass transfer*. Oxford: Pergamon Press.

Cowell, N. D., & Evans, H. L. (1961). Studies in canning processes. Lag factors and slopes of tangents to heat penetration curves for canned foods heated by conduction. *Food Technol.*, *15*, 407–412.

Cox, P. W., Bakalis, S, Ismail, H., Forester, R., Parker, D. J. & Fryer, P. J. (2003). Visualization of three-dimensional flows in rotating cans using positron emission particle tracking (PEPT). *J. Food Eng.*, *60*(3), 229–240.

Croft, D. R., & Lilley, D. G. (1977). *Heat transfer calculations using finite difference equations*. London, UK: Applied Science Publishers.

Cuesta, F. J., & Lamua, M. (1995). Asymptotic modelling of the transient regime in heat conduction in solids with general geometry. *J. Food Eng.*, *24*(3), 295–320.

Datta, A. K. (1985). Numerical modelling of natural convection and conduction heat transfer in canned foods with application to on-line process control. PhD thesis, Univ. Florida.

Datta, A. K., & Liu, J. (1992). Thermal time distributions for microwave and conventional heating of food. *Food & Bioproducts Processing, Trans. IChemE*, *70*, C2, 83–90.

Datta, A. K., & Teixeira, A. A. (1987). Numerical modeling of natural convection heating in canned liquid foods. *Trans. American Society Mechanical Engineers*, *30*(5), 1542–1551.

Datta, A. K., & Teixeira, A. A. (1988). Numerically predicted transient temperature and velocity profiles during natural convection heating in canned liquid foods. *J. Food Sci.*, *53*(1), 191–195.

Datta, A. K., Teixeira, A. A., & Chau, K. V. (1984). Implication of analytical heat conduction equation in thermal processing of canned foods. *ASAE Paper* No. 84–6518, St Joseph, MI: American Society of Agricultural Engineers.

De Baerdemaeker, J., Singh, R. P., & Segerlind, L. J. (1977). Modelling heat transfer in foods using the finite-element method. *J. Food Proc. Eng.*, *1*, 37–50.

Decareau, R. V. (Ed.) (1985). *Microwaves in the food processing industry*. Orlando, FL: Academic Press.

Decareau, R. V., & Peterson, R. A. (1986). *Microwave processing and engineering*. Chichester (UK), Ellis Horwood.

Deniston, M. F. (1984). Heat-transfer coefficients to liquids with food particles in axially rotating cans. MS thesis, Univ. California, Davis.

Deniston, M. F., Hassan, B. H., & Merson, R. L. (1987). Heat-transfer coefficients to liquids with food particles in axially rotating cans. *J. Food Sci.*, *52*, 962–966.

DeSalvo, G. J., & Gorman, R. W. (1989). ANSYS – Engineering Analysis System – User's Manual, Swanson Analysis Systems, Inc., Swansee, USA.

Dincer, I., Varlik, C., & Gun, H. (1993). Heat transfer rate variation in a canned food during sterilization. *Int. Comm. on Heat & Mass Transfer, 20*, 301–309.

Duquenoy, A. (1980). Heat transfer to canned liquids. In P. Linko, *et al.* (Eds.), *Food process engineering*, Vol. 1, (pp. 483–489). London, UK: Applied Science Publishers.

Duquenoy, A. (1984). Heat transfer in cans: conduction and convection. In B. M. McKenna (Ed.), *Engineering and food*, Vol. 1, (pp. 209–216). London, UK: Elsevier Applied Science Publishers.

Ede, A. J. (1967). Advances in free convection. In *Advances in heat transfer*, Vol. 4, (pp. 1–30). New York Academic Press.

Eisner, M. (1988). *Introduction to the technique of rotary sterilization*. Milwaukee, WI.: author.

Emond, S. P., & Tucker, G. S. (2001). Optimising the in-container mixing of a tomato product during end-over-end rotary thermal processing. In J. Welti-Chanes, G. V. Barbosa-Cánovas & J. M. Aguilera (Eds.), *International Conference Engineering and Food, ICEF8*, Vol. 1, (pp. 669–663). Lancaster PA: Technomic Pub. Co., Inc.

Engelman, M. S., & Sani, R. L. (1983). Finite-element simulation of an in-package pasteurization process. *Numerical Heat Transfer, 6*, 41–54.

Erdogdu, F., Balaban, M. O., & Chau, K. V. (1998a). Modeling of heat conduction in elliptical cross section: 1. Development and testing of the model. *J. Food Eng. 38*(2), 223–239.

Erdogdu, F., Balaban, M. O., & Chau, K. V. (1998b). Modeling of heat conduction in elliptical cross section: 2. Adaptation to thermal processing of shrimp. *J. Food Eng. 38*(2), 241–258.

Erdogdu, F., Balaban, M. O., & Chau, K. V. (2001a). A numerical method for conduction heat transfer in oval shapes. In J. Welti-Chanes, G. V. Barbosa-Cánovas, & J. M. Aguilera (Eds.), *Engineering and food, ICEF8*, Vol. 2, (pp. 1799–1803). Lancaster, PA: Technomic Pub. Co.

Erdogdu, F., Balaban, M. O., & Chau, K. V. (2001b). Modeling of heat conduction in elliptical cross sections. In J. Welti-Chanes, G. V. Barbosa-Cánovas, & J. M. Aguilera (Eds.), *Engineering and food, ICEF8*, Vol. 2, (pp. 1864–1868). Lancaster, PA: Technomic Pub. Co.

Erdogdu, F., & Turhan, M. (2006), Analysis of dimensionless ratios of regular geometries for infinite geometry assumptions in conduction heat transfer problems. *J. Food Eng., 77*(4), 818–824.

Eszes F., & Rajkó. R. (2004). Modelling heat penetration curves in thermal processes. In P. Richardson (Ed.), *Improving the thermal processing of food* (pp. 307–333). Cambridge, UK: Woodhead Publishing Ltd.

Evans, H. L. (1958). Studies in canning processes. II. Effects of the variation in temperature of the thermal properties of foods. *Food Technol., 12*(6), 276–282.

Evans, H. L., & Board, P. W. (1954). Studies in canning processes. I. Effect of headspace on heat penetration in products heating by conduction. *Food Technol., 8*, 258–263.

Evans, L. B., & Stefany, N. E. (1966). An experimental study of transient heat transfer to liquids in cylindrical enclosures. *Chem. Eng. Prog. Symp. Ser., 62*, (64), 209–215.

Evans, L. B., Reid, R. C., & Drake, E. M. (1968). Transient natural convection in a vertical cylinder. *AIChE J., 14*, 251–259.

Fagan, M. J. (1992) *Finite element analysis – Theory and practice*. Harlow, UK: Longman Scientific and Technical.

Fernandez, C. L., Rao, M. A., Rajavasireddy, S. P., & Sastry, S. K. (1988). Particulate heat transfer to canned snap beans in a steritort. *J. Food Proc. Preserv., 10*, 183–198.

Flambert, C. M. F., & Deltour, J. (1971). Calculation of the temperatures of foods in cylindrical cans heating by conduction. *Ind. Alim. Agric., 88*(9), 1279–1287 [in French].

Flambert, C. M. F., & Deltour, J. (1973a). Calculation of the lethality of conduction heat treatments applied to food in cylindrical cans. *Ind. Alim. Agric., 90*(1), 5–10 [in French].

Flambert, C. M. F., & Deltour, J. (1973b). Exact temperature and lethality calculation for sterilizing process. In Leniger, H. (Ed.), *1st int. congress on heat and mass transfer in food processing*. The Netherlands: Wageningen.

Fourier, J. B. J. (1822). *Théorie analytique de la chaleur*. France: Paris. [English translation by A. Freeman, 1955, New York: Dover Publications Inc.]

Fujita, H. (1952). On a certain non-linear problem of thermal conduction in a can. *Bull. Jap. Soc. Sci. Fish., 17*, 393–400 [in Japanese].

George, R. M. (1990). A literature survey of the thermal diffusivity of food products. *Technical Bulletin* No. 73, Chipping Campden, Glos., UK: Campden & Chorleywood Food Research Association.

George, R. M. (1993). Making (micro)waves. *Food Processing, 62*(5), 23–28.

George, R. M., & Campbell, G. M. (1994). The use of metallic packaging to improve heating uniformity and process validation during microwave sterilization. In R. Field (Ed.), *Food Process Engineering*, IChemE Symposium, (pp. 219–225), Univ. Bath, UK.

Ghani, A. G., Farid, M. M., Chen, X. D., & Richards, P. (1999a). Numerical simulation of natural convection heating of canned food by computational fluid dynamics. *J. Food Eng., 41*(1), 55–64.

Ghani, A. G., Farid, M. M., Chen, X. D., & Richards, P. (1999b). An investigation of deactivation of bacteria in a canned liquid food during sterilization using computational fluid dynamics (CFD). *J. Food Eng., 42*(4), 207–214.

Ghani, A. G., Farid, M. M., Chen, X. D., & Richards, P. (2001a). Thermal sterilization of canned food in a 3-D pouch using computational fluid dynamics. *J. Food Eng., 48*(2), 147–156.

Ghani, A. G. A., Farid, M. M., Chen, X. D. & Watson, C. (2001b). Numerical simulation of transient two-dimensional profiles of temperature and flow of liquid food in a can during sterilization. In J. Welti-Chanes, G. V. Barbosa-Cánovas, & J. M. Aguilera (Eds.), *Eighth International Conference Engineering and Food*, ICEF8, Vol. 2 (pp. 1832–1837). Lancaster, PA: Technomic Pub. Co.

Ghani, A. G., Farid, M. M., Chen, X. D., & Richards, P. (2002a). Heat transfer in a 3-D pouch using computational fluid dynamics. *J. PHOENICS, 12*(3), 293–305.

Ghani, G. A., Farid, M. M., & Chen, X. D. (2002b). Numerical simulation of transient temperature and velocity profiles in a horizontal can during sterilization using computational fluid dynamics. *J. Food Eng., 51*(1), 77–83.

Ghani, A. G., Farid, M. M., & Chen, X. D. (2002c). Theoretical and experimental investigation of the thermal inactivation of *Bacillus stearothermophilus* in food pouches. *J. Food Eng. 51*(3), 221–228.

Ghani, A. G., Farid, M. M., & Zarouk, S. J. (2003). The effect of can rotation on sterilization of liquid food using computational fluid dynamics. *J. Food Eng., 57*(1), 9–16.

Gillespy, T. G. (1951). Estimation of the sterilizing values of processes as applied to canned foods. I – Packs heating by conduction. *J. Sci. Food Agric., 2*, 107–125.

Gillespy, T. G. (1953). Estimation of the sterilizing values of processes as applied to canned foods. II – Packs heating by conduction: complex processing conditions and value of coming-up time of retort. *J. Sci. Food Agric., 4*, 553–565.

Gouvaris, A. K., & Scholefield, J. (1988). Comparisons of a computer evaluation with a standard evaluation of retort pouch thermal processing. *Int. J. Food Sci. Technol., 23*, 601–606.

Guerrei, G. (1993). Pasteurization of beverages: unsteady state heat transfer model and process analysis. *Food & Bioproducts Processing, Trans. IChemE, 71*, C2, 67–76.

Hallström, B., Skjöldebrand, C., & Trägärdh, C. (1988). *HeattTransfer and food products*, London: Elsevier Applied Science.

Hammitt, F. G., & Chu, P. T. (1962). Transient internal natural convection heating and cooling of closed, vertical, cylindrical vessels. *Amer. Soc. Mech. Eng.* Paper 62-WA-309, pp. 2–12.

Hassan, B. H. (1984). Heat-transfer coefficients from particles in liquid in axially rotating cans. PhD thesis, Univ. California, Davis.

Hayakawa, K. (1964). Development of formulas for calculating the theoretical temperature history and sterilizing value in a cylindrical can of thermally conductive food during heating. PhD thesis, Rutgers State Univ., New Brunswick, NJ.

Hayakawa, K. (1969). Estimating central temperatures of canned food during the initial heating and cooling period of heat process. *Food Technol.*, *23*(11), 1473–1477.

Hayakawa, K. (1970). Experimental formulas for accurate estimation of transient temperature of food and their application to thermal process evaluation. *Food Technol.*, *24*(12), 1407–1418.

Hayakawa, K. (1971). Estimating food temperatures during various processing or handling treatments. *J. Food Sci.*, *36*, 378–385.

Hayakawa, K. (1974). Response charts for estimating temperatures in cylindrical cans of solid food subjected to time variable processing temperatures. *J. Food Sci.*, *39*, 1090–1098.

Hayakawa, K. (1975). Roots of characteristic equations for heat conduction or mass diffusion in an infinite cylinder or sphere. *Lebensm.-Wiss. u. -Technol.*, *8*, 231–233.

Hayakawa, K., & Ball, C. O. (1968). A note on theoretical heating curve of a cylindrical can of thermally conductive food. *Can. Inst. Food Sci. Technol. J.*, *1*(2), 54–60.

Hayakawa, K., & Ball, C. O. (1969a). Charts for calculating average temperature of thermally conductive food in a cylindrical can during heat processing. *Can. Inst. Food Sci. Technol. J.*, *2*(10), 12–19.

Hayakawa, K., & Ball, C. O. (1969b). A note on theoretical cooling curves of a cylindrical can of thermally conductive food. *Can. Inst. Food Sci. Technol. J.*, *2*(3), 115–119.

Hayakawa, K., & Ball, C. O. (1971). Theoretical formulas for temperatures in cans of solid food and for evaluating various heat processes. *J. Food Sci.*, *36*(2), 306–310.

Hayakawa, K. & Giannoni-Succar, E. B. (1996). Modified Duhamel's theorem for variable. *J. Food Eng.*, *27*(2), 409–422.

Hayakawa, K., & Succar, J. (1983). A method for determining the apparent thermal diffusivity of spherical foods. *Lebensm. -Wiss. u. -Technol.*, *16*, 373–375.

Hayakawa, K., & Villalobos, G. (1989). Formulas for estimating Smith *et al.* parameters to determine mass average temperature of irregular shaped bodies. *J. Food Proc. Eng.*, *11*, 237–256.

Hayakawa, K., Giannoni-Succar, E. B., Huang, F., & Zhou, L. (1997). Use of empirical temperature response function for modified Duhamel's theorem application. *J. Food Eng.*, *34*(3), 331–353.

Heldman, D. R., & Singh, R. P. (1980). *Food process engineering*, Section 3.4.6. New York:Van Nostrand Reinhold/AVI.

Hendrickx, M. (1988). The application of transmission line matrix (TLM) modelling to food engineering. Doctoral thesis, Univ. Leuven, Belgium.

Hendrickx, M., Silva, C., Oliveira, F., & Tobback, P. (1992). Optimizing thermal processes of conduction heated foods: generalized equations for optimal processing temperatures. In "*Food Engineering in a Computer Climate*" (pp. 271–276), pub. Institution of Chemical Engineers, Rugby, U.K.

Hendrickx, M., Silva, C., Oliveira, F., & Tobback, P. (1993). Generalized semi-empirical formulas for optimal sterilization temperatures of conduction-heated foods with infinite surface heat-transfer coefficients. *J. Food Eng.*, *19*(2), 141–158.

Hicks, E. W. (1951). On the evaluation of canning processes. *Food Technol.*, *5*(4), 132–142.

Hiddink, J. (1975). Natural convection heating of liquids, with reference to sterilization of canned foods. Doctoral thesis *Agric. Research Report No. 839*. Netherlands Wageningen: Agricultural. University.

Hiddink, J., Schenk, J., & Bruin, S. (1976), Natural convection heating of liquids in closed containers. *Appl. Sci. Res.*, *32*, 217–235.

Holdsworth, S. D., & Overington, W. J. G. (1975) Calculation of the sterilizing value of food canning processes. *Technical Bulletin No. 28*. Chipping Campden, Glos., UK: Campden & Chorleywood Food Research Association.

Hughes, J. P., Jones, T. E. R. & James, P. W. (2003). Numerical simulation and experimental visualization of the isothermal flow of liquid containing a headspace bubble inside a closed cylinder during off-axis rotation. *Food and Bioproducts Processing, Trans IChemE, 81*(C2), 119–128.

Hurwicz, H., & Tischer, R. G. (1952). Heat processing of beef. I. A theoretical consideration of the distribution of temperature with time and in space during processing. *Food Res.*, *17*, 380–392.

Ikegami, Y. (1977). Heat penetration in canned foods containing solids and liquid. *Canners' J.*, *56*(7), 548–552.

Ikegami, Y. (1978). Heat transfer and food surface temperature change with time. *Canners' J.*, *57*(7), 593–596 [in Japanese].

Ingersoll, L. R., Zobel, O. J., & Ingersoll, A. C. (1953). *Heat conduction*. London: Thames and Hudson.

Iwata, G. (1940). Temperature–time relation for oval cans during heating. *Bull. Jap. Soc. Sci. Fish.*, *9*, 117–120 [in Japanese].

Jakobsen, F. (1954). Notes on process evaluation. *Food Res.*, *19*, 66–79.

Jaluria, Y. (1980). *Natural convection heat and mass transfer*. Oxford: Pergamon Press.

James, P. W., Hughes, J. P., & Jones, P. R. (2001). Numerical and experimental simulation of the flow in rotating containers. NAFEMS Int. J. CFD Case Studies, 3, 31–60.

Johns, W. R (1992a). Simulation of food processes with uncertain data. In W. Hamm & S. D. Holdsworth (Eds.), *Food engineering in a computer climate"* (pp. 1–24). Rugby, U.K: Institution of Chemical Engineers.

Johns, W. R. (1992b). Simulation of food processes with uncertain data. *Food & Bioproducts Processing, Trans. IChemE, 70*, C2, 59–68.

Jones, D. E. A. (1931). Heat penetration by convection. *Food Technol. (UK)*, *1*, 63–65.

Jones, D. E. A. (1932). The cooling of foodstuffs. *Food Technol. (UK)*, *1*, 214–216.

Jowitt, R., & Mynott, A. R. (1974). Some factors affecting natural convective heat transfer to canned foods. *Dechema Monographien*, *77*, 153–164.

Jun, S., & Sastry, S. (2005). Modeling and optimisation of ohmic heating of foods inside a flexible package. *J. Food Proc. Eng.*, *28*(4), 417–436.

Kent, M. (1987) *Electric and dielectric properties of food materials*. London: Science and Technology Publishers.

Kim, K. H., & Teixeira, A. A. (1997). Predicting internal temperature response to conduction-heating of odd-shaped solids. *J. Food Process Eng.*, *20*(1), 51–63.

Kumar, A., & Battacharya, M. (1991). Transient temperature and velocity profiles in a canned non-Newtonian liquid food during sterilization in a still-cook retort. *Int. J. Heat Mass Transfer*, *34*(4/5), 1083–1096.

Kumar, A., Bhattacharya, M., & Blaylock, J. (1990). Numerical simulation of natural convection heating of canned thick viscous liquid food products. *J. Food Sci.*, *55*, 1403–1411, 1420.

Langstroth, G. O. (1931). Thermal investigations on fish muscle. *Control Can. Biol. Fish.*, *6* (NS), 377–389.

Lanoiseuille, J.-L., Canadau Y., & Debray, E. (1995). Predicting internal temperature of canned foods during thermal processing using a linear recursive model. *J. Food Sci., 60*(4), 833–835, 840.

Larkin, J. W., & Steffe, J. F. (1982). Error analysis in estimating thermal diffusivity from heat penetration data. *J. Food Proc. Eng., 6*, 135–158.

Lekwauwa, A. N., & Hayakawa, K. (1986). Computerized model for the prediction of thermal responses of packaged solid–liquid food mixture undergoing thermal processes. *J. Food Sci., 51*(4), 1042–1049, 1056.

Lenz, M. K., & Lund, D. B. (1977). The lethality–Fourier number method: experimental verification of a model for calculating temperature profiles and lethality in conduction-heating canned foods. *J. Food Sci., 42*(4), 989–996, 1001.

Lenz, M. K., & Lund, D. B. (1978). The lethality–Fourier number method. Heating rate variations and lethality confidence intervals for forced-convection heated foods in containers. *J. Food Proc. Eng., 2*(2), 227–271.

Leonhardt, G. F. (1976a). Estimation of the central temperature of thermally conductive food in cylindrical and rectangular cans during heat processing. *J. Food Sci., 41*, 685–690.

Leonhardt, G. F. (1976b). Estimation of the average temperature of thermally conductive food in cylindrical and rectangular cans during heat processing. *J. Food Sci., 41*, 691–695.

Lewis, M. J. (1987). *Physical properties of food processing system*. Chichester, UK: Ellis Horwood.

Luikov, A. V. (1968). *Analytical heat diffusion theory*. New York: Academic Press.

Lund, D. B., & Norback, J. P. (1983). Heat transfer and selected topics. In I. Saguy (Ed.), *Computer-aided techniques in food technology* (pp. 137–148). New York: Marcel Dekker.

Maesmans, G., Hendrickx, M., DeCordt, S., Fransis, A., & Tobback, P. (1992). Fluid-to-particle heat-transfer coefficient determination of heterogeneous foods: a review. *J. Food Proc. Preserv., 16*, 29–69.

Magnus, W., Oberhettinger, F., & Soni, R. P. (1966). *Formulas and theorems for the special functions of mathematical physics*. Berlin: Springer-Verlag.

Manson, J. E., Zahradnik, J. W., & Stumbo, C. R. (1970). Evaluation of lethality and nutrient retentions of conduction-heating food in rectangular containers. *Food Technol., 24*(11), 1297–1301.

Manson, J. E., Stumbo, C. R., & Zhradnik, J. W. (1974). Evaluation of thermal processes for conduction heating foods in pear-shaped containers. *J. Food Sci., 39*, 276–281.

Márquez, C. A., De Michelis, A., Salvadori, V. O., & Mascheroni, R. H. (1998). Application of transfer functions to the thermal processing of particulate foods enclosed in a liquid medium. *J. Food Eng., 38*(2), 189–205.

Martens M., Scheerlinck, N., De Belie, N., & De Baerdemaecker, J. (1991). Numerical model for the combined simulation of heat transfer and enzyme inactivation kinetics in cylindrical vegetables. *J. Food Eng. 47*(3), 185–193.

Masilov, V. A., & Medvedev, O. K. (1967). Physical constants for tomato products. *Pisch. Tekhnol., 4*, 78–79 [in Russian].

McGinnis, D. S. (1986.) Prediction of transient conduction heat transfer in foods packaged in flexible retort pouches. *Can. Inst. Food Sci. Technol. J., 19*(4), 148–157.

Merson, R. L. and Stoforos, N. G. (1990). Motion of spherical particles in axially rotating cans. Effect of liquid-particle heat transfer. In W. E. L. Spiess., & H. Schubert (Eds.), *Engineering and food* Vol. 2, (pp. 60–69). London: Elsevier Applied Science.

Merson, R. L., Leonard, S. J., Meija, E., & Heil, J. (1981). Temperature distributions and liquid-side heat-transfer coefficients in model liquid foods in cans undergoing flame sterilization heating. *J. Food Proc. Eng.*, *4*, 85–97.

Merson, R. L., Singh, R. P., & Carroad, P. A. (1978). An evaluation of Ball's formula method of thermal process calculations. *Food Technol.* *32*(3): 66–76.

Metaxas, A. C. and Meredith, R. J. (1983). *Industrial microwave heating*. London: Peregrinus.

Mickley, H. S., Sherwood, T. K., & Reed, C. E. (1957.) *Applied mathematics in chemical engineering* (2nd Edition). New York: McGraw-Hill.

Miglioli, L., Massini, R., Padrelli, T., & Cassara, A. (1983). Mechanism of heat transfer by natural convection: heating of water inside a cylindrical container. *Industria Conserve*, *58*, 158–163 [in Italian].

Miles, C. A., Beek, G. van, & Veerkamp, C. H. (1983). Calculation of the thermophysical properties of foods. In R. Jowitt *et al.* (Eds.), *Physical properties of foods* (pp. 269–312). London: Applied Science Publishers.

Minkowycz, W. J., Sparrow, E. M., Schneider, G. E., & Fletcher, R. H. (1988). *Handbook of numerical heat transfer*. New York:Wiley.

Mirespassi, T. J. (1965). Heat transfer tables for time-variable boundary temperatures in slabs. *Brit. Chem. Eng.*, *10*(11), 764–769.

Mohamed, I. O. (2003). Computer simulation of food sterilization using an alternating direction implicit finite difference method. *J. Food Eng.*, *60*(3), 301–306.

Mudgett, R. E. (1986a). Electrical properties of foods. In M. A. Rao, & S. S. H. Rizvi (Eds.), *Engineering properties of foods* (pp. 329–390). New York: Marcel Dekker.

Mudgett, R. E. (1986b). Microwave properties of heating characteristics of foods. *Food Technol.*, *40*, 84–93.

Nahor, H. B., Scheerlinck, N., Verniest, R., De Baerdemaeker, J., & Nicolai, B. M. (2001). Optimal experimental design for the parameter estimation of conduction heated foods. *J. Food Eng,*. *48*(2), 109–119.

Naveh, D. (1982). Analysis of transient conduction heat transfer in thermal processing of foods using the finite-element method. PhD thesis, Univ. of Minnesota, Minneapolis.

Naveh, D., & Kopelman, I. J. (1980). Effects of some processing parameters on heat transfer coefficients in a rotating autoclave. *J. Food Proc. Preserv.*, *4*(4), 67–77.

Naveh, D., Kopelman, I. J., & Pflug, I. J. (1983a). The finite element method in thermal processing of foods. *J. Food Sci.*, *48*, 1086–1093.

Naveh, D., Kopelman, I. J., Zechman, L., & Pflug, I. J. (1983b). Transient cooling of conduction heating products during sterilization: temperature histories. *J. Food Proc. Preserv.*, *7*, 259–273.

Newman, M., & Holdsworth, S. D. (1989). Methods of thermal process calculation for food particles. *Technical Memorandum No. 321 (revised)*. Chipping Campden, Glos., UK: Campden & Chorleywood Food Research Association.

Nicolai, B. M., & De Baerdemaeker, J. (1992). Simulation of heat transfer in foods with stochastic initial and boundary conditions. *Food and Bioproducts Processing Trans. IChemE*, *70*, Part C, 78–82.

Nicolai, B. M., & De Baerdemaeker, J. (1996). Sensitivity analysis with respect to the surface heat-transfer coefficient as applied to thermal process calculations. *J. Food Eng.*, *28*(1), 21–33.

Nicolaï, B. M., & De Baerdemaeker, J. (1997). Finite element perturbation analysis of non-linear heat conduction problems with random field parameters. *Int. J. Numerical Methods for Heat and Fluid Flow*, *7*(5), 525–544.

Nicolai, B. M., Verboven, P., & Sheerlinck, N. (2001). Modelling and simulation of thermal processes. In P. Richardson (Ed.), *Thermal technologies in food processing* (pp. 92–112). Cambridge: Woodhead Publishing.

Nicolaï, B. M., Verboven, P., Scheerlinck, N., & De Baerdemaeker, J. (1998). Numerical analysis of the propagation of random parameter fluctuations in time and space during thermal processes. *J. Food Eng., 38*, 259–278.

Nicolai, B. M., Van Den Broek, P., Skellekens, M., De Roek, G., Martens, T., & De Baerdemaeker, J. (1995). Finite element analysis of heat conduction in lasagne during thermal processing. *Int. J. Food Sci. Technol., 30*(3), 347–363.

Noh, B. S., Heil, J. R., & Patino, H. (1986). Heat transfer study on flame pasteurization of liquids in aluminum cans. *J. Food Sci., 51*(3), 715–719.

Norohna, J., Hendrickx, M., Van Loey, A., & Tobback, P. (1995). New semi-empirical approach to handle time-variable boundary conditions during sterilization of non-conductive heating foods. *J. Food Eng., 24*(2), 249–268.

Ohlsson, T., & Bengtsson, N. E. (1975). Dielectric food data for microwave sterilization processing. *J. Microwave Power, 10*, 93–108.

Okada, M. (1940a). Cooling of canned foods. *Bull. Jap. Soc. Sci. Fish., 9*, 64–68 [in Japanese].

Okada, M. (1940b). Natural convection in can-shaped space, referred to the laws of similtude. *Bull. Jap. Soc. Sci. Fish. 8*, 325–327 [in Japanese].

Okada, M. (1940c). Cooling of rectangular cans of food. *Bull. Jap. Soc. Sci. Fish., 9*, 208–213 [in Japanese].

Okos, M. R. (Ed.). (1986). *Physical and chemical properties of food*. St Joseph, MI.: American Society of Agricultural Engineers.

Olivares, M., Guzman, J. A., & Solar, I. (1986). Thermal diffusivity of non-homogeneous food. *J. Food Proc. Preserv., 10*, 57–67.

Olson, F. C. W. (1947). Recent Japanese researches on canning technology. *Food Technol., 1*, 553–555.

Olson, F. C. W., & Jackson, J. M. (1942). Heating curves: theory and practical applications. *Ind. Eng. Chem., 34*(3), 337–341.

Olson, F. C. W., & Schultz, O. T. (1942). Temperature in solids during heating and cooling. *Ind. Eng. Chem., 34*(7), 874–877.

Ozisik, M. N. (1980). *Heat Conduction*. Scranton, PA.: International Text Book Co.

Palazoglu. K. (2006). Influence of convection heat-transfer coefficient on the heating rate of materials with different thermal diffusivities. *J. Food Eng., 73*(3), 290–296.

Patkai, G., Kormendy, I., & Erdelyi, M. (1990). Outline of a system for the selection of the optimum sterilization process for canned foods. Part II. The determination of heat-transfer coefficients and conductivities in some industrial equipments for canned products. *Acta Alimentaria, 19*(4), 305–320.

Paulus, K., & Ojo, A. (1974). Heat transfer during flame sterilization. *Proc. IVth Congress Food Sci. Technol.*, Vol. 4, pp. 443–448.

Peralta Rodriguez, R. D. (1982). Heat transfer to simulated canned liquid foods undergoing sterilization. PhD thesis, Univ. California, Davis.

Peralta Rodriguez, R. D. (1987a). Heat Transfer in Flame Sterilization of Liquid Food Simulants. *Technical Bulletin No. 62*. Chipping Campden, Glos., UK: Campden & Chorleywood Food Research Association.

Peralta Rodriguez, R. D. (1987b). Microcomputer routines for calculations of Bessel functions and roots of characteristic equations for heat or mass diffusion in bodies with cylindrical geometries. *Lebensm. -Wiss. u. -Technol., 20*, 309–310.

Peralta Rodriguez, R. D., & Merson, R. L. (1982). Heat transfer and chemical kinetics during flame sterilization. *Food Proc. Eng., AIChE Symp. Series, No. 218* (78), 58–67.

Peralta Rodriguez, R. D., & Merson, R. L. (1983). Experimental verification of a heat transfer model for simulated liquid foods undergoing flame sterilization. *J. Food Sci., 48*, 726–733.

Perez-Martin, R. I., Banga, J. R., & Gallard, J. M. (1990). Simulation of thermal processes in tuna can manufacture. In W. E. L. Spiess & H. Schubert (Eds.), *Engineering and food* (pp. 848–856), London: Elsevier Applied Science.

Perry, J. H., & Green, M. (Eds.) (1984). *Perry's chemical engineers' handbook* (6th Edition). New York: McGraw-Hill.

Peterson, W. R., & Adams, J. P. (1983). Water velocity effect on heat penetration parameters during institutional size retort pouch processing. *J. Food Sci., 49*, 28–31.

Plazl, I., Lakner, M., & Koloini, T. (2006). Modelling of temperature distributions in canned tomato based dip during industrial pasteurization. *J. Food Eng., 75*(3), 400–406.

Pornchaloempong, P., Balaban, M. O., & Chau, K. V. (2001). Simulation of conduction heating in conical shapes. In J. Welti-Chanes, G. V., Barbosa-Cánovas, & J. M. Aguilera (Eds.), *Engineering and food, ICEF8, Vol.1* (pp. 671–675). Lancaster, PA: Technomic Pub. Co.

Pornchaloempong, P., Balaban, M. O., Teixeira, A. A., & Chau, K. V. (2003). Numerical simulation of conduction heating in conically shaped bodies. *J. Food Proc. Eng., 25*, 539–555.

Price, R. B., & Bhowmik, S. R. (1994). Heat transfer in canned foods undergoing agitation. *J. Food Eng., 23*(4), 621–629.

Puri, V. M., & Anantheswaran, R. C. (1993). The finite-element method in food processing: a review. *J. Food Eng., 19*(3), 247–274.

Quarini, J., & Scott, G. (1997). Transient natural convection in enclosures filled with non-Newtonian fluids. In R. Jowitt (Ed.), *Engineering and food*, ICEF7 Part 2, §K, (pp. 49–53). Sheffield, UK: Sheffield Academic Press.

Quast, D. G., & Siozawa, Y. Y. (1974). Heat transfer rates during heating of axially rotated cans. *Proc. IVth Int. Congress Food Sci. and Technol.*, Vol. 4, (pp. 458–468).

Ramaswamy, H. S., Lo, K. V., & Tung, M. A. (1982). Simplified equations for transient temperatures in conduction foods with convective heat transfer at the surface. *J. Food Sci., 47*(6), 2042–2065.

Ramaswamy, H. S., Tung, M. A., & Stark, R. (1983). A method to measure surface heat transfer from steam/air mixtures in batch retorts. *J. Food Sci., 48*, 900–904.

Rao, M. A., & Anantheswaran, R. C. (1988). Convective heat transfer in fluid foods in cans. *Adv. Food Res., 32*, 39–84.

Richardson, P. S., & Holdsworth, S. D. (1989). Mathematical modelling and control of sterilization processes. In R. W. Field & J. A. Howell(Eds.), *Process engineering in the food industry* (pp. 169–187). London: Elsevier Applied Science.

Riedel, L. (1947). The theory of the heat sterilization of canned foods. *Mitt. Kältetech. Inst. Karlsruhe*, No. 1, 3–40 [in German].

Rowley, A. T. (2001). Radio frequency heating. In P. Richardson (Ed.), *Thermal technologies in food processing* (pp. 163–177). Cambridge, UK: Woodhead Publishing Ltd.

Ruckenstein, E. (1971). Transport equations in transfer co-ordinates. *Chem. Eng. Sci., 26*, 1795–1802.

Rumsey, T. R. (1984). Modeling heat transfer in cans containing liquid and particulates. *AMSE Paper* No. 84–6515. St Joseph, MI.: American Society of Mechanical Engineers.

Ryynänen, S. (1995). The electromagnetic properties of food materials. *J. Food Eng.*, *26*(4), 409–429.

Sablani, S. S., & Ramaswamy, H. S. (1993). Fluid/particle heat-transfer coefficients in cans during end-over-end processing. *Lebens. –Wiss. u. –Technol.*, *26*(6), 498–501.

Sablani, S. S., Ramaswamy, H. S., & Prasher, S. O. (1995). A neural network approach for thermal processing applications. *J. Food Processing Preserv.*, *19*, 283–301.

Sablani, S. S., & Ramaswamy, H. S. (1996). Particle heat-transfer coefficients under various retort operation conditions with end-over-end rotation. *J. Food Process Eng.*, *19*(4), 403–424.

Sablani, S. S., & Ramaswamy, H. S. (1997a). Heat transfer to particles in cans with end-over-end rotation: influence of particle size and concentration(%v/v). *J. Food Process Eng.*, *20*(4), 265–283.

Sablani, S. S., Ramaswamy, H. S., Sreekanth, S., & Prasher, S. O. (1997b). Neural network modeling of heat transfer to liquid particle mixtures in cans subjected to end-over-end processing. *Food Research Int.*, *30*(2), 105–116.

Sablani, S. S., Ramaswamy, H. S., & Mujumdar, A. S. (1997). Dimensionless correlations or convective heat transfer to liquid and particles in cans subjected to end-over-end rotation. *J. Food Eng.*, *34*(4), 453–472.

Sablani, S. S., & Ramaswamy H. S. (1998). Multi-particle mixing behavior and its role in heat transfer during end-over-end agitation of cans. *J. Food Eng.*, *38*(2), 141–152.

Salvadori, V. O., Mascheroni, R. H., Sanz, P. D., & Dominguez Alsonso, M. (1994a). Application of the *z*-transfer functions to multidimensional heat transfer problems. *Latin Amer. Appl. Res.*, *24*, 137–147.

Salvadori, V. O., Sanz, P. D., Dominguez Alsonso, M., & Mascheroni, R. H. (1994b). Application of z-transfer functions to heat and mass transfer problems; their calculation by numerical methods. *J. Food Eng.*, *23*(3), 293–307.

Sargent, R. W. H., Perkins, J. D., & Thomas S. (1982). SPEEDUP: Simulation program for the economic evaluation and design of unified processes. In M. E. Lesley (Ed.), *Computer-aided Process Plant Design*. Houston: Guelph.

Sastry, S. K. (1984). Convective heat-transfer coefficients for canned mushrooms processed in still retorts. *ASME Paper* No. 84–6517. St Joseph, MI.: American Society of Mechanical Engineers.

Sastry, S. K., Beelman, R. B., & Speroni, J. J. (1985). A three-dimensional finite element model for thermally induced changes in foods: application to degradation of agaritine in canned mushrooms. *J. Food Sci.*, *50*, 1293–1299, 1326.

Scheerlinck, N., & Nicolaï, B. M. (1999). User's manual. Leuven, Belgium: Katholieke Univ., Laboratory of Postharvest Technology.

Schultz, O. T., & Olson, F. C. W. (1938). Thermal processing of canned foods in tin containers. I. Variation of heating rate with can size for products heating by convection. *Food Res.*, *3*, 647–651.

Segerlind, L. J. (1984). *Applied finite element analysis*, Wiley, New York.

Sheen, S., Tong, C.-H., Fu, Y., & Lund, D. B. (1993). Lethality of thermal processes for food in anomalous-shaped plastic containers. *J. Food Eng.*, *20*(3), 199–213.

Shiga, I. (1970). Temperatures in canned foods during processing. *Food Preserv. Quarterly*, *30*(3), 56–58.

Shin, S., & Bhowmik, S. R. (1990). Computer simulation to evaluate thermal processing of food in cylindrical plastic cans. *J. Food Eng.*, *12*(2), 117–131.

Shin, S., & Bhowmik, S. R. (1993). Determination of thermophysical properties of plastic cans used for thermal sterilization of foods. *Lebens. -Wiss. u. –Technol.*, *26*(5), 476–479.

Silva, C. L. M., Hendrickx, M., Oliveira, F., & Tobback, P. (1992). Optimum sterilization temperatures for conduction heating food considering finite surface heat transfer coefficients. *J. Food Sci.*, *57*(3), 743–748.

Silva, C. L. M., Oliveira, F. A. R., & Hendrickx, M. (1994). Quality optimization of conduction heating foods sterilized in different packages. *Int. J. Food Sci. Technol.*, *29*, 515–530.

Simpson, R., Almonacid, S., & Teixeira, A. (2003). Bigelow's General Method Revisited: Development of a New Calcualtion Technique. *J. Food Sci.*, *68*(4), 1324–1333.

Simpson, R., Almonacid, S., & Mitchell, M. (2004). Mathematical model development, experimental validations and process optimization: retortable pouches packed with seafood in cone fustrum shape. *J. Food Eng.*, *63*(2), 153–162.

Simpson, R., Aris, I., & Torres, J. A. (1989). Sterilization of conduction heated foods in oval-shaped containers. *J. Food Sci.*, *54*(5), 1327–1331, 1363.

Singh, R. P. (1982) Thermal diffusivity in thermal processing. *Food Technol.*, *36*(2), 87–91.

Singh, R. P., & Segerlind, L. J. (1974). The finite element method in food engineering. *ASAE Paper No. 74–6015*. St Joseph, MI.: American Society of Agricultural Engineers.

Skinner, R. H. (1979). The linear second-order system as an empirical model of can-centre temperature history. *Proc. Int. Meeting Food Microbiol. and Technol.* (pp. 309–322). Parma, Italy.

Smith, G. D. (1974). *Numerical solution of partial differential equations*, London: Oxford University Press.

Smith, R. E. (1966). Analysis of transient heat transfer from anomalous shapes with heterogeneous properties. PhD thesis, Oklahoma State Univ., Stillwater.

Smith, R. E., Nelson, G. L., & Henrickson, R. L. (1967). Analyses on transient heat transfer from anomalous shapes. *Trans. ASAE.*, *10*(2), 236–245.

Smith, R. E., Nelson, G. L., & Henrickson, R. L. (1968). Applications of geometry analysis of anomalous shapes to problems in transient heat transfer. *Trans. ASAE.*, *11*(2), 296–303.

Soule, C. L., & Merson, R. L. (1985). Heat-transfer coefficients to Newtonian liquids in axially rotated cans. *J. Food Proc. Eng.*, *8*(1), 33–46.

Stevens, P. M. (1972). Lethality calculations, including effects of product movement, for convection heating and for broken-heating foods in still-cook retorts. PhD thesis, Univ. Massachusetts, Amherst.

Stoforos, N. G. (1988). Heat transfer in axially rotating canned liquid/particulate food systems. PhD thesis, Dept. Agric. Eng., Univ. California, Davis.

Stoforos, N. G., & Merson, R. L. (1990). Estimating heat-transfer coefficients in liquid/particulate canned foods using only liquid temperature data. *J. Food Sci.*, *55*(2), 478–483.

Stoforos, N. G., & Merson, R. L. (1991). Measurements of heat-transfer coefficients in rotating liquid/particle systems. *Biotechnol. Progress*, *7*, 267–271.

Stoforos, N. G., & Merson, R. L. (1992). Physical property and rotational speed effects on heat transfer in axially rotating liquid/particle canned foods. *J. Food Sci.*, *57*(3), 749–754.

Stoforos, N. G., & Merson, R. L. (1995). A solution to the equations governing heat transfer in agitating liquid/particle canned foods. *J. Food Process Eng.*, *18*(2), 165–186.

Storofos, N. G., Noronha, J., Hendrickx, M., & Tobback, P. (1997). Inverse superposition for calculating food product temperatures during in-container thermal processing. *J. Food Sci., 62*(2), 219–224, 248.

Taggert, R., & Farrow, F. D. (1941). Heat penetration into canned foods. Part 1. *Food, 16*, 325–330.

Taggert, R., & Farrow, F. D. (1942). Heat penetration into canned foods. Part 2. *Food, 17*, 13–17.

Tan, C.-S., & Ling, A. C. (1988). Effect of non-uniform heat transfer through can surfaces on process lethality of conduction heating foods. *Can. Inst. Food Sci. J., 21*(4), 378–385.

Tani, S. (1938a). Heat conduction in canned foods. *Bull. Jap. Soc. Sci. Fish., 8*, 79–83 [in Japanese].

Tani, S. (1938b). Natural convection in can-shaped space. *Bull. Jap. Soc. Sci. Fish., 8*, 76–78 [in Japanese].

Tatom, J. W., & Carlson, W. O. (1966). Transient turbulent free convection in closed containers. *Proc. 3rd Int. Heat Transfer Conf. (ASME)*, Vol. 2, pp. 163–171.

Tattiyakul, J., Rao, M. A., & Datta, A. K. (2001). Simulation of heat transfer to a canned corn starch dispersion subjected to axial rotation. *Chem Eng. & Process., 40*, 391–99.

Tattiyakul, J., Rao, M. A., & Datta, A. K. (2002a). Heat transfer to a canned starch dispersion under intermittent agitation. *J. Food Eng., 54*(4), 321–329.

Tattiyakul, J., Rao, M. A., & Datta, A. K. (2002b). Heat transfer to three canned fluids of different thermo-rheological behaviour under intermittent agitation. *Food and Bioproducts Processing, Trans. IChemE, 80*(C1), 20–27.

Teixeira, A. A., Dixon, J. R., Zahradnik, J. W., & Zinsmeister, G. E. (1969). Computer determination of spore survival distributions in the thermal processing of conduction-heated foods. *Food Technology, 23*(3), 352–354.

Teixeira, A. A., Tucker, G. S., Balaban, M. O., & Bichier, J. (1992). Innovations in conduction-heating models for on-line retort control of canned foods with any *j*-value. In *Advances in food engineering* P. R. Singh & M. A. Wirakartakusumah (Eds.), (pp. 293–308). Boca Raton, FL: CRC Press.

Teixeira Neto, R. O. (1982). Heat transfer rates to liquid foods during flame-sterilization. *J. Food Sci., 47*, 476–481.

Thijssen, H. A. C., & Kochen, L. H. P. J. M. (1980). Short-cut method for the calculation of sterilizing conditions for packed foods yielding optimum quality retention at variable temperature of heating and cooling medium. In *Food process engineering*, Vol. 1 (pp. 122–136). London:Applied Science Publishers.

Thijssen, H. A. C., Kerkhoff, P. J. A. M., & Liefkens, A. A. A. (1978). Short-cut method for the calculation of sterilization conditions yielding optimum quality retention for conduction-type heating of packaged foods. *J. Food Sci., 43*(4), 1096–1101.

Thompson, G. E. (1919). Temperature-time relations in canned foods during sterilization. *J. Ind. Eng. Chem., 11*, 657–664.

Thompson, G. E. (1922). Heat flow in a finite cylinder having variable surface geometry. *Phys. Rev.* (2nd Series), *20*, 601–606.

Thorne, S. (1989). Computer prediction of temperatures in solid foods during heating or cooling. In S. Thorne (Ed.), *Developments in food preservation* (pp. 305–324). London: Applied Science Publishers.

Thorpe, R. H., & Atherton, D. (1969). Processing studies on canned new potatoes. In J. Woodman (Ed.), *Potatoes for canning, AMDEC Annual progress report 1969*, Chipping Campden, Glos., UK: Campden & Chorleywood Food Research Association.

Tucker, G., & Badley, E. (1990). CTemp: *Centre temperature prediction program for heat sterilization processes* (User's Guide). Chipping Campden, Glos., UK: Campden & Chorleywood Food Research Association.

Tucker, G., & Clark, P. (1989). Computer modelling for the control of sterilization processes. *Technical Memorandum No. 529*, Chipping Campden, Glos. UK: Campden & Chorleywood Food Research Association.

Tucker, G., & Clark, P. (1990). Modelling the cooling phase of heat sterilization processes using heat-transfer coefficients. *Int. J. Food Sci. Technol.*, *25*(6), 668–681.

Tucker, G. S. (2004). Improving rotary thermal processing. In P. Richardson (Ed.), *Improving the thermal processing of foods* (pp. 124–137). Cambridge: Woodhead Publishing.

Tucker, G. S., & Holdsworth, S. D. (1991a). Optimization of quality factors for foods thermally processed in rectangular containers. *Technical Memorandum No. 627*. Chipping, Glos., UK: Campden & Chorleywood Food Research Association.

Tucker, G. S., & Holdsworth, S. D. (1991b). Mathematical modelling of sterilization and cooking process for heat preserved foods - application of a new heat transfer model. *Food & Bioproducts Processing, Trans. IChemE*, *69*, Cl, 5–12.

Tucker, G. S., Kassim, H. O., Johns, W. R., & Best, R. J. (1992). Zone modelling: Part I – Application to thermal processing of homogeneous material in simple geometry, *Technical Memo. No. 654*. Chipping Campden, Glos., UK: Campden & Chorleywood Food Research Association.

Tung, M. A., Morello, G. F., & Ramaswamy, H. S. (1989). Food properties, heat transfer conditions and sterilization considerations in retort processes. In R. P. Singh & A.G Medina (Eds.), *Food properties and computer-aided engineering of food processing systems*, pp. 49–71. New York: Kluwer Academic Publishers.

Turhan, M., & Erdogdu, F. (2003). Error associated with assuming a finite regular geometry as an infinite one for modeling of transient heat and mass transfer processes. *J. Food Eng.*, *59*(2/3), 291–296.

Turhan, M., & Erdogdu, F. (2004). Errors based on location and volume average changes in infinite geometry assumptions in heat and mass transfer processes. *J. Food Eng.*, *64*(2), 199–206.

Uno, J., & Hayakawa, K. (1979). Nonsymmetric heat conduction in an infinite slab. *J. Food Sci.*, *44*(2), 396–403.

Uno, J., & Hayakawa, K. (1980). A method for estimating thermal diffusivity of heat conduction food in a cylindrical can. *J. Food Sci.*, *45*, 692–695.

Varma, M. N., & Kannan, A. (2006). CFD studies on natural convection heating of canned foods in conical and cylindrical containers. *J. Food Eng.*, *77*(4), 1024–1036.

Verboven, P., De Baerdemaeker, J., & Nicolaï, B. M. (2004). Using computational fluid dynamics to optimise thermal processes. In P. Richardson (Ed.), *Improving the thermal processing of foods* (pp. 82–102). Cambridge, UK: Woodhead Publishing.

Videv, K. (1972). Mathematical model of the sterilization process of canned foods during convection heating. Doctoral thesis, Institute of Food and Flavour Industries, Plovdiv, Bulgaria [in Bulgarian].

Wadsworth, J. I., & Spardaro, J. J. (1970). Transient temperature distribution in whole sweet potato roots during immersion heating. *Food Technol.*, *24*(8), 913–928.

Wang, P. Y., Draudt, H. N., & Heldman, D. R. (1972). Temperature histories of conduction-heating foods exposed to step changes in ambient temperatures. *Trans. ASAE*, *15*(6), 1165–1167.

Wang, Y., Lau, M. H., Tang, G., & Mao. R. (2004). Kinetics of chemical marker M-1 formation in whey protein gels for developing sterilization processes based on dielectric heating. *J. Food Eng., 64*(1), 111–118.

Welt, B. A., Teixeira, A. A., Chau, K. V., Balaban, M. O., & Hintenlang, D. E. (1997). Explicit finite difference methods of heat transfer simulation and thermal process design. *J. Food Sci., 62*(2), 230–236,

Williamson, E. D., & Adams, L. H. (1919). Temperature distribution in solids during heating and cooling. *Phys. Rev.* (2nd series), *14*(2), 99–114.

Yang, W. H., & Rao, M. A. (1998). Transient natural convection heat transfer to starch dispersion in a cylindrical container: numerical solution and result. *J. Food Eng., 36*(4), 395–415.

Zechman, L. G. (1983). Natural convection heating of liquids in metal containers. MS thesis, Univ. Minnesota, Minneapolis.

3
Kinetics of Thermal Processing

3.1. Introduction

3.1.1. General Effects of Thermal Processing

The successful thermal processing of packaged food products requires sufficient heat to inactivate micro-organisms, both those which cause spoilage and those which cause food poisoning. For this purpose it is necessary to know how heat-resistant micro-organisms are in order to establish a time and temperature for achieving the objective. In the canning trade this time and temperature relationship is known as the *process* and should not be confused with the sequence of engineering operations.

The heat applied to the food product not only inactivates micro-organisms but also facilitates the cooking of the product to give an acceptable texture and destroys active enzymes. In addition, it has the effect of destroying the nutrients, color and other quality attributes. The balance between excessive heating and under processing, known as *optimization*, will be discussed later in Chapter 7.

In this chapter the thermal kinetics of microbial inactivation, enzyme inactivation and quality changes will be discussed.

3.1.2. The Nature of Microbial Behavior

The structure, chemical composition and behavior of bacteria have been the subjects of extensive investigation and the reader wishing for a comprehensive treatment of the subject should consult advanced microbiology textbooks. More specific treatments are given in the standard canning texts, such as Ball and Olson (1957), Hersom and Hulland (1980) and Stumbo (1973).

In canning the primary concern is with bacteria, which are unicellular organisms less than 3 μm in size. These are either spherical (cocci), cylindrical (rods) or helical in shape and can multiply by binary fission. Bacteria are divided into families and genera, the members of the latter being sub-divided into species. Two of the more important genera are *Bacillus* and *Clostridium*, whose species have the ability to form endospores, which are more commonly referred to as *spores*. These are durable cells that can exist with little or no metabolic feed, are extremely resistant to chemical and physical treatments, and can accomplish rapid but controlled release of the dormant condition through activation and

TABLE 3.1. Some typical heat resistance data.

Organism	Conditions for inactivation
Vegetative cells	10 min at 80°C
Yeast ascospores	5 min at 60°C
Fungi	30–60 min at 88°C
Thermophilic organisms	
Bacillus stearothermophilus	4 min at 121.1°C
Clostridium thermosaccharolyticum	3–4 min at 121.1°C
Mesophilic organisms	
Clostridium botulinum spores	3 min at 121.1°C
Clostridium botulinum toxins Types A & B	0.1–1 min at 121.1°C
Clostridium sporogenes	1.5 min at 121.1°C
Bacillus subtilis	0.6 min at 121.1°C

germination, under anaerobic conditions. In the case of *Clostridium botulinum*, this can lead to the production of a deadly toxin and cause the disease of botulism, which can be fatal. The purified toxins of *Clostridium botulinum* can be destroyed by heating at 80°C for 10 min; the presence of a food substrate may modify the conditions required (Hersom & Hulland 1980).

Bacterial endospores have a very variable and usually high heat resistance, those of *Clostridium botulinum* being among the most resistant for pathogenic micro-organisms. Table 3.1 shows some heat resistance data. More extensive heat resistance data are given in Appendix A.

3.1.3. Other Factors Affecting Heat Resistance

3.1.3.1. Water Activity

It should be noted that dry heat resistance data are generally much higher than those for wet conditions; however, with conventional canning this condition rarely arises. It can exist under adverse conditions of mixing powders into formulated products. Heat resistance for oily and fatty products is also higher. This emphasizes the point that it is necessary to determine the heat resistance of bacteria in the presence of food substrate.

3.1.3.2. pH

The acidity of the substrate or medium in which micro-organisms are present is an important factor in determining the degree of heating required. For products with a pH less than 4.5 (high-acid products), a relatively mild pasteurization process is all that is necessary to stabilize the product. This category includes acid fruits and acidified products. Some spore formers may cause spoilage, for example, *Bacillus coagulans, Clostridium butyricum* and *B. licheniformis*, as well as ascospores of *Byssoclamys fulva* and *Byssoclamys nivea,* which are often present in soft fruit such as strawberries. For pH values greater than 4.5, so-called low-acid products, it is necessary to apply a time–temperature regime sufficient to inactivate spores of *C. botulinum*. In the industry the process is referred to as a botulinum process.

Ito and Chen (1978) have reviewed the evidence for the pH required to prevent the growth of *C. botulinum* and concluded that a pH of 4.6 or less is the necessary requirement. The widely used and accepted pH value of 4.5 falls within this criterion.

It is possible to identify at least four groups of products:

Group 1: Low-acid products (pH 5.0 and above) – meat products, marine products, milk, some soups and most vegetables.
Group 2: Medium-acid products (pH 5.0–4.5) – meat and vegetable mixtures, specialty products, including pasta, soup, and pears.
Group 3: Acid products (pH 4.5–3.7) – tomatoes, pears, figs, pineapple and other fruits.
Group 4: High-acid products (pH 3.7 and below) – pickles, grapefruit, citrus juices and rhubarb.

Table 3.2 gives a range of typical pH values. It should be noted that the classification is not rigorous, but only intended for guidance (Hersom & Hulland 1980). The pH of some marginal fruit products can vary with growing conditions, variety and season, and similarly formulated products can have a wide range of pH. It is important to check the products in marginal cases to ensure that an adequate heating regime is applied. When necessary, especially in the case of new product formulations, microbiological advice should be sought from a competent canning organization.

The US Food and Drug Administration currently defines low-acid canned foods as having a pH greater than 4.6 and a water activity greater than 0.85.

3.1.3.3. Other Factors

A number of other factors affect the heat resistance, including lipids and oily materials, dielectric constant, ionic species, ionic strength, oxygen level, organic acids and antibiotics. Details of these will be found in Hersom and Hulland (1980) and Vas (1970).

3.1.4. Measuring Heat Resistance

This is a specialized technique, and the reader should consult standard microbiological texts for details. Brown (1991) has tabulated the various methods which are available.

The two most popular ways of measuring heat resistance in the range 60–135°C are the end-point method and the multiple-point method. The end-point method involves preparing a number of replicate containers containing a known number of spores and heating them for successively longer times until no survivors are obtained by culturing each container. The multiple-point method involves heating a batch of spore suspension continuously and withdrawing samples at selected intervals, followed by determination of the number of survivors.

TABLE 3.2. Range of pH values for various foods.

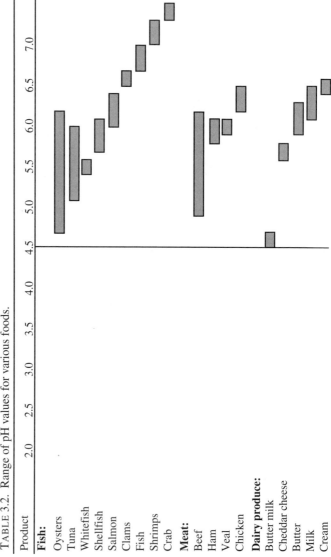

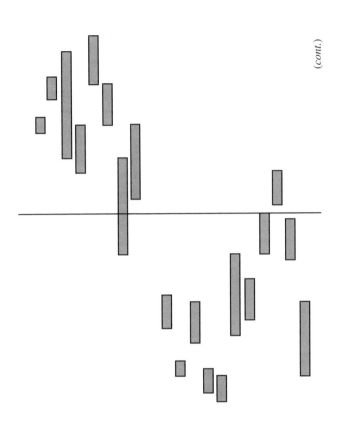

(cont.)

Soups:
Bean
Beefbroth
Chicken noodle
Duck
Mushroom
Pea
Tomato
Vegetable
Juices:
Apple
Cranberry
Grapefruit
Lemon
Lime
Orange
Pineapple
Tomato
Carrot
Vegetable
Vinegar

TABLE 3.2. (Continued)

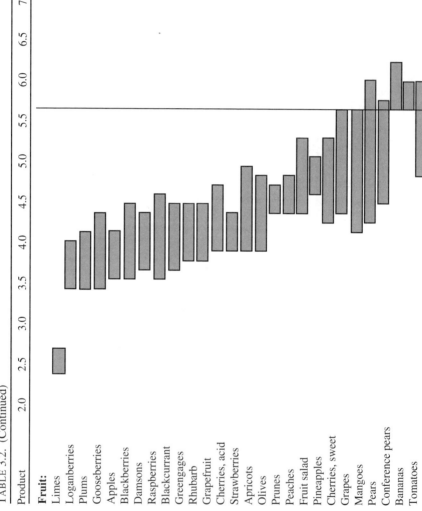

Vegetables:

Sugar beet
Carrots
Pumpkin
Celery
Beetroot
Beans/tomato sauce
Turnip
Water melon
Runner beans
Cauliflower
Mixed vegetables
Potatoes
Spinach
Buter beans
Asparagus
Garden peas
Processed peas
Brussells sprouts
Broccoli
Honeydew melon

Simplified equations for calculating the lethality of the heating and cooling phases for capillary tube techniques have been presented by Dickerson (1969) and a programmed heating technique has been developed by Hayakawa (1969). A computerized method of estimating microbial inactivation parameters is described by Hayakawa *et al.* (1981).

One of the major problems of heating spore suspensions is allowing for the time it takes for the heat to be transferred through the container walls. For temperatures in the range 100–120°C it is relatively easy to correct for the "come-up" time; however, for higher temperatures, where the heating times can be as short as the wall penetration lag times, the problem is more difficult. Consequently, the standard capillary tube method becomes less accurate.

A thermoresistometer may be used for determining the heat resistance of spores (Stumbo 1973, 1984; Brown *et al.* 1988; Brown 1992; Gaze & Brown 1990). This has the advantages of carefully controlling the residence time and direct exposure of the microbial suspension to the steam heating medium. Brown *et al.* (1988) have produced some of the most reliable data available for heat-resistant spores, especially *Clostridium botulinum*, for the temperature range 120–150°C.

Rodrigo *et al.* (1993) have also used a thermoresistometer to determine the kinetics of *Clostridium sporogenes PA 3679*. Indirect heating methods have also been applied to determining the TDT of micro-organisms (Hass *et al.* 1994).

Flow methods have also been devised, although again these tend to be used for the higher-temperature processing region (Swartzel 1984, 1989; Swartzel & Jones 1985; Srimani *et al.* 1990; Cole & Jones 1990). Computerized methods of experimental design for the determination of D-values for inactivating bacterial spores have been recommended by Hachigan (1989).

Teixeira (2002, 2004) has discussed dynamic methods of determining heat resistance including Paired Equivalent Isothermal Exposures (PEIE), which make use of end-point reaction data and temperature/time data from continuous sterilization experiments. The method has been shown to be superior to estimating resistance from traditional survivor curves, especially at higher processing temperatures.

3.1.5. The Statistical Nature of Microbial Death

Bigelow (1921) and Ball (1923) developed the theory of thermal process evaluation on the basis of the direct killing or inactivation of micro-organisms. However, other workers, among them Stumbo (1973), Gillespy (1946), and Jakobsen (1954), showed that there was a statistical basis for the thermal inactivation of micro-organisms.

The theory was based on the concept that spores die at all temperatures; however, the higher the temperature the more likely it is that death will occur. Every spore has an inherent probability of escaping death, which does not change with time. This probability defines the heat resistance of a micro-organism. If P is the probability of escaping death for unit time, then the possibility of escaping for t units of time is P^t.

3.1.5.1. Spores of Equal Heat Resistance

If N spores of equal heat resistance are considered, then the number of survivors S expected after time t is given by

$$S = NP^t. \tag{3.1}$$

Taking logs gives

$$\log S = \log N + t \log P.$$

This indicates that bacterial death is of a logarithmic order, and a plot of $\log S$ versus t will give a straight line, with slope $d(\log S)/dt = \log P$ (a constant). Since P takes values from 0 to 1, i.e. $0 < P < 1$, then $\log P$ is negative.
   If $\log P = -1/D$ then

$$S = N \times 10^{-t/D}. \tag{3.2}$$

D is known as the *decimal reduction time*, usually measured in minutes (see Section 3.2.2 for further discussion).
   Substituting for S from equation (3.1), we have

$$P^t = 10^{-t/D}. \tag{3.3}$$

If $t = D$, then $P^t = P^D = 0.1$, i.e. a 10% survival would be expected, and if $t = 2D$, then $P^D = 0.01$ and a 1% survival would result.
   The survival equation may also be written in exponential terms, i.e. if $\log_e P = \ln P = -k$, then

$$S = Ne^{-kt}. \tag{3.4}$$

3.1.5.2. Spores of Unequal Heat Resistance

For a collection of spores of varying heat resistance the survival equation becomes

$$S = N_1 e^{-k't} + N_2 e^{-k''t} + N_3 e^{-k'''t} + \cdots, \tag{3.5}$$

where $\Sigma N_i = 1$, $k' = \ln P'$, etc.
   Assuming that $\ln k$ is normally distributed among the population of spores, then

$$S = (1 + akt)^{-1/a}$$

or

$$\ln S = -(1/a) \ln(1 + akt), \tag{3.6}$$

where a is the variance of $\ln k$, i.e. $(\mathrm{var}\, k)/k^2$.
   The above equations may be written in base 10 notation as follows:

$$S = 10^{-Kt}$$

$$S = (1 + aKt/M)^{-1/a} \tag{3.7}$$

$$\log S = -(1/a) \log(1 + aKt/M)$$

where $M = \log e = 0.4343$. Whereas equation (3.1) indicated that the survivor relationship between $\log S$ and t was rectilinear, equation (3.6) or (3.8) shows that the curve is concave upwards. This is observed in practice for many microbial systems. A more extensive treatment of the topic is given by Aiba and Toda (1967).

More recently a technique has been described that uses the number of active microbial survivors in the food. This is based on the principle that the number and concentration of active survivors can be assessed from the number of spoiled heat treated out of the initial sample. The method has been validated using spores of *Clostridium sporogenes* in pea purée (Körmendy & Mohácsi-Farkas 2000).

3.1.6. Practical Aspects

The practical implications of this logarithmic order of death are as follows: It is impossible to give a time and temperature that will guarantee to kill every spore in a can of food, or indeed every spore in every can of a batch. Increasing the process increases the likelihood that an individual will be sterilized.

In order to produce canned foods which are adequately processed to prevent them being a public health hazard, it is necessary to establish a very low probability of spore survival. For low-acid foods the aim is to attain a probability of spore survival of one in 10^{12} or better. This corresponds to a $12D$ process, and is known as a *minimum process*. Processes established on this basis do not always destroy heat-resistant thermophiles; consequently, a process which deals adequately with pathogenic organisms is called a commercial process and the operation of applying the process is called *commercial sterilization*. These processes are adequate provided the cans are stored at temperatures below $35°C$, since above this temperature residual thermophiles may grow and produce a spoiled can. Cans destined for hot climates may require increased processes.

The level of initial infection, N, of a can is very important, as can be seen from equation (3.2), since the greater the value of N the higher the value of the survival S for given values of t and D. In practice, this means that raw materials must be examined carefully and every precaution taken to reduce the spore loading to a minimum by using hygienic handling conditions.

3.2. Methods of Representing Kinetic Changes

3.2.1. Basic Kinetic Equations

The most general equation for studying the kinetics of reactions, whether microbial inactivation, enzyme inactivation, the degradation of a heat-labile component, or the effects of cooking, is given by

$$-dc/dt = k_n c^n, \tag{3.8}$$

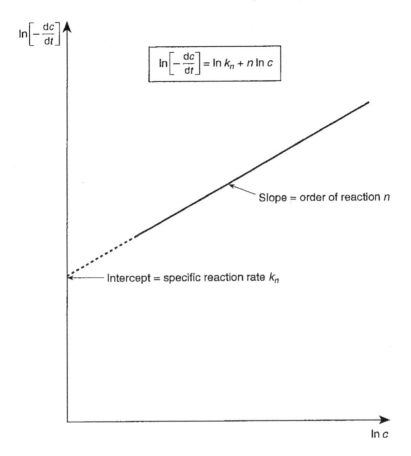

FIGURE 3.1. Determination of reaction order n from equation (3.9).

where c is the concentration of a reacting species at any time t, k_n is the specific reaction rate, with units $[\text{concentration}]^{1-n}[\text{time}]^{-1}$, and n the order of the reaction. The concentration of reactants can be either the number of micro-organisms or the concentration of a compound in grams per liter or other appropriate molar units.

When studying the kinetics of a thermal treatment it is essential to determine the order of the reaction, so that an appropriate kinetic equation can be obtained for subsequent use. This is done by obtaining the rate term $-dc/dt$, plotting c against t and then using the logarithmic form of equation (3.8), i.e.

$$\ln(-dc/dt) = \ln k_n + n \ln c. \tag{3.9}$$

A typical plot is shown in Figure 3.1, from which the reaction order n can be found from the slope and the value of k from the intercept on the y-axis. A least-squares statistical approach should be used for the accurate determination of the constants and their statistical significance.

3.2.1.1. Zero-Order Reactions ($n = 0$)

The reaction rate is independent of the concentration, and the reaction rate constant, k_0, can be found from the equation

$$-dc/dt = k_0. \tag{3.10}$$

Integration gives

$$c_0 - c = k_0 t, \tag{3.11}$$

where c_0 and c are the concentrations at times 0 and t respectively. The units of k_0 are [concentration] [time]$^{-1}$.

Zero-order reactions are encountered infrequently in thermal processing; some examples are the caramelization of sugars (Buera *et al.* 1987a, b), and vitamin C degradation in model meat systems of different water activities (Laing *et al.* 1978).

3.2.1.2. First-Order Reactions ($n = 1$)

The reaction rate is proportional to the concentration of the reacting species or the number of micro-organisms, and the reaction constant, k_1, can be found from the equation

$$-dc/dt = k_1 c. \tag{3.12}$$

Integration gives

$$\ln(c/c_0) = -k_1 t \tag{3.13}$$

or

$$\ln(c_1/c_2) = k_1(t_2 - t_1), \tag{3.14}$$

where c_1 and c_2 are the concentrations at times t_1 and t_2 respectively. The units of k_1 are [time]$^{-1}$ and are independent of concentration.

The majority of reactions, including microbial inactivation, encountered in thermal processing obey, or are assumed to obey, first-order (or pseudo-first-order) kinetics. In this text the subscript giving the reaction order will be omitted, thus in the various forms of the first-order reaction equation the reaction rate constant will be signified by k and equation (3.13) takes the form

$$\ln(c/c_0) = -kt$$

or

$$c/c_0 = e^{-kt}. \tag{3.15}$$

3.2.1.3. Second-Order Reactions ($n = 2$)

The reaction rate is proportional to the square of the concentration in the case of a single reactant and the product of the concentrations for two reactants. For the

TABLE 3.3. Reaction orders for the effect of heat on some quality factors for green peas and white beans (Van Loey *et al.* 1995).

Product	Attribute	Reaction order n
Peas (*Pisum sativum*)	Color	1.31 ± 0.18
	Hardness	1.13 ± 0.20
Beans (*Phaseolus vulgaris*)	Hardness	0.36 ± 0.17
	Appearance	0.44 ± 0.16

single reactant case, the reaction constant, k_2, can be found from the equation

$$-dc/dt = k_2 c^2. \qquad (3.16)$$

Integration gives

$$c^{-1} - c_0^{-1} = k_2 t. \qquad (3.17)$$

The units of k_2 are [concentration]$^{-1}$[time]$^{-1}$. Second-order reactions are infrequently encountered in thermal processing; however, the destruction of thiamin in milk, was shown to be second-order by Horak and Kessler (1981).

3.2.1.4. Other Orders

Van Loey *et al.* (1995) have shown that reaction orders may be fractional or not whole integers for some types of degradation reaction, for example the color and texture of peas and the texture and appearance of beans. Table 3.3 shows the n-values obtained from experiments on peas and beans. These values reveal the care that should be taken in establishing the reaction order, so that the correct equations may be used for determining the effect of thermal processing on the heat-labile components in the food products.

Readers requiring further information on reaction kinetics should consult standard textbooks and monographs, such as Atkins (1994), Moore (1972, 1983), Steinfeld and Fransisco (1989) and Stevens (1961).

3.2.2. Decimal Reduction Time

For microbial inactivation the number of viable organisms, N_0 at time zero and N at time t, is used as a measure of the concentration. Equation (3.15) thus becomes

$$\ln(N/N_0) = -kt, \qquad (3.18)$$

which can also be expressed as

$$\log_{10}(N/N_0) = -kt/2.303 \qquad (3.19a)$$

or

$$N/N_0 = e^{-kt}. \qquad (3.19b)$$

The standard convention in thermobacteriology has been to define a *decimal reduction time*, D, which is the heating time in minutes to give an expected inactivation of 90% of organisms, or 10% survival (Katzin *et al.* 1942). This can be determined from heat resistance experiments by plotting log N against t and using the equation

$$\log N = \log N_0 - t/D. \tag{3.20}$$

The term $\log(N/N_0)$ is also known as the germ reduction value, G (see Machmerth 1983; Swartzel 1986).

From equation (3.19b), the interrelation between the kinetic factors results in equation (3.21):

$$D = 2.3/60k, \tag{3.21}$$

where D is in minutes and k in seconds. Note that the interrelation between kinetics factors (D and k in equation (3.21)) has been derived for a first-order kinetics and is not valid for different kinetics orders.

A typical microbial inactivation curve, known as the *survivor curve* or *thermal death curve*, is a plot of log N against time, and is shown in Figure 3.2. D is the time, in minutes, to traverse one log cycle and the slope is $-1/D$.

The logarithmic relationship between microbial inactivation and heating time was originally established by Esty and Meyer (1922), before the D-value concept was used, for spores of various strains of *Clostridium botulinum*.

3.2.3. More Complex Inactivation Models

Figure 3.3 shows some of the different types of survivor curve often encountered in experimental work. It can be seen that it is often difficult to determine the initial number of spores accurately from the non-rectilinear nature of these curves. Curves that show a concave downwards behavior are considered to be due to the clumping of organisms; the true inactivation level can be obtained from the part of the curve following the initial lag period. Curves that are concave upwards are due to organisms of different heat resistance (see Section 3.1.5.2). The true rate of inactivation can be obtained from the latter part of the curve where the most heat resistance is located. A review of the different types of curve, including shoulder, sigmoidal, upward concave and biphasic types, has been presented by Cerf (1977). One of the commonest types of curve encountered shows an initial shoulder, considered to be due to activation of organisms, followed by a constant period of inactivation, and then a residual tailing.

A more complex model, which includes heat activation, injury, preliminary inactivation of the less heat-resistant fraction and inactivation of the remainder, has been developed (Rodriguez *et al.* 1988; Teixeira *et al.* 1990). The model has three terms and allows for the accurate prediction of the thermal inactivation behavior of microbial populations. The first term describes the rapid decay of the least resistant fraction of the organisms:

$$N_A = N_0 e^{-k_a t}, \tag{3.22}$$

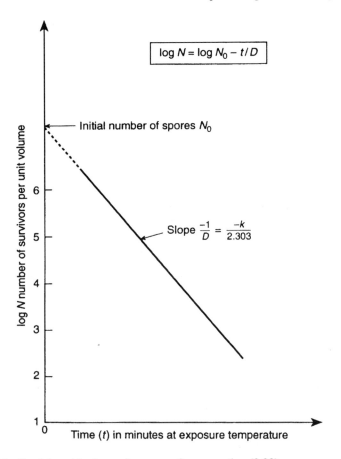

$$\log N = \log N_0 - t/D$$

Initial number of spores N_0

Slope $\dfrac{-1}{D} = \dfrac{-k}{2.303}$

log N number of survivors per unit volume

Time (t) in minutes at exposure temperature

FIGURE 3.2. Semi-logarithmic survivor curve from equation (3.20).

where k_a is the reaction rate for activation. The second term is a combination of thermal activation and subsequent destruction of dormant spores initially present:

$$N_B = N_0(1 - e^{-k_a t})e^{-k_d t} \qquad (3.23)$$

where k_d is the reaction rate constant for destruction. This is the term which generates the "shoulder" of the inactivation curve. It represents a pulse rising rapidly from an initial value of zero, peaking and then decaying slowly towards zero. The third term describes the inactivation of initially active spores:

$$N_C = N_0 e^{-k_{d'} t}. \qquad (3.24)$$

The total number of survivors can be obtained by summation

$$N_t = N_A + N_B + N_C.$$

Rodriguez *et al.* (1988), using spores of *Bacillus subtilis*, showed that a two-term model, $N_B + N_C$, was adequate for predicting the spore inactivation, especially

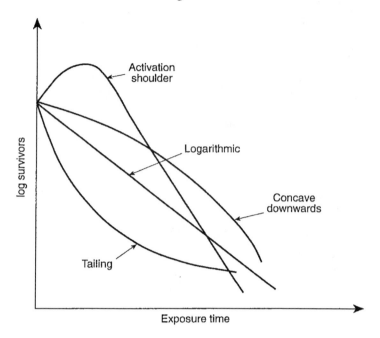

FIGURE 3.3. Examples of different types of spore survivor curves

for shorter times and higher temperatures. The values for the rate constants, k_a, k_d, and $k_{d'}$, for spores of *Clostridium botulinum* are 1.4×10^{-2}, 1.6×10^{-3} and 0.16 s^{-1} respectively.

Abraham *et al.* (1990) showed that a mathematical model based on one proposed by Shull *et al.* (1963) was applicable to experimental survivor curves that were basically nonlinear. The model, represented by the following equation, contains two terms, one for activation of dormant spores (specific reaction rate k_A) and the other for inactivation (specific reaction rate k_B):

$$N/N_0 = \phi e^{-k_D t} + (1 - \phi)e^{-k_A t}, \qquad (3.25)$$

where

$$\phi = [1 - k_A/(k_D - k_A]N_D/N_A$$

and N_D and N_A are the number of dormant and activated spores respectively.

Empirical formulae have been developed (Hayakawa *et al.* 1969; Hayakawa 1982) for estimating nonlinear survivor curves of thermally vulnerable factors. The method proposed was simpler than the traditional method of Stumbo (1973); it was based on programmed heating and included cases where the thermal death time (TDT) curve was a linear relation with heating time. Teixeira (2004) has used this model for determining the heat resistance factors for survivor curves showing "shoulders" and "tails," considered to be the result of activation and deactivation of sub-populations of microbial species. This uses as many terms as are required to fit the survival curve, using a number of terms as in equation (3.25a).

$$N(t) = N_{10}e^{-k_1t} + N_{20}e^{-k_2t} + N_{30}e^{-k_3t} + \cdots + N_0e^{-k_nt} \qquad (3.25a)$$

3.2.4. Temperature Dependence of Death Rate

The fact that the death rate varies with temperature is very important in all aspects of thermal processing, especially in establishing adequate processes. This section will examine the most important of these models and their applicability in design, simulation and modeling situations.

Thermal inactivation of mixed microbial populations has been studied theoretically by Körmendy et al. (1998). This showed that the nth-order time constant and survivor ratio are independent of the initial microbial concentration.

Körmendy and Körmendy (1997) have shown that when the semi-logarithmic thermal inactivation models are non-linear, the Bigelow model can be used for calculating thermal processes if $t\{T, N\} = g\{T\}$. $F\{N\}$, where t is the heat treatment time at constant temperature; T is the temperature, and N is the number of surviving micro-organisms at time t. The function $g\{T\}$ depends only on T and the function $F\{N\}$ depends only on N. A simple graphical method is presented for determining whether semilogarithmic curves correspond to a $g\{T\} \cdot f\{N\}$ type function.

An alternative concept to the kinetic theory of microbial inactivation is the statistical approach, i.e., using a cumulative form of the temporal distribution of lethal events. These are capable of handling non-linear survival curves with upward and downward concavity as well as "shoulders" andd "tails." This is briefly discussed in Section 3.1.5.1., where the early work of the 1940s is discussed.

A number of new approaches have been examined:

(a) *Weibull Distribution* (Peleg & Cole 1998; Peleg & Penchina 2000; Peleg et al. 2002; Peleg 2003a, b, c; Peleg 2004; Van Boekel 2002; Corradini et al. 2005).
    This model is essentially a power-law distribution of the form $\log_{10} S(t) = -b(T) \cdot t^{n(T)}$ where $S(t)$ is the survival ratio of micro-organisms with time t, i.e., Nt/N_0, $b(T)$, the scale or steepness factor, is a temperature (T) dependent coefficient (between 0.1–6.0), and $n(T)$, the shape or curvature factor, is a temperature dependent coefficient (approximately 0.4). When $n(t)$ is unity the equation resolves to the linear Arrhenius model. When $n(t) < 1$, the curve is upward concave and when $N(t) > 1$ it is downward concave. The equation has been applied satisfactorily to the interpretation of the inactivation *Clostridium botulinum* spores (Campanella & Peleg 2001).

(b) *Log-Logistic Model* (Campanella et al. 2001; Peleg et al. 2002; Anderson et al. 1996; Davies et al. 2000)
    This empirical model is represented by the following equation:
    $b(T) = \log_e\{1 + \exp[k(T - -T_c)]\}$, where k and T_c are constants

(c) *Prentice Model* (Prentice 1976; Kilsby et al. 2000; Mycock 2002)

This is a seven-parameter model, which has been used successfully to model the kinetic behavior of *C. Botulinum* 213B and *Salmonella Bedford.* Mycock (2002) showed that the heat resistance data for *Clostridium botulinum* 213B was found to be modeled by the Prentice function and *Salmonella Bedford* 286 by a log normal distribution, both using a log time scale. The z-value in °C (Celsius degrees) for *C. botulinum* was found to be dependent on temperature, viz., 8.7 at 110°C; 10.6 at 121°C and 13.2 at 135°C. It is interesting that the z-value increases considerably with increasing temperature, above the standard processing temperature of 121.1°C. This could have implications on the processing times required for processes at higher temperatures.

(d) *Other Models* Geeraerd *et al.* (2004) have discussed a number of other models, including a Biphasic model and a Sigmoidal model. For the latter a freeware for Microsoft® Excel known as GinaFiT (Geeraerd *et al.* 2004 Inactivation Model Fitting Tool) has been developed for the Sigmoidal model. A discontinuous linear model (Corradini *et al.* 2005), which considers that no inactivation takes place below a specified temperature and that the increase in the rate parameter above this temperature is linear. The capabilities in predicting the survival patterns for *Clostridium botulinum* and *Bacillus sporothermodurans* are shown to be superior to the Arrhenius model.

3.2.4.1. Thermal Death Time Method: D–z Model

Some of the earliest work on the heat resistance of micro-organisms was done by W. D. Bigelow of the National Canners' Association, Washington, USA (see Bigelow & Esty 1920; Bigelow 1921), who first used the term 'thermal death time' in connection with the minimum time required for the total destruction of a microbial population. When the temperature was plotted against the log time a straight-line relationship was observed. Subsequent to this work the D-value – obtained from the slope, $-1/D$, of the log survivor curve against time – was used to quantify heat resistance (see Figure 3.2). When the D-value is plotted against the temperature a linear relationship is observed; this is generally referred to as the *thermal death time curve*. The slope of the log D versus temperature curve, i.e. $d(\log D)/dT$, is $-1/z$, where z is the temperature change necessary to alter the TDT by one log-cycle. From Figure 3.4 it can be seen that for any two values of D and the corresponding temperatures T, z can be determined from the relationship

$$\log(D_1/D_2) = (T_2 - T_1)/z. \tag{3.26}$$

An important and unique value of D measured at a reference temperature is designated D_{ref} or D_r; this is defined by

$$\log(D/D_{\text{ref}}) = -(T - T_{\text{ref}})/z. \tag{3.27}$$

The standard reference temperature was 250° F, but its exact Celsius equivalent, 121.1°C, is now in more common usage, though less satisfactory since the ".1"

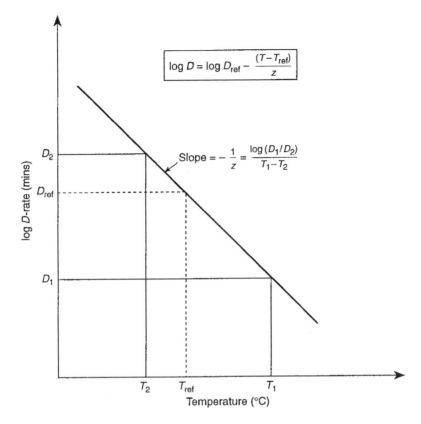

$$\log D = \log D_{ref} - \frac{(T - T_{ref})}{z}$$

$$\text{Slope} = -\frac{1}{z} = \frac{\log (D_1/D_2)}{T_1 - T_2}$$

FIGURE 3.4. Thermal death time curve to obtain z-value.

represents a recurring decimal. The use of 120°C or 121°C has been rejected on the grounds that it could lead to misunderstanding and errors in using existing F-values. A more theoretical treatment of the use of reference temperatures is given by Holdsworth and Overington (1975), who suggested the use of an absolute standard of 400 K and processing times in seconds.

The ratio of D_{ref}/D is known as the *lethal rate* L; this is a quantitative measure of the rate of inactivation of organisms at a given temperature:

$$L = (D_{ref}/D) = 10^{(T - T_{ref})/z}. \tag{3.28}$$

An important property of lethal rates is that they are additive, so it is possible to quantify the process over the temperature range used. The integrated lethal rate is known as the F-value (see Section 4.1.2).

3.2.4.2. Arrhenius Kinetic Approach: k–E Model

One of the most important approaches to modeling the effect of temperature T (in kelvin) on the specific reaction rate $k(\text{s}^{-1})$ was produced by Svante Arrhenius in a paper discussing the kinetics of the inversion of sucrose (Arrhenius 1889). He suggested an equilibrium existed between inert and active molecules and that the inversion reaction only took place with molecules that were energetically activated by heat. Using the van't Hoff isochore and applying it to the equilibrium, the following classical equation was obtained:

$$\ln k = \ln A - E/RT$$

or

$$k = A e^{-E/RT}. \tag{3.29}$$

From the activation energy theory A is a constant known as the pre-exponential factor, collision number or frequency factor (s^{-1}), E is the activation energy (kJ mol^{-1}) and R is the molar gas constant (8.135 J mol^{-1} K^{-1} or 1.987 cal mol^{-1} K^{-1}). The activation energy can be conceived as the minimum energy that molecules must possess before reaction can occur, and the exponential term $e^{-E/RT}$ is the fraction of the molecules that collectively have minimum energy. The fundamental basis for this theory is extremely well expounded in the text by Glasstone et al. (1941). Other work that is relevant to the subject can be found in the papers of Karplus et al. (1965) and Truhlar (1978).

The values of A and E may be determined from a plot of $\ln k$ versus $1/T$. The values of A for unimolecular chemical reactions vary from 10^{14} to 10^{20} s^{-1}, for bimolecular reactions from 10^4 to 10^{11} s^{-1}, with intermediate values for chain reactions.

When applied to the thermal inactivation of microbial spores, equation (3.29) is written in a form which eliminates the factor A by using a reference temperature, T_{ref}, and the corresponding specific reaction rate constant, k_{ref}:

$$\ln k = \ln k_{ref} - (E/R)[1/T - 1/T_{ref}]. \tag{3.30}$$

For heat-resistant bacterial spores A is extremely large, of the order of $10^{30} - 10^{60}$ s^{-1}, and E can be up to 500 kJ mol^{-1} (see Appendix 1 for a table of values). For Clostridium botulinum spores, Simpson and Williams (1974) concluded that the most appropriate value for k was obtained by using $A = 2 \times 10^{60}$ s^{-1} and $E = 310.11 \times 10^3$ Jmol^{-1} K^{-1} for temperatures in the range $373 < T < 423$ K. Casolari (1979) has commented critically on the meaning of such high A-values. He concludes that equilibrium thermodynamics is neither a fruitful field nor a satisfying approach to determining reaction rates in rapidly increasing temperature fields. The factors A and E are assumed in the theory to be independent of temperature; however, this may not always be the case. Casolari (1981) has proposed a modified version of the Arrhenius equation to deal with nonlinear forms of inactivation curves. This includes a term showing the dependence of k on the concentration of cells, and this has been tested by Malmborg (1983).

3.2.4.3. Absolute Reaction Rate Theory

The Arrhenius model can be developed further using the theory of absolute reaction rates and a thermodynamic evaluation of the constants. Thus the variation of k with temperature T can be expressed by the equation

$$k = (k_B/h_P)T \, e^{(\Delta S^*/R)} e^{(\Delta H^*/RT)}, \qquad (3.31)$$

where k_B is the Boltzmann constant ($1.381 \times 10^{-23} \, \text{J K}^{-1}$), h_P the Planck constant ($6.626 \times 10^{-34} \, \text{J s}$), ΔS^* the entropy of activation ($\text{kJ kmol}^{-1} \, \text{K}^{-1}$) and ΔH^* the enthalpy of activation (kJ kmol^{-1}).

Comparing equation (3.31) with (3.29), the pre-exponential factor, A, and the activation energy, E, are

$$A = (k_B T/h_P)e^{(\Delta S^*/R)}$$

$$E = \Delta H^*.$$

The theory is based on the fact that spores will die when they are activated to a sufficiently high level by increasing the probability of collision. The rate constant, k, increases with temperature and makes the probability of death greater.

For many systems, including microbial inactivation by heat, an empirical relationship between ΔS^* and ΔH^* has been observed. This is known as the *compensation law* because, as can be seen from the following equation, one factor compensates the other:

$$\Delta S^* = a \Delta H^* + b. \qquad (3.32)$$

When $T = a^{-1}$, the two thermodynamic entities are equal; this temperature is known as the *compensation* or *isokinetic* temperature. The model can be used, within experimental limits, for the prediction of entropies of activation from enthalpies. The reader requiring further information should consult the papers by Banks *et al.* (1972), Kemeny and Rosenberg (1973), Boon (1973), and Harris (1973).

Warren (1973) developed a more extensive model which accounts for the effects of temperature and pH on the thermal death rate of spores. The validity of the model was tested using the data of Gillespy (1947, 1948) for the inactivation kinetics of *Clostridium thermosaccharolyticum* and that of Xezones and Hutchings (1965) for *C. botulinum*. The predicted results were in good agreement with the experimental ones. The model is sufficiently important to justify further refinement and use as a predictive tool.

Other models have also been developed for the combined effect of pH and temperature on microbial inactivation:

(a) *The Davey – linear Arrhenius model* D-LA (Davey 1993; Davey *et al.* 1995; Chiruta *et al.* 1997; Davey 1994; Davey & Cerf 1996; Davey *et al.* 2001) and this is given by equation (3.33):

$$\ln k = C_0 + C_1 T^{-1} + C_2 \text{pH} + C_3 \text{pH}^2. \qquad (3.33)$$

For *C. botulinum*, in spaghetti/tomato sauce, the values of the constants were $C_0 = 105.23$, $C_1 = -3.704 \times 10^4$, $C_2 = -2.3967$ and $C_3 = 0.1695$, using the data of Xezones and Hutchings (1965).

This equation has been extended by Gaillard *et al.* (1998) and Khoo *et al.* (2003) to include a second temperature dependent term:

$$\ln k = C_0 + C_1 T^{-1} + C_2 T^{-2} + C_3 pH + C_4 pH^2. \tag{3.34}$$

This shows that the value of k is a maximum when $pH_{max} = -C_3/2C_4$.

(b) *Square-root model* is a special form of the Ratowsky-Belehradek model (Belehradek 1926; McMeekin *et al.* 1993; Khoo 2003)

This is represented by multiplicative equation:

$$(k)^{1/2} = b(T - T^*) \times (pH - pH^*). \tag{3.35}$$

where the * terms are conceptual values. This is a three-term model which includes T, pH, and TpH and is more complex to apply than the classical log-linear and D-LA models.

(c) *nth order polynomials—noP* These have been widely used to model bacterial growth (McMeekin *et al.* 1993). They take the form:

$$\ln k = \alpha_0 + \alpha_1 T^1 + \alpha_2 T^2 + \alpha_3 T^3 + \alpha_4 pH + \alpha_5 pH^2 + \alpha_6 pH^3 + X \tag{3.36}$$

where α_n are the coefficients of the cubical model and X are any interactive terms.

(d) *The Mafart Model* The model has been used to extend the generalized sterilization chart for combined temperature and pressure (Davey 1993b). Gaillard *et al.* (1998) and Mafart (1999a, b) have presented an extended model, the Mafart model, for the prediction of the heat resistance of *Bacillus cereus* spores, which include water activity as well as pH.

3.2.4.4. Quotient Indicator Method

Another method used to express the temperature dependence of specific reaction rate is the *quotient indicator*, Q. This is the ratio of two specific reaction rates measured at different temperatures, such as T and $T = 10°C$:

$$Q_{10} + k_{T+10}/k_T. \tag{3.37}$$

The Q_{10} ratio has values of 1.8–4 for chemical reactions, 8–20 for inactivation of micro-organisms, and 10–100 for denaturation of proteins. The higher the Q_{10}-value the more sensitive the inactivation rate is to temperature change.

The Q_{10} method can be compared with the D–z method, and can be converted using the formula

$$Q_{10} = D_T/D_{T+10} = 10^{10/z}. \tag{3.38}$$

For *C. botulinum*, $z = 10°C$, $Q_{10} = 10$.

The Q_{10} is extensively used in milk pasteurization and sterilization practice (see Burton 1988), and for biochemical and sensory changes induced by heating; it is rarely used in conventional canning technology.

3.2.4.5. Equivalence of the Models

For first order kinetics the thermal death time D–z model as expressed in equation (3.27) can be directly compared with the Arrhenius k–E model equation (3.30) to give the following identities:

$$D = \ln 10/(60\text{k}) = 2.303/(60\text{k}), \qquad (3.39)$$

where k is in reciprocal seconds and D is in minutes; and

$$E = 2.303RTT_{\text{ref}}/z, \qquad (3.40)$$

where E is in kilojoules per mole and z in Kelvin or degrees Celsius. Equation (3.40) is often expressed in the approximate form:

$$E = 2.303RT_{\text{ref}}^2/z, \qquad (3.41)$$

which is satisfactory when processing temperatures are near to the reference temperature 121.1°C. It also implies that the z-value is a function of temperature. Datta (1993) examined the errors associated with using $T_{\text{ref}} = 121$°C, and showed that errors of the order of 1.3% at 116°C, 8% at 80°C and 62% at 151°C were inherent. Nunes *et al.* (1993) used equation (3.40) to determine the most appropriate T_{ref} to minimize the errors. For a z-value of 10°C and values of E from 320 to 350 kJ mol^{-1}, T_{ref} was between 140 and 150°C. Hallström *et al.* (1988) have examined the errors in Arrhenius kinetic factors as a function of processing temperature. Several workers, among them Haralumpu *et al.* (1985), Cohen and Saguy (1985), and Arabshahi and Lund (1985), have dealt with the problems of statistical fit of kinetic data using the two models. Three least-square regression models – a separate linear model, a multiple linear model and a nonlinear model – have been examined and the third found to be the best fit. The two models, however, are essentially different; the D-value is proportional to the temperature, whereas k is proportional to $1/T$. Despite the differences, experimental results for *Bacillus stearothermophilus* (Gillespy 1947; Navani *et al.* 1970; Jonsson 1977; Perkin *et al.* 1977; Ocio *et al.* 1994) fit both models with a high degree of correlation. For the heat inactivation of *Saccharomyces uvarum*, Manji and Van De Voort (1985) found that the data fitted the Arrhenius model better than the D–z model. Jones (1968) showed significant differences between the two approaches for calculating process lethalities; however, the data used were inconsistent and invalidated the results (Cowell 1968). Ramaswamy *et al.* (1989), extending the work of Lund (1975) and Cleland and Robertson (1985), concluded that over a limited range of processing temperatures, 115–135°C, near to the reference temperature 121.1°C, there is little to choose between the two approaches. The D–z model predicts greater lethality values than the k–E model for temperatures below T_{ref} and vice versa.

TABLE 3.4. Statistical accuracy of some D- and z-values.

Reference	Statistical parameter	D-value	z-values
Lewis (1956)	Coefficient of variation	10%	–
Hicks (1961)	Standard error of the mean	–	2%
Stumbo et al. (1950)	Coefficient of variation	–	2%
Kaplan et al. (1954)	Coefficient of variation	–	9%
Kaplan et al. (1954)	Mean value [PA 3679]	–	10.4°C
Lenz and Lund (1977)	Standard deviation	10% (0.10D)	1.11°C
Patino and Heil (1985)	Standard deviation	3.6% (0.03–0.06D)	0.28–0.55°C

The practical and theoretical aspects of alternatives to the $D - z$ concept have been discussed by Brown (1992b) and Casadei and Gaze (1994). The different methods of determining heat resistance are reviewed, including, indirect, particle, mixing, direct steam and electrical methods. The conclusion was reached that in relation to non-logarithmic survivor curves, the $D - z$ approach was less appropriate for determining thermal processes. The authors recommended the use of the logistic model or the Gompertz exponential models.

Sapru et al. (1993) have compared a number of predictive models, applicable mainly to higher processing temperatures. Cunha et al. (1997) have shown that for isothermal heating the Bigelow $D - z$ model gives results in good agreement with the Arrhenius $E_A - k$ model; however, this is not the case for non-isothermal heating, where the initial temperature plays an important role.

3.2.4.6. Errors in Experimental Kinetic Factors

In the previous section the differences between two kinetic models were discussed, as well as the errors arising from these approaches. It is, however, necessary to look at the inherent error in the experimental results. Hicks (1961) considered that the following factors contributed to the variations encountered in experimental results: the nature of the organism; the history of the spore prior to use; the conditions of sporulation; the nature of the substrate in which the spores are heated; and the substrate used for detecting survivors. Discussion of these factors is beyond the scope of this book, and the practicing technologist should seek information from microbiological sources. It is, however, important for the process engineer and the food technologist to appreciate the sources of error in the data that they have to use.

The errors found in some of the experimental kinetic factors and their effect on process evaluation have been widely discussed – see, for example, Lewis (1956), Hicks (1961), Lund (1978, 1983), Patino and Heil (1985), and Datta (1993). Table 3.4 gives some data on the accuracy of experimental D- and z-values.

3.2.4.7. Conclusion

Pflug (1982), an international authority on food sterilization, recommends the TDT method as an objective tool for use in the laboratory, in the manufacturing plant, and in validation and monitoring of sterilization processes. For the purposes of this book it should be recognized that discrepancies exist; however, provided the processing temperatures are not excessive, the problem can be minimized. Care should be taken to examine the original experimental microbiological data to make sure they are consistent with the application envisaged. The situation is more complex at the higher processing temperatures encountered in continuous-flow sterilization for aseptic packaging applications (Holdsworth 1992). This also applies to high-temperature, short-time (HTST) processing of packaged foods, where it is practicable.

3.3. Kinetics of Food Quality Factor Retention

3.3.1. Introduction

When food is heated for the purpose of destroying micro-organisms, several types of chemical and physico-chemical reactions also occur, some desirable but often excessive (e.g. enzyme destruction, cooking, texture softening), others less desirable but often inevitable to some degree (e.g. nutrient destruction and loss of organoleptic quality, color, texture and flavor). These reactions are essentially chemical and less heat-dependent than microbial destruction. The art of canning depends on being able to choose processes which are microbiologically safe but result in the least loss of quality. In this section some of the kinetic factors for these changes will be discussed.

3.3.2. Kinetic Representation

Both the D–z and k–E methods, discussed in Section 3.2, have been used to express the temperature dependence of reaction rate.

From equation (3.12), the basic kinetic equation for a first-order reaction is

$$-dc/dt = kc, \tag{3.42}$$

where c is the concentration of the species being studied and dc/dt is the rate of change of concentration with time. This yields the familiar relation

$$c_1/c_2 = e^{-k_c(t_1-t_2)}$$

or

$$\ln(c_1/c_2) = k_c(t_2 - t_1), \tag{3.43}$$

where c_1 and c_2 are the concentrations at times t_1 and t_2, respectively, and k_c refers to a chemical constituent or physical attribute.

The temperature dependence is given by the Arrhenius equation

$$k_c = A e^{-E/RT}. \tag{3.44}$$

The D–z concept is also used for heat-vulnerable factors. These can be determined using the equations given in section 3.2.2. The D_{ref}-value, often designated

TABLE 3.5. Some comparative z-values for heat-vulnerable constituents.

Vulnerable factor	z-value (°C)	Tables
Bacterial spores	7–12	Table A.1
Vegetative cells	4–8	Table A.1
Vitamins	25–30	Table B.1
Proteins	15–37	Table B.2
Enzymes	10–50	Table B.3
Overall sensory quality	25–45	Table B.4
Texture and cooking	17–47	Table B.5
Colour degradation	17–57	Table B.6

as D_c, is taken at the reference temperature T_{ref}, and represents 90% destruction of a thermolabile component; the corresponding z-value is designated z_c. Equation (3.45) is used to determine D-values for thermolabile components:

$$\log(D/D_{ref})_c = -(T - T_{ref})/z_c, \tag{3.45}$$

where the subscript c refers to a chemical compound or physical attribute.

Adams (1978) compared experimental results for the thermal inactivation of horseradish peroxidase with the theoretical predictions of the D–z, k–E and absolute reaction rate models, but found little difference between them. This is probably because there is less temperature sensitivity with these compounds than with bacterial spores. Cohen and Saguy (1985) have statistically evaluated the applicability of the Arrhenius model to food quality losses.

3.3.3. Kinetic Factors

For chemical constituents of food the rates of degradation reactions are slower than for the destruction of microbial species. Table 3.5 shows these differences in general terms. For more detailed information the reader should consult Villota and Hawkes (1986, 1992).

3.3.4. Experimental Procedures

The main experimental methods used to determine the kinetics of the effect of temperature on quality attributes and heat-vulnerable components of food are as follows:

3.3.4.1. Steady-State Methods

These use blended samples of the food, or food and an added component, which are heated to a constant temperature for constant periods of time, the time being increased for each successive experiment. The reaction mixture is then analyzed for the component being studied at each period of time, and the results studied to determine the kinetic factors, using either the D–z or the k–E model. The main problem with this method is that a considerable period of time is required to analyze the large number of samples required. For further discussion of the technique and error analysis, see Lenz and Lund (1980) and Hill and Grieger-Block (1980).

3.3.4.2. Unsteady-State Methods

With these methods the samples are submitted to a variable time–temperature heating profile, either similar to those encountered in commercial canning or to a programmed sequence. The kinetic factors are determined using a trial-and-error method, which can be computerized and the average retention of a component or components obtained. For further details, the reader is referred to the work of Hayakawa (1969), Nasri et al. (1993), and Nunes et al. (1991).

A linear gradient technique has been developed by Cunha et al. (2000) for the kinetics of the acid hydrolysis of sucrose. A theoretical equation was developed for a linearly rising temperature gradient and an analytical solution was derived using exponential integrals. Values of $k_{77°C}$ varied from 0.10 to 1.47.

The large differences in the values of kinetic factors found in the literature (see Appendix B for some tabulated data) often result from differing experimental methods (Sadeghi & Swartzel 1990; Swartzel 1982; Nunes et al. 1993). Considerable care needs to be taken in obtaining reproducible and satisfactory results that may be used with confidence in optimizing process conditions.

3.3.5. Specific Components

3.3.5.1. Vitamin Destruction

The destruction of nutrients and vitamins during the processing of canned foods is detrimental to their nutritional value. It is necessary to know the kinetics of destruction in order to determine the conditions of processing necessary for minimizing this effect (Lund 1977). There has been a continuous trend in thermal processing to reduce process times and increase temperatures in order to reduce nutrient degradation. Similarly, attempts have been made to utilize thin-profile heat-stable plastic pouches and containers.

Greenwood et al. (1944) showed that the rate of thiamine destruction during processing meat doubled with a 10°C increase in temperature compared with a tenfold increase in the rate of inactivation of heat-resistant bacteria. The most extensively quoted data on thiamine destruction in puréed meats and vegetables was reported by Feliciotti and Esselen (1957), who determined D-values, specific reaction rates and half-life values, for temperatures ranging from 108 to 150°C at pH values from 3.5 to 7. Hayakawa (1977) has compared experimentally obtained results for thiamine degradation in canned puréed meats and puréed vegetables with computer predictions using kinetic models and heat transfer data. He concluded that theoretical predictions could be used as a good indicator of the effect of differing processes on nutrient retention. Mauri et al. (1989) have reviewed the existing data on thiamin degradation in foods and model solutions of high water activity.

Davey and Cerf (1996) have studied the effect of combined temperature and pH on the kinetics of destruction of vitamins B_1 and C and shown that the model represented by equation (3.33) can be used to predict the degree of denaturation satisfactorily.

Some kinetic data for the thermal destruction of nutrients are given in Table B.1 (Appendix B).

3.3.5.2. Protein Denaturation

Relatively little information is available on the kinetics of protein destruction in thermally processed foods; mostly it relates to milk proteins, and haemoglobin in meat products. Some kinetic data are presented in Table B.2.

3.3.5.3. Enzyme Inactivation

Enzymes in many food products control the deteriorative reactions in fresh produce and it is necessary, when handling raw materials such as vegetables, to inactivate the enzymes prior to processing. This is one of the objectives of the heat treatment known as *blanching*; for a recent excellent review, see Selman (1989), and for earlier work on the nature of thermal destruction of enzymes and their regeneration, see Leeson (1957). Apart from this, many other products require adequate enzyme inactivation to stabilize them, especially at higher processing temperatures, where the process times for bacterial destruction may be less than those for enzyme inactivation (Adams 1978). Some kinetic data for thermal destruction of enzymes are given in Table B.3.

3.3.5.4. Quality Attributes

The effect of thermal processing on sensory and other factors (e.g. color, texture, flavor and visual appearance) has received only limited treatment despite the importance of the subject. Overall sensory effects have been studied by several workers, among them Cho and Pyun (1994), Mansfield (1974), Hayakawa et al. (1977), and Ohlsson (1980). Cooking effects have been more extensively covered because of the importance of mouth-feel and chewing requirements (Paulus 1984; Rao & Lund 1986). Some kinetic data are given in Table B.4 for overall sensory quality, in Table B.5 for texture and softening, and in Table B.6 for color degradation. Ávila and Silva (1999) have produced an extensive list of kinetic parameters for color degradation in fruits and vegetables.

3.3.6. Summary

Despite the adverse effects of heat on thermally processed foods, many products are commercially in high demand and produce a significant contribution to the diet. Many products are esteemed as brand image products in their own right, without reference to the initial raw material or product. There is, however, a considerable need to be able to quantify the effects of heat on food components in order to be able to predict the effect of different thermal processes. Much of the difficulty in experimental work comes from the variability of the composition of raw materials and prepared products ready for canning and the reliability of the experimental techniques. It is essential to have statistically well-designed experiments in order to account for the variability. This is an important area for future research, especially considering the need to conserve energy and natural resources.

References

Abraham, G., Debray, E., Candau, Y., & Piar, G. (1990). Mathematical model of thermal destruction of *Bacillus stearothermophilus* spores. *Appl. Environ. Microbiology*, *56*(10), 3073–3080.

Adams, J. B. (1978). The inactivation and regeneration of peroxidase in relation to HTST processing of vegetables. *J. Food Technol.*, *13*, 281–293.

Aiba, S., & Toda, K. (1967). Thermal death rate of bacterial spores. *Process Biochem.*, *2*(2), 35–40.

Anderson, W. F., McClure, P. J., Baird-Parker, A. C. & Cole, M. B. (1996). The application of a log-logistic model to describe the thermal inactivation of *C. boulinum 213B* at temperatures below 121.1°C. *J. Appl. Bact.*, *80*, 283–290.

Arabshahi, A., & Lund, D. B. (1985). Considerations in calculating kinetic parameters from experimental data. *J. Food Process Eng.*, *7*, 239–251.

Arrhenius, S. A. (1889). Influence of temperature on the rate of inversion of sucrose. *Z. Physik. Chem.*, *4*, 226.

Atkins, P. W. (1994). *Physical chemistry* (5th Edition). London: Oxford University Press.

Ávila, I. M. L. B., & Silva, C. L. M. (1999). Modelling kinetics of thermal degradation of colour in peach puree. *J. Food Eng.*, *39*(2), 161–166.

Ball, C. O. (1923). Thermal process time for canned food. *Bull. Nat. Res. Council*, *7*(37), 9–76.

Ball, C. O., & Olson, F. C. W. (1957). *Sterilization in food technology – Theory, practice and calculations*. New York: McGraw-Hill.

Banks, B. E. C., Damjanovic, V., & Vernon, C. A. (1972). The so-called thermodynamic compensation law and thermal death. *Nature*, *240*(5377), 147–148.

Belehradek, J. (1926). Influence of temperature on biological processes. *Nature, 118*, 117–118.

Bigelow, W. D. (1921). The logarithmic nature of the thermal death time curves. *J. Infectious Diseases*, *29*, 528–536.

Bigelow, W. D., & Esty, J. R. (1920). Thermal death point in relation to time of typical thermophilic organisms. *J. Infectious Diseases*, *27*, 602–617.

Boon, M. R. (1973). Thermodynamic compensation rule. *Nature*, *243*, 401.

Brown, K. L. (1991). Principles of heat preservation. In J. A. G. Rees & J. Bettison (Eds.) *Processing and packaging of heat preserved foods*. Glasgow and London: Blackie.

Brown, K. L. (1992a). Heat resistance of bacterial spores. PhD thesis, University of Nottingham, UK.

Brown, K. L. (1992b) Practical and Theoretical Aspects of the death kinetics of micro-organisms – Alternatives to D and z concept. *Technical Bulletin No. 92*. Chipping Campden, Glos. UK: Campden & Chorleywood Food Research Association.

Brown, K. L., Gaze, J. E., McClement, R. H., & Withers, P. (1988). Construction of a computer-controlled thermoresistometer for the determination of the heat resistance of bacterial spores over the temperature range 100–150°C. *Int. J. Food Sci. Technol.*, *23*, 361–371.

Buera, M. P., Chirife, J., Resnik, S. L., & Lozano, R. D. (1987a). Nonenzymatic browning in liquid model systems of high water activity: kinetics of colour changes due to caramelization of various single sugars. *J. Food Sci.*, *52*(4), 1059–1062, 1073.

Buera, M. P., Chirife, J., Resnik, S. L. & Wetzler, G. (1987b). Nonenzymatic browning in liquid model systems of high water activity: kinetics of colour changes due to Maillard's reaction between different single sugars and glycine and comparison with caramelization browning. *J. Food Sci.*, *52*(4), 1063–1067.

Burton, H. (1988). *UHT processing of milk and milk products*. London: Elsevier Applied Science.

Campanella, O. H. & Peleg, M. (2001). Theoretical comparison of a new and traditional method to calculate *C botulinum* survival during thermal inactivation. *J. Sci. Food Agric.*, *81*, 1069–1076.

Casadei, M. A., & Gaze, J. (1994). Practical and theoretical aspects of the death kinetics of micro-organisms: alternatives to the D and z concept. *Technical Memo. No. 706*. Chipping Campden, Glos. UK: Campden & Chorleywood Food Research Association.

Casolari, A. (1979). Uncertainties as to the kinetics of heat inactivation of micro-organisms. *Proceedings international meeting – Food microbiology and technology* (pp. 231–238). Parma, Italy.

Casolari, A. (1981). A model describing microbiological inactivation and growth kinetics. *J. Theor. Biol.*, *88*, 1–34.

Cerf, O. (1977). A review: Tailing of survival curves of bacterial spores. *J. Appl. Bact.*, *13*(6), 851–857.

Chiruta, J., Davey, K. R., & Thomas, C. J. (1997). Thermal inactivation kinetics for three vegetative bacteria as influenced by combined temperature and pH in liquid medium. *Food and Bioproducts Processing, Trans. IChemE.*, *75*(C3), 174–180.

Cho, H.-Y., & Pyun, Y.-R. (1994). Prediction of kinetic parameters for food quality changes using equivalent time at standard temperature. In *Developments in food engineering Part 1* (pp. 453–455). Glasgow and London: Blackie.

Cleland, A. C., & Robertson, G. L. (1985). Determination of thermal processes to ensure commercial sterility of canned foods. In S. Thorne (Ed.), *Developments in food preservation – 3*. London: Applied Science Publishers.

Cohen, E., & Saguy, I. (1985). Statistical evaluation of Arrhenius model and its applicability to food quality losses. *J. Food Process. Preserv.*, *9*, 273–290.

Cole, M. B., & Jones, M. V. (1990). A submerged-coil heating apparatus for investigating thermal inactivation of micro-organisms. *Letters Appl. Microbiol.*, *11*, 233–235.

Corradini, M. G., Normand, M. D., & Peleg, M. (2005). Calculating the efficacy of heat sterilization processes. *J. Food Eng.*, *67*(1/2), 59–69.

Cowell, N. D. (1968). Methods of thermal process evaluation. *J. Food Technol.*, *3*, 303–304.

Cunha, L. M., & Oliviera, F. A. R. (2000). Optimal experimental design for estimating the kinetic parameters of processes described by the first-order Arrhenius model undergoing linearly increasing temperature profiles. *J. Food Eng.* *46*(1), 53–60.

Cunha, L. M., Oliveira, F. A. R., Brandão, T. R. S., & Oliveira, J. C. (1997). Optimal experimental design for estimating the kinetic parameters of the Bigelow model. *J. Food Eng.*, *33*(1/2), 111–128.

Datta, A. K. (1993), Error estimates for kinetic parameters used in the food literature. *J. Food Eng.*, *18*(2), 181–199.

Davey, K. (1993a). Linear-Arrhenius models for bacterial growth and death and vitamin denaturations. *J. Ind. Microbiol., 12*, 172–179.

Davey, K. R. (1993b). Extension of the generalized sterilization chart for combined temperature and pH. *Lebens. -Wiss. u. -Technol., 26*(5), 476–479.

Davey, K. R. (1994). Modelling the combined effect of temperature and pH on the rate coefficient for bacterial growth. *Int. J. Food Microbiol., 23*, 295–303.

Davey, K., Hall, R. F., & Thomas, C. J. (1995). Experimental and model studies of the combined effect of temperature and pH on the thermal sterilization of vegetative bacteria in liquid. *Food & Bioproducts Process. Trans. IChemE, 73*(C3), 127–132.

Davey K. R., & Cerf, O. (1996). Predicting concomitant denaturation of vitamin as influenced by combined process temperature and pH in Batch and continuous flow sterilization of liquids. *Food & Bioproducts Processing, Trans. Inst. ChemE., 74*(C4), 200–206.

Davey, K. R. Thomas, C. J., & Cerf, O., (2001). Thermal death of bacteria. *J. Appl. Microbiol. 90*(1), 148–150.

Dickerson, R. W. Jr (1969). Simplified equations for calculating lethality of the heating and cooling phases of thermal inactivation determinations. *Food Technol., 23*(3), 108–111.

Esty, J. R., & Meyer, K. F. (1922). The heat resistance of spores of *B. botulinus* in canned foods. *J. Infectious Diseases, 31*, 650–663.

Feliciotti, E., & Esselen, W. B. (1957). Thermal destruction rate of thiamine in pureed meat and vegetables. *Food Technol., 11*, 77–84.

Gaillard, S., Leguérinel, I., & Mafart, P. (1998). Model for combined effects of temperature, pH and water activity on thermal inactivation of *Bacillus cereus* spores. *J. Food Sci., 63*(5), 887–889.

Gaze, J. E., & Brown, K. L. (1990). Comparative heat resistance studies on spores of *Clostridium botulinum, Clostridium sporogenes*, and *Bacillus stearothermophilus*: thermoresistometer studies and Bacillus datafile. *Technical Memorandum No. 568.* Chipping Campden, Glos., UK: Campden & Chorleywood Food Research Association.

Geeraerd, A. H., Valdramidis, V. P., Bernaerts, K., & Van Impe, J. F. (2004). Evaluating microbial inactivation models for thermal processing. In P. Richardson (Ed.), *Improving the thermal processing of food* (pp. 427–453). Cambridge, UK: Woodhead Publishing Ltd.

Gillespy, T. G. (1946). The heat resistance of spores of thermophilic bacteria I. *Annual Report 1946* (pp. 40–49). Chipping Campden, Glos., UK: Campden & Chorleywood Food Research Association.

Gillespy, T. G. (1947). The heat resistance of spores of thermophilic bacteria II. Thermophilic anaerobes. *Annual Report 1947* (pp. 40–51), Chipping Campden, Glos., UK: Campden & Chorleywood Food Research Association.

Gillespy, T. G. (1948). The heat resistance of spores of thermophilic bacteria III. *Annual Report 1948* (pp. 34–43). Chipping Campden, Glos., UK: Campden & Chorleywood Food Research Association.

Glasstone, S., Laidler, K. J., & Eyring, H. (1941). *The theory of rate processes* (pp. 1–27). McGraw-Hill, New York and London.

Greenwood, D. A., Kraybill, H. R., Feaster, J. F., & Jackson, J. M. (1944). Vitamin retention in processed meat. *Ind. Eng. Chem., 36*, 922–927.

Haas, J., Bultermann, R. B., & Schubert, H. (1994). Determination of the thermal death kinetics of bacterial spores by indirect heating methods. In T. Yano, R. Matsuno, & K. Nakamura (Eds.), *Developments in food engineering ICEF 6 Part 2* (pp. 698–700).

Hachigian, J. (1989). An experimental design for determination of D-values describing inactivation kinetics of bacterial spores; design parameters selected for using computer simulation. *J. Food Sci.*, *54*(3), 720–726.

Hallström, B., Skjöldebrand, C., & Trägärdh, C. (1988). *Heat transfer and food Products*. London: Elsevier Applied Science.

Haralumpu, S. G., Saguy, I., & Karel, M. (1985). Estimation of Arrhenius model parameters using three least-squares methods. *J. Food. Process. Preserv.*, *9*, 129–143.

Harris, P. S. (1973). Compensation effect and experimental error. *Nature*, *243*, 401–402.

Hayakawa, K. (1969). New parameters for calculating mass average sterilizing value to estimate nutrients in thermally conductive food. *Can. Inst. Food Sci. Technol. J.*, *2*, 165–172.

Hayakawa, K. (1977). Review of computerized prediction of nutrients in thermally processed canned food. *J. Assoc. Offic. Anal. Chem.*, *60*(6), 1243–1247.

Hayakawa, K. (1982). Empirical formulae for estimating nonlinear survivor curves of thermally vulnerable factors. *Can. Inst. Food Sci. Technol. J.*, *15*(2), 116–119.

Hayakawa, K., Schnell, P. G., & Kleyn, D. H. (1969). Estimating thermal death time characteristics of thermally vulnerable factors by programmed heating of sample solution or suspension. *Food Technol.*, *23*, 1090–1095.

Hayakawa, K., Timbers, G. E., & Steir, E. F. (1977). Influence of heat treatment on the quality of vegetables: organoleptic quality. *J. Food Sci.*, *42*, 1286–1289.

Hayakawa, K., Matsuda, N., Komaki, K., & Matsunawa, K. (1981). Computerized estimation of reaction kinetic parameters for thermal inactivation of micro-organisms. *Lebensm.-Wiss. u. -Technol.*, *14*, 70–78.

Hersom, A. C., & Hulland, E. D. (1980). *Canned foods. Thermal processing and microbiology* (7th Edition). Edinburgh: Churchill Livingstone.

Hicks, E. W. (1961). Uncertainties in canning process calculations. *Food Res.*, *32*(6), 218–223.

Hill, C. G. Jr., & Griecer-Block, R. A. (1980). Kinetic data: generation, interpretation and use. *Food Technol.*, *34*(2), 56–66.

Holdsworth, S. D. (1992). *Aseptic processing and packaging of food products*. London: Elsevier Applied Science.

Holdsworth, S. D., & Overington, W. J. G. (1975). Calculation of the sterilizing value of food canning processes. *Technical Bulletin No. 28*. Chipping Campden, Glos., UK: Campden & Chorleywood Food Research Association.

Horak, F. P. & Kessler, H. G. (1981). Thermal destruction of thiamine – a second order reaction. *Z. Lebensm. Untersuch. Forsch.*, *172*, 1–6.

Ito, K. A. & Chen, J. K. (1978). Effect of pH on growth of *Clostridium botulinum* in foods. *Food Technol.*, *32*(6), 71–72, 76.

Jakobsen, F. (1954). Notes on process evaluation. *Food Res.*, *19*, 66–79.

Jones, M. C. (1968). The temperature dependence of the lethal rate in sterilization calculations. *J. Food Technol.*, *3*, 31–38.

Jonsson, U., Snygg, B. G., Harnulv, B. G., & Zachrisson, T. (1977). Testing two models for the temperature dependence of the heat inactivation of *Bacillus stearothermophilus* spores. *J. Food Sci.*, *42*, 1251–1252, 1263.

Kaplan, A. M., Reynolds, H., & Lichtenstein, H. (1954). Significance of the variations in observed slopes of thermal death time curves for putrefactive anaerobes. *Food Res.*, *19*, 173–179.

Karplus, M., Porter, R. N., & Sharma, R. D. (1965). Exchange reactions with activation energy. *J. Chem. Phys.*, *43*, 3259.

Katzin, L. I., Sandholzer, L. A., & Strong, M. E. (1942). Application of the decimal reduction principle to a study of the heat resistance of coliform bacteria to pasteurization. *J. Bacteriol.*, *45*, 265–272.

Kemeny, G., & Rosenberg, B. (1973). Compensation law in thermodynamics and thermal death. *Nature*, *243*, 400–401.

Khoo, K. Y., Davey, K. R. & Thomas, C. J. (2003). Assessment of four model forms for predicting thermal inactivation kinetics of *Escherichia coli* in liquid as affected by combined exposure time, liquid temperature and pH. *Food & Bioproducts Processing, Trans. I ChemE.*, *81*(C2), 129–137.

Kilsby, D. C., Davies, K. W., McClure, P. J., Adair, C & Anderson, W. A. (2000). Bacterial thermal death kinetics based on probability distributions; the heat destruction of *C. Botulinum* and *Salmonella Bedford*. *J. Food Protection*, *63*(9), 1197–1203.

Körmendy, I., & Körmendy, L. (1997). Considerations for calculating heat inactivation processes when semilogarithmic thermal inactivation models are non-linear. *J. Food Eng.*, *34*(1), 33–40.

Körmendy, I., Körmendy, L., & Ferenczy, A. (1998). Thermal inactivation kinetics of mixed microbial populations. A hypothesis paper. *J. Food Eng.*, *38*(4), 439–453.

Körmendy, I., & Mohácsi-Farkas, C. (2000). Heat treatment and defective units' ratio: surviving active spores in pea puree. *J. Food Eng.*, *45*(4), 225–230.

Laing, B. M., Schuelter, D. L., & Labuza, T. P. (1978). Degradation kinetics of ascorbic acid at high temperatures. *J. Food Sci.*, *43*(5), 1440–1443.

Leeson, J. A. (1957). The inactivation of enzymes by heat: a review of the evidence for the regeneration of enzymes. *Scientific Bulletin No. 2*. Chipping Campden, Glos., UK: Campden & Chorleywood Food Research Association.

Lenz, M. K. & Lund, D. B. (1977). The lethality-Fourier number method. Confidence intervals for calculated lethality and mass-average retention of conduction-heating canned foods. *J. Food Sci.*, *42*, 1002–1007.

Lenz, M. K., & Lund, D. B. (1980). Experimental procedures for determining destruction kinetics of food components. *Food Technol.*, *34*(2), 51–55.

Lewis, J. C. (1956). The estimation of decimal reduction times. *Appl. Microbiol.*, *4*, 211–214.

Lund, D. B. (1975). Effects of blanching, pasteurization and sterilization on nutrients. In R. E. Harris & E. Karamas (Eds.) *Nutritional evaluation of food processing* (2nd Edition). Westport, CT: AVI.

Lund, D. B. (1977). Design of thermal processes for maximizing nutrient retention. *Food Technol.*, *31*(2), 71–75.

Lund, D. B. (1978). Statistical analysis of thermal process calculations. *Food Technol.*, *32*(3), 76–78, 83.

Lund, D. B. (1983). Influence of variations of reaction kinetic parameters and in heat transfer parameters on process lethality. In T. Motohiro & K. Hayakawa (Eds.), *Heat sterilization of food* (pp. 165–171). Tokyo: Koseicha-Koseikaku.

Machmerth, W. (1983). *Theoretical basis and optimization of thermal sterilization*. Leipzig: J. A. Barth.

Mafart, P. (1999a). Modelling the heat destruction of micro-organisms. *Sci. des Aliments*, *19*(2), 131–146.

Mafart, P., & Leguérinel, I. (1999b). Modeling combined effects of temperature and pH on heat resistance of spores by a linear-Bigelow equation. *J. Food Sci.*, *63*(1), 6–8.

Malmborg, A. (1983). Application of theories for the kinetics of inactivation of micro-organisms. In J. V. McLoughlin & B. M. McKenna (Eds.), *Proc. 6th international congress – Food science and technology.* Dublin: Boole.

Manji, B., & Van De Voort, F. R. (1985). Comparison of two models for process holding time calculations, convection system. *J. Food Protection*, *48*(4), 359–363.

Mansfield, T. (1974). *A brief study of cooking.* San José, CA: Food & Machinery Corporation.

Mauri, L. M., Alzamora, S. M., Chirife, J., & Tomio, M. J. (1989). Review: kinetic parameters for thiamine degradation in foods and model solutions of high water activity. *Int. J. Food Sci. Technol.*, *24*, 1–30.

McMeekin T. A., Olley, J. N., Ross, T., & Ratkowsky, D. A. (1993). *Predictive microbiology – Theory and application.* Taunton, UK: Research Studio Press.

Moore, W. J. (1972). *Physical Chemistry.* London: Longmans.

Moore, W. J. (1983). *Basic Physical Chemistry.* Englewood Cliffs, NJ: Prentice Hall.

Mycock, G. (2002). Heat process evaluation and non-linear kinetics. In Tucker, G. S. (Ed.), *Second international symposium on thermal processing – Thermal processing: validation challenges* (Session 2:1). Chipping Campden UK: Campden & Chorleywood Food Research Association.

Nasri, H., Simpson, R., Bouzas, J., & Torres, J. A. (1993). An unsteady-state method to determine kinetic parameters for heat inactivation of quality factors: conduction-heated foods. *J. Food Eng.*, *19*, 291–301.

Navani, S. K., Scholefield, J., & Kibby, M. R. (1970). A digital computer program for the statistical analysis of heat resistance data applied to *Bacillus stearothermophilus* spores. *J. Appl. Bact.*, *33*(4), 609–620.

Nunes, R. V., Rhim, J. W., & Swartzel, K. R. (1991). Kinetic parameter evaluation with linearly increasing temperature profiles: integral methods. *J. Food Sci.*, *56*, 1433–1437.

Nunes, R. V., Swartzel, K. R., & Ollis, D. F. (1993). Thermal evaluation of food process: the role of reference temperature. *J. Food Eng.*, *20*, 1–15.

Ocio, M. J., Rhim, P. S., Alvarruiz, A. & Martinez, A. (1994). Comparison of TDT and Arrhenius models for rate constant inactivation predictions of *Bacillus stearothermophilus* heated in mushroom-alginate substrate. *Letters in Applied Microbiology*, *19*, 114–117.

Ohlsson, T. (1980). Temperature dependence of sensory quality changes during thermal processing. *J. Food Sci.*, *45*, 836–839, 847.

Patino, H., & Heil, J. R. (1985). A statistical approach to error analysis in thermal process calculations. *J. Food Sci.*, *50*(4), 1110–1114.

Paulus, K. O. (1984). Cooking kinetics: models and knowledge. In B. M. Mckenna (Ed.), *Engineering and food Vol. 1*, (pp. 99–108). London, Elsevier Applied Science.

Peleg, M., & Cole M. B. (1998). Reinterpretation of microbial survival curve. *Critical Rev. Food Sci. & Nutrition*, *38*(5), 353–380.

Peleg, M., & Penchina, C. M. (2000). Modeling microbial survival during exposure to a lethal agent with varying intensity. *Critical Rev. Food Sci. & Nutrition*, *40*(2), 159–172.

Peleg, M., Engel, R., Gonzáles-Martinez, C., & Corradini, M. G. (2002). Non Arrhenius and Non- WLF kinetics in food systems. *J. Sci. Food Agric.*, *82*, 1346–1355.

Peleg, M. (2003a). Microbiological survivor curves: Interpretation, mathematical modeling and utilization. *Comments on Theoretical Biology*, *8*, 357–387.

Peleg, M. (2003b). Calculation of the non-isothermal inactivation patterns of microbes having sigmoidal isotherms semi-logarithmic survival curves. *Critical Rev. Food Sci & Nutrition*, *43*(6), 353–380.

Peleg, M. (2003c). Modeling applied to processes: The case of thermal preservation. In P. Zeuthen & L. Bøgh-Sørensen (Eds.), *Food preservation techniques* (pp. 507–522). Cambridge, UK: Woodhead Publishing.

Peleg, M. (2004). Analyzing the effectiveness of microbial inactivation in thermal processing. In P. Richardson (Ed.), *Improving the thermal processing of foods* (pp. 411–426). Cambridge: Woodhead Publishing.

Perkin, A. G., Burton, H., Underwood, H. M., & Davies, F. L. (1977). Thermal death kinetics of *B. stearothermophilus* spores at ultra-high temperatures. II. Effect of heating period on experimental results. *J. Food Technol., 12,* 131–138.

Pflug, I. J. (1982). *Microbiology and engineering of sterilization processes* (5th Edition). Minneapolis: University of Minnesota Press.

Prentice, R. L. (1976). A generalization of the probit and logit methods for dose response curves. *Biometrics, 32,* 761–768.

Ramaswamy, H. S., Van De Voort, F. R., & Ghazala, S. (1989). An analysis of TDT and Arrhenius methods for handling process and kinetic data. *J. Food Sci., 54*(5), 1322–1326.

Rao, M. A., & Lund, D. B. (1986). Kinetics of thermal softening of foods–a review. *J. Food Proc & Preserv., 10,* 311–329.

Rodrigo, M., Martinez, A., Sanchez, T., Peris, M. J., & Safon, J. (1993). Kinetics of *Clostridium sporogenes PA 3679* spore destruction using computer-controlled thermoresistometer. *J. Food Sci., 58*(3), 649–652.

Rodriguez, A. C., Smerage, G. H., Teixeira, A. A., & Busta, F. F. (1988). Kinetic effects of lethal temperatures on population dynamics of bacterial spores. *Trans. ASAE, 31*(5), 1594–1601.

Sadeghi, F., & Swartzel, K. R. (1990). Generating kinetic data for use in design and evaluation of high temperature processing systems. *J. Food Sci., 55*(3), 851–853.

Sapru, V., Smerage, G. H., Teixeira, A. A., & Lindsay, J. A. (1993). Comparison of predictive models for bacterial spore population resources to sterilization temperature. *J. Food Sci., 58*(1), 223–228.

Selman, J. D. (1989). The blanching process. In S. Thorne (Ed.), *Developments in food preservation–4* (pp. 205–249). London: Elsevier Applied Science.

Shull, J. J., Cargo, G. T., & Ernst, R. R. (1963). Kinetics of heat activation and thermal death of microbial spores. *Appl. Microbiol., 11,* 485–487.

Simpson, S. G., & Williams, M. C. (1974). An analysis of high temperature/short time sterilization during laminar flow. *J. Food Sci., 39,* 1047–1054.

Srimani, B., Stahl, R., & Loncin, M. (1990). Death rates of bacterial spores at high temperatures. *Lebensm.-Wiss. u. -Technol., 13,* 186–189.

Steinfeld, J. I., & Fransico, J. S. (1989). *Chemical kinetics and dynamics*, New York: Prentice Hall.

Stevens, B. (1961). *Chemical kinetics*, London: Chapman & Hall.

Stumbo, C. R. (1948). A technique for studying resistance of bacterial spores to temperatures in the higher range. *Food Technol., 2,* 228–240.

Stumbo, C. R. (1973). *Thermobacteriology in food processing* (2nd Edition). New York: Academic Press.

Stumbo, C. R., Murphy, J. R., & Cochran, J. (1950). Nature of thermal death time curves for P. A. 3679 and *Clostridium botulinum. Food Technol., 4,* 321–326.

Swartzel, K. R. (1982). Arrhenius kinetics applied to product constituent losses in ultrahigh temperature processing. *J. Food Sci., 47,* 1886–1891.

Swartzel, K. R. (1984). A continuous flow procedure for kinetic data generation. *J. Food Sci.*, *49*, 803–806.

Swartzel, K. R. (1986). Equivalent-point method of thermal evaluation of continuous flow systems. *J. Agric. Food Chem.*, *34*, 396–401.

Swartzel, K. R. (1989). Non-isothermal kinetic data generation for food constituents. In R. P. Singh & A. G. Medina (Eds.), *Food properties and computer-aided engineering of food processing systems* (pp. 99–103). Dordrecht: Kluwer Academic Publishers.

Swartzel, K. R., & Jones, V. A. (1985). System design and calibration of continuous flow apparatus for kinetic studies. *J. Food Sci.*, *50*, 1203–1204, 1207.

Teixeira, A. A., Sapru, V., Smerage, G. H., & Rodriguez, A. C. (1990). Thermal sterilization model with complex reaction kinetics. In W. E. L. Spiess and H. Schubert (Eds.), *Engineering and food Vol. 2* (pp. 158–166). London: Elsevier Applied Science.

Teixeira, A. A. (2002). Origins of the 12-D process. In Tucker, G. S. (Ed.), *Second International Symposium on thermal processing – Thermal processing: validation challenges* (Session 1:2). Chipping Campden, Glos., UK: Campden & Chorleywood Food Research Association.

Teixeira, A. A. (2004). Dynamic methods for estimating thermal resistance parameters. In G. S. Tucker (Ed.), *Third international symposium on thermal processing – Process and package innovation for convenience foods* (Session 3.4.). Chipping Campden UK: Campden & Chorleywood Food Research Association.

Truhlar, D. G. (1978). Interpretation of the activation energy. *J. Chem. Educ.*, *55*(5), 310.

Van Boekel, M. J. A. S. (2002). On the use of the Weibull model to describe thermal inactivation of microbial vegetative cells. *Int. J. Food Microbiology*, *72*, 159–172.

Van Loey, A., Fransis, A., Hendrickx, M., Maesmans, G., & Tobback, P. (1995). Kinetics of quality changes in green peas and white beans during thermal processing. *J. Food Eng.*, *24*, 361–377.

Vas, K. (1970). Problems of thermal processing. *J. Appl. Bact.*, *33*, 157–166.

Villota, R., & Hawkes, J. G. (1986). Kinetics of nutrients and organoleptic changes in foods during processing. In M. R. Okos (Ed.), *Physical and chemical properties of food*. St Joseph, MI: American Society of Agricultural Engineers.

Villota, R., & Hawkes, J. G. (1992). Reaction kinetics in food systems. In D. R. Heldman and D. B. Lund (Eds.), *Handbook of food engineering*. New York: Marcel Dekker.

Warren, D. S. (1973). A physico-chemical model for the death rate of a micro-organism. *J. Food Technol.*, *8*(3), 247–257.

Xezones, H., & Hutchings, I. J. (1965). Thermal resistance of *Clostridium botulinum* (62A) as affected by fundamental constituents. *Food Technol.*, *19*(6), 113–119.

4
Sterilization, Pasteurization and Cooking Criteria

4.1. Sterilization Value

4.1.1. Definitions

The need for a criterion against which to judge the efficacy of a process is paramount in thermal processing technology. First it is necessary to decide on a target organism, which in the case of sterilization is usually *Clostridium botulinum*, with a z-value of 10°C, and then it is necessary to know the temperature history to which the package has been submitted or the temperature distribution obtained at the point of slowest heating in the food product during the process.

The unit of processing value is known as the F-value, and was originally developed from the pioneering work of Bigelow *et al.* (1920) at the National Canners' Association in Washington, and developed subsequently by Ball (Ball & Olson 1957). It is a system that was adopted by the canning industry in the USA and later the UK and is in universal use in other countries, which were influenced by the pioneering American work.

4.1.2. Lethal Rates

A measure of the lethal effect of heat on micro-organisms at a given temperature has already been discussed in Section 3.2.4.1 – the lethal rate L, in minutes, which is defined by the equation

$$L = 10^{(T - T_{\text{ref}})/z}, \tag{4.1}$$

and is the ratio of the D-value at a particular temperature, T, to the D-value at the reference temperature T_{ref}. For a constant temperature T the F-value is equal to the L-value; however, if the temperature varies from an initially ambient temperature to the retort temperature, or the reverse on cooling, then it is necessary to integrate the achieved lethal rates at the various temperatures (Figure 4.1). The basic equation for calculating F-values for processes is given by:

$$F = \int_0^t L\,dt, \tag{4.2}$$

123

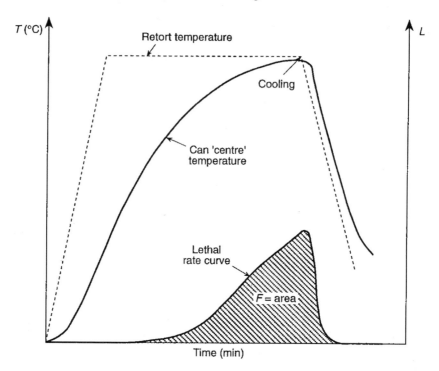

FIGURE 4.1. Graph of temperature (T) and lethal rate (L) against time.

or, in the classical form,

$$F_r = \int_0^t 10^{\frac{T - T_{ref}}{z}} dt \qquad (4.3)$$

The value of $T - T_{ref}$ is always negative when the processing temperature, and hence the internal temperature, T, is less than the reference temperature.

   If the relationship between temperature and time at the point of slowest heating is in a mathematical form, then the equation can be analytically integrated. However, it is more usual, from heat penetration experiments, to know the time–temperature relationship in numerical form. The equation may then be integrated by a mathematical technique such as the trapezium rule. An important property of lethal rates is that they are additive and use can be made of lethal rate tables (Table 4.1), calculated using equation (4.1), to determine F-values. Table 4.2 lists lethal rate tables available for different temperature systems and different reference temperatures; however, it is easier to construct a table using equation (4.1) for the particular conditions being used.

4.1.3. Reference Temperatures

While it has become established practice to use a reference temperature in degrees Celsius, it was the practice at first to use the Fahrenheit system.

TABLE 4.1. Lethal rate tables ($T_{ref} = 121.1°C$ and $z = 10°C$).

T(°C)	0.0	0.1	0.2	0.3	0.4	0.5	0.6	0.7	0.8	0.9
90	0.001	0.001	0.001	0.001	0.001	0.001	0.001	0.001	0.001	0.001
91	0.001	0.001	0.001	0.001	0.001	0.001	0.001	0.001	0.001	0.001
92	0.001	0.001	0.001	0.001	0.001	0.001	0.001	0.001	0.001	0.002
93	0.002	0.002	0.002	0.002	0.002	0.002	0.002	0.002	0.002	0.002
94	0.002	0.002	0.002	0.002	0.002	0.002	0.002	0.002	0.002	0.002
95	0.002	0.003	0.003	0.003	0.003	0.003	0.003	0.003	0.003	0.003
96	0.003	0.003	0.003	0.003	0.003	0.003	0.004	0.004	0.004	0.004
97	0.004	0.004	0.004	0.004	0.004	0.004	0.004	0.005	0.005	0.005
98	0.005	0.005	0.005	0.005	0.005	0.005	0.006	0.006	0.006	0.006
99	0.006	0.006	0.006	0.007	0.007	0.007	0.007	0.007	0.007	0.008
100	0.008	0.008	0.008	0.008	0.009	0.009	0.009	0.009	0.009	0.010
101	0.010	0.010	0.010	0.010	0.011	0.011	0.011	0.011	0.012	0.012
102	0.012	0.013	0.013	0.013	0.013	0.014	0.014	0.014	0.015	0.015
103	0.015	0.016	0.016	0.017	0.017	0.017	0.018	0.018	0.019	0.019
104	0.019	0.020	0.020	0.021	0.021	0.022	0.022	0.023	0.023	0.024
105	0.025	0.025	0.026	0.026	0.027	0.028	0.028	0.028	0.029	0.030
106	0.031	0.032	0.032	0.033	0.034	0.035	0.035	0.036	0.037	0.038
107	0.039	0.040	0.041	0.042	0.043	0.044	0.045	0.046	0.047	0.048
108	0.049	0.050	0.051	0.052	0.054	0.055	0.056	0.058	0.059	0.060
109	0.062	0.063	0.065	0.066	0.068	0.069	0.071	0.072	0.074	0.076
110	0.078	0.079	0.081	0.083	0.085	0.087	0.089	0.091	0.093	0.095
111	0.098	0.100	0.102	0.105	0.107	0.110	0.112	0.115	0.117	0.120
112	0.123	0.126	0.129	0.132	0.135	0.148	0.141	0.145	0.148	0.151
113	0.155	0.158	0.162	0.166	0.170	0.174	0.178	0.182	0.186	0.191
114	0.195	0.200	0.204	0.209	0.214	0.219	0.224	0.229	0.234	0.240
115	0.245	0.251	0.257	0.263	0.269	0.275	0.282	0.288	0.295	0.302
116	0.309	0.316	0.324	0.331	0.339	0.347	0.355	0.363	0.372	0.380
117	0.389	0.398	0.407	0.417	0.427	0.437	0.447	0.457	0.468	0.479
118	0.490	0.501	0.513	0.525	0.537	0.550	0.562	0.575	0.589	0.603
119	0.617	0.631	0.646	0.661	0.676	0.692	0.708	0.724	0.741	0.759
120	0.776	0.794	0.813	0.832	0.851	0.871	0.891	0.912	0.933	0.955
121	0.977	1.000	1.023	1.047	1.072	1.096	1.122	1.148	1.175	1.202
122	1.230	1.259	1.288	1.318	1.349	1.380	1.413	1.445	1.479	1.514
123	1.549	1.585	1.622	1.660	1.698	1.738	1.778	1.820	1.682	1.905
124	1.950	1.995	2.042	2.089	2.138	2.188	2.239	2.291	2.344	2.399
125	2.455	2.512	2.570	2.630	2.692	2.754	2.818	2.884	2.952	3.020
126	3.090	3.162	3.236	3.311	3.388	3.467	3.548	3.631	3.715	3.802
127	3.890	3.981	4.074	4.169	4.266	4.365	4.467	4.571	4.677	4.786
128	4.898	5.012	5.129	5.248	5.370	5.495	5.623	5.754	5.888	6.026
129	6.166	6.310	6.457	6.607	6.761	6.918	7.079	7.244	7.413	7.586
130	7.762	7.943	8.128	8.318	8.511	8.710	8.913	9.120	9.333	9.550

Further values may be calculated from the formula $L = 10^{(T-121.1)/z}$.

Thus much of the literature in canning uses a reference temperature of 250°F; subsequently the industry has moved to use the direct Celsius equivalent of 121.1°C or, less frequently, 121.11°C, rather than 120°C or 121°C. In time the processes of standardization will probably settle on one of the more rational values.

TABLE 4.2. Some tables of lethal rates.

T_{ref}	z-value	Temperature range	Smallest ΔT	Reference
250°F	18°F	191–218°F	0.25°F	Patashnik (1953)
250°F	21°F	181–205°F	0.50°F	Patashnik (1953)
250°F	17–23°F	180–260°F	2.00°F	Patashnik (1953)
250°F	10–26°F	180–260°F	2.00°F	Ball and Olson (1957)
250°F	18°F	200–264°F	0.10°F	Sleeth (1978)
250°F	18°F	190–270°F	0.10°F	Lopez (1987)
250°F	12–23°F	200–264°F	0.25°F	Stumbo (1973)
250°F	12–23°F	200–264°F	0.25°F	Hersom and Hulland (1980)
120°C	10°C	100–121.9°C	0.10°C	Brennan et al. (1976)
120°C	10°C	90–155°C	0.10°C	Shapton et al. (1971)
121.0°C	10°C	90–130°C	1.00°C	Lewis (1987)
121.1°C	10–26°C	90–125°C	0.10°C	Cheftel and Thomas (1963)
121.1°C	10°C	90–130°C	0.10°C	See Table 4.1
121.11°C	10°C	100–121.9°C	0.01°C	Brennan et al. (1976)
121.11°C	7–12°C	90–124°C	0.10°C	Hersom and Hulland (1980)
85°C*	10°C	65–94°C	0.10°C	Shapton et al. (1971)

* For pasteurization purposes.

It is usual to be more specific in defining F-values by using adscripts z and T_{ref}, so that the commonest value is $F_{121.1}^{10}$ or F_{250}^{18}, designated F_0. Thus equation (4.3) becomes

$$F_0 = \int_0^t 10^{(T-121.1)/10} dt \equiv \int_0^t 10^{(T-250)/18} dt. \qquad (4.4)$$

The system of using F-values with adscripts defining the conditions is particularly useful and avoids confusion. It can be extended to other processes, such as the heat destruction of yeast spores in beer, a pasteurization process, using $F_{212}^{12.5}$ or F_{100}^{7}. It is, however, necessary when using this system to state at what point the value has been determined. An alternative system uses subscripts alone, e.g. F_c when the F-value has been determined at the center point in the food mass. This is also known as the point of slowest heating or critical point. When this system is not used it may be assumed, unless stated to the contrary, that the F-value has been determined from time–temperature data obtained at this unique point.

The original unit of process was 1 min at 250°F and, although the system still exists, there is a continuing change to 1 min at 121.1°C. Whichever system is used, the F-value is not affected, provided that z-values and temperatures are in consistent units. If a process has an F-value of 4 min, the process is said to be equivalent in sterilizing value to 4 min at 121.1°C, assuming instant attainment of this temperature at the start and instant cooling to sub-lethal temperatures at the end. This is very useful, since it means that different processes can be compared for their lethal effects. For every F-value there is a range of combinations of time and temperature that will produce the equivalent sterilizing value. In practice, there are limits to the suitability of a process; for sterilizing processes, temperatures are generally in the range

115–130°C (240–270°F), and for pasteurization processes a temperature of 70–100°C (160–212°F) is the normal value.

Like the lethal rate, the F-value is additive; thus the various stages may be evaluated separately, and their contributions to the total F-value for a complete heating-holding-cooling cycle may be obtained from the equation

$$F_{total} = F_{heating} + F_{holding} + F_{cooling}. \tag{4.5}$$

4.1.4. A Processing Point of View to Derive F Value

When defining a closed system (canned food, retortable pouches, a particle in a moving system, etc.) and doing a survivor balance, we obtain equation (4.7):

In general, for an open system in non steady state condition the survivor balance can be expressed as

$$[QN]_i - [QN]_o + M\left[\frac{dN}{dt}\right]_I = \left[\frac{dMN}{dt}\right]_S. \tag{4.6}$$

Applying equation (4.6) for the particular case of a closed system, the above general survivor balance is reduced to

$$\left[\frac{dN}{dt}\right]_I = \left[\frac{dN}{dt}\right]_S \tag{4.7}$$

Where

N: Micro-organisms concentration
t: Time
i, o: input and output
I: Inactivation
S: System

Considering first-order kinetics for a micro-organism's inactivation and replacing into equation (4.7):

$$-kN = \left[\frac{dN}{dt}\right]_S \tag{4.8}$$

Separating variables and integrating, taking into account the D value definition:

$$-\int_0^D k\, dt = \int_{N_o}^{\frac{N_o}{10}} \frac{dN}{N} \quad \text{and, therefore,} \quad k = \frac{\ln 10}{D}. \tag{4.9}$$

Given that D values can be expressed as a function of temperature according to equation (3.26),

$$D = D_r 10^{\frac{T_r - T}{z}}. \tag{4.10}$$

Replacing equation (4.10) into equation (4.9) and then into equation (4.8), we obtain

$$-\frac{\ln 10}{D_r 10^{\frac{T_r-T}{z}}} N = \left[\frac{dN}{dt}\right]_s. \tag{4.11}$$

Where

T_r: Reference temperature
D_r: Decimal reduction time at reference temperature
z: temperature change necessary to alter TDT by one log-cycle
Integrating, equation (4.11), from N_0 to $N_0/10^x$ for micro-organisms (where x represents the number of decimal reductions) and between 0 through t for time:

$$x \cdot D_r = \int_0^t 10^{\frac{T-T_r}{z}} dt \tag{4.12}$$

Where the product xD_r was denominated as F_r, then

$$F_r = \int_0^t 10^{\frac{T-T_r}{z}} dt. \tag{4.13}$$

In the case of $T_r = 121.1°C$ (250°F), F_r has been denominated as F_o.

$$F_r = F_o \text{ at } T = 250°F \tag{4.14}$$

$$F_o = \int_0^t 10^{\frac{T-121.1}{z}} dt \tag{4.15}$$

Note that equation (4.15) was derived for closed systems and first order inactivation kinetics.

4.1.5. Integrated F-Values, F_s

Stumbo (1948, 1949) considered that in order to evaluate a process properly it was necessary to consider the probability of survival of micro-organisms throughout the pack and not just at the center. In this method (Stumbo 1953), the pack was divided into iso-F regions, i.e. regions that received the same thermal process, which formed a nest of concentric shells of geometry corresponding to that of the pack. Using the F-value for any particular region λ, denoted by F_λ, and for the center F_c, Stumbo showed that there was a linear relation between $(F_\lambda - F_c)$ and v, the fraction of the total pack volume enclosed by λ, for values of from 0 to 0.4. This is adequate since the region for which the probability of survival is significant is generally less than $\lambda = 0.15$. The relation is given by

$$F_\lambda = F_c + mv, \tag{4.16}$$

where m is the slope of the line.

Using F_s for the total integrated sterilizing value, also known as the mass-average value, for cylindrically shaped cans, the following equation was obtained:

$$F_s = F_c + 1.084D_r + \log[(F_\lambda - F_c)/D_r], \tag{4.17}$$

where D_r is the D-value of the target organism at T_{ref}. This equation sums up all the probabilities from the center of the can to $\upsilon = 0.19$ of the can volume. In practice, F_λ is the value of F_0 at a point whose j-value (Chapter 5) is $j_c/2$, and the value of 1.084 is $\log \upsilon + \log 2.303$.

The F_s-value can be obtained experimentally from the number of organisms present at the start of the process n_0 and the number surviving at the end of the process n_t, using the equation

$$F_s = D_r \log(n_0/n_t) \tag{4.18}$$

The method is described fully with examples of application in Stumbo (1973), and extensions to the subject have been reported by Stumbo and Longley (1966), Manson and Zahradnik (1967), and Skinner and Urwin (1971). A critique of the method is given by Holdsworth and Overington (1975). An application to nutrient retention is reported by Jen et al. (1971), which gives a slightly different formula than equation (4.17) and is discussed in Section 4.2. Newman and Holdsworth (1989) have derived equations for other geometries, in relation to food particulate sterilization; however, these are also applicable to containers.

In 1951 two researchers, Gillespy (1951) and Hicks (1951), independently published a mathematical derivation for the integrated sterilizing value. In the first of these the F_s-value is given by the equation

$$F_s = F_c + D_r \log(2.303 f_h q / D), \tag{4.19}$$

where f_h is the reciprocal of the slope to the heating curve and q is the slope of the curve of $(F_\lambda - F_c)$ against υ. This equation is similar to Stumbo's inasmuch as it shows the dependence of F_s on the D_r-value.

In the second, using a different technique involving the first-order approximation of the heat transfer equation for both heating and cooling, the following equation was obtained:

$$F = f_h \exp[b(T_R - T_{\text{ref}})]E_i(-Ab\,e^{-\gamma t'}) - E_i(-Ab), \tag{4.20}$$

where A and γ relate to the heat transfer equation (see Chapter 2) and t' is the heating time at temperature T_R, b the inactivation rate for a specified microorganism and media, and $E_i(x)$ is the exponential integral (see Chapter 6). The total number of survivors in the whole can, n_f, is obtained by integrating c_f, the concentration of spores at the end of the process, with respect to the volume element dV:

$$n_f = \int_0^\upsilon c_f \, dV. \tag{4.21}$$

4.1.6. F-Values for Cans of Differing Sizes

An important relation between the F-value and the size of containers was obtained by Gillespy (1951). If two series of equal volumes of the same food are packed into A2 and A10 cans, the number of cans will be approximately in the ratio 11:2. If the cans receive processes of equal F-values, then the degrees of sterility with respect to spores will be the same for each pack, but the degrees of sterility with respect to cans will differ. Therefore a higher F-value would have to be attained in the A10 cans to give the same degree of sterility as in the A2 pack.

Given the original mean number of spores, N, in each of the two series of equal volumes of the same food packed in cans of volume V_1 and V_2, then the mean spore survivals m_1 and m_2 would be given by

$$m_1 = N 10^{-F_1/D_{ref}}$$

$$m_2 = N 10^{-F_2/D_{ref}}$$

Since $m_1/m_2 = V_1/V_2$, then

$$F_1 - D_{ref} \log V_1 = F_2 - D_{ref} \log V_2$$

$$F_2 = F_1 + D_{ref} \log(V_2/V_1). \tag{4.22}$$

This can be used to determine the required F-value for a new can size, when the F-value is known for a given can size.

4.1.7. Arrhenius Approach

The Arrhenius equation for microbial death given in Chapter 3 is also used for deriving a measure of the lethal effect of heat. The approach is widely used in continuous sterilization (Holdsworth 1992), but less so in canning process calculations.

The basic equation relates the microbial reduction to the temperature coefficient of microbial destruction:

$$\ln(n_0/n_t) = \int_0^t k \, dt, \tag{4.23}$$

where n_0 and n_t represent the concentrations of micro-organisms at times 0 and t; using the Arrhenius equation, this becomes

$$\ln(n_0/n_t) = A \int_0^t e^{-E/RT} \, dt. \tag{4.24}$$

One of the difficulties of using this equation is that the values of A are extremely large and difficult to handle. This is overcome by using a reference temperature, so that the equation becomes

$$\ln(n_0/n_t) = \int_0^t e^{-(E/RT - E/RT_{ref})} \, dt \tag{4.25}$$

The F-value is obtained from

$$F = D_{ref} \log(n_0/n_t).$$

Deindoerfer and Humphrey (1959a, b) used the symbol ∇ to represent the term $\ln(n_0/n_t)$, while Nunes and Swartzel (1990) used the letter G for $A^{-1} \ln(n_0/n_t)$, where A is the pre-exponential factor of the Arrhenius equation (4.25).

Maesmans *et al.* (1995), have examined Nunes' and Swarztel's method, known as the equivalent point method, and concluded that for time-variable process conditions the concept is less satisfactory than for isothermal heating used in continuous sterilization.

4.2. Cooking Values

4.2.1. Historical Perspective

The application of heat during the processing operation causes not only microbial destruction but also nutrient degradation, texture changes (usually softening) and enzyme inactivation. The final quality of the canned product thus depends on the amount of heat it has received. Biochemical reactions proceed at a substantially slower rate than microbial inactivation, and this is reflected in the higher z-values, which are given in Table 4.3.

The degree of heat treatment with respect to these factors can be represented by C-value equations similar to the F-value equations, beginning with

$$C = \int_0^t 10^{(T-T_{\text{ref}})/z_c} \, dt, \qquad (4.26)$$

where z_c is the thermal destruction rate analogous to the z-factor for microbial inactivation. This concept originated from the work of Mansfield (1962), at FMC, San Jose, CA, and is now standard nomenclature. The E-value of Ball and Olson (1957) does not seem to have been used extensively and has been superseded by the C-value concept. Reichert (1977) used E for enzyme inactivation to distinguish it from the C-value for cooking. Definitions of C_0 vary, but the general reference temperature is 100°C or 212°F and the z_c is appropriate to the heat vulnerable factor being considered. It is probably best if the C-value is quoted with the adscripts depicting the z_c-value (super) and T_{ref} (sub), i.e. $C_{T_{\text{ref}}}^{z_c}$.

TABLE 4.3. Values of z for heat-vulnerable components.

Heat-vulnerable component	Approximate range of z-value (°C)	Detailed tables
Bacterial species	7–12	Table A.1
Vegetative cells	4–8	Table A.1
Enzymes	10–50	Table B.1
Vitamins	25–30	Table B.2
Proteins	15–37	Table B.3
Sensory factors		
Overall	25–47	Table B.4
Texture and softening	25–47	Table B.5
Colour	25–47	Table B.6

The detailed tables in the Appendices are extended versions of those given in Holdsworth (1992).

TABLE 4.4. Variation of F- and C-values with temperature and reference temperature.

Temperature (°C)	$F_{121.1}^{10}$ (min)	$C_{100}^{33.1}$ (min)	$C_{121}^{33.1}$ (min)
100	7.76×10^{-3}	1.00	0.23
110	7.76×10^{-2}	2.15	0.46
121.1	1.00	5.05	1.00
130	7.76	10.00	1.85
140	77.60	21.50	3.71
150	776.00	46.40	7.42

Table 4.4 shows how the F-value and the C-value vary with temperature. It can be seen that the sterilizing values rise by a factor of 10 for every 10°C rise in temperature.

While point values, F_c are applicable to microbial inactivation, it is the total integrated or mass-average values that are most appropriate for cooking; thus we can define a C_s-value by

$$C_s = D_{\text{ref}} \log(c_0/c) \tag{4.27}$$

where c_0 and c are the concentrations of heat-vulnerable species at times 0 and t, respectively, and D_{ref} is the D-value corresponding to the z_c value, and

$$\log(c_0/c) = \frac{1}{D_{\text{ref}}} \int_0^t 10^{(T-T_{\text{ref}})/z_c} \, dt. \tag{4.28}$$

Jen et al. (1971) developed the Stumbo equations (4.17) and (4.18) to take into account the nature of the distribution of heat-vulnerable components, and indicated that Stumbo's distribution equation $(F_\lambda - F_c) = m\upsilon$ does not apply to the outer portions of the can; consequently, a new relation equation (4.29) was developed, which encompassed the whole volume,

$$\ln(1 - \upsilon) = -m(F_\lambda - F_c). \tag{4.29}$$

Using this equation in the F_s equation,

$$C_s \equiv F_s = F_c + D_r \log \left[\frac{D_r + A(F_\lambda - F_c)}{D_r} \right] \tag{4.30}$$

where A depends on the geometry. Jen et al. (1971) indicated that A was 10.9 for a cylindrical can, while Newman and Holdsworth (1989) obtained values of $A = 9.28$ for a spherical object, 10.73 for a cylindrical object and 11.74 for a cube-shaped object.

The art of canning consists of selecting times and temperatures which achieve the desired purpose of microbial destruction and adequate cooking without undue loss of nutrients. The C-value concept, combined with optimization theory, helps to put this on a more scientific basis. The application of this concept is dealt with in Chapter 7.

4.2.2. Origin and Rationale of Cooking Value

Cooking value was derived from the F_o definition. To have a clear understanding of it usefulness as a quality indicator, we shall first take a close look at the derivation of the F_o value (see 4.1.4).

In the same manner, with the same constraints and rationale as was derived F_o (equation 4.15), if a quality factor has first order inactivation kinetics, it is possible to obtain the following expression for a quality factor in a closed system:

$$x \cdot D_r^c = \int_0^t 10^{\frac{T-T_r}{z_c}} dt \tag{4.31}$$

Where D_r value is the reference D value for the target attribute and z_c is the corresponding z value for the target attribute. As was the case for F_r, we can define:

$$x \cdot D_r^c = C_r \tag{4.32}$$

Then:

$$C_r = \int_0^t 10^{\frac{T-T_r}{z_c}} dt \text{ or, in its common form: } C_o = \int_0^t 10^{\frac{T-100}{z_c}} dt. \tag{4.33}$$

One alternative, a practical use of the aforementioned equation, is the calculation of the cooking value on the surface. But, as mentioned in the beginning it will be necessary to have a z_c value and, in addition, a corresponding value for D_r. Without knowing D_r the obtained value for C_r is not interpretable and understandable. Depending on the target attribute (D_r), C_r will have different meanings.

Clearly, the calculation of cooking value (C_r) at the cold spot is not important because it is reflecting the minimum cooking value in the whole food product. To calculate the cooking value, besides the temperature history, the only requirement is z_c value. According to its definition z value represents the temperature dependency but has no relation to the thermal resistance of a given attribute. On the other hand, D value has a direct relation with the thermal resistance of the target attribute, and it is not required to calculate the cooking value. Then, the intricate problem will be how to interpret the obtained cooking value. Clearly, it will have different meanings depending on the target attribute. According to the appendices B.4 and B.5, D_{121} values vary widely from 0.45 through 2350 min.

How, for example, can one interpret a cooking value of 30 min ($T_r = 100°C$)? Choosing real values for quality factors from the appendix (pea purée and green beans) with the same z_c (32.5°C) but with different D_r (4 and 115 min at 121°C), the following results were obtained: In the case of the less resistant attribute, we obtained 0.8 decimal reductions and surface retention of 15.84%, and for the most resistant attribute, 0.028 decimal reductions and a surface retention of 93.8%.

Another critical aspect of the utilization of cooking value is the fact that z_c presents a wide range among different target attributes. It would seem difficult to accept a universal value of 33.1°C for z_c. According to appendices B.4 and

B.5, z_c ranges from 2.66 through 109.7°C. A quite small difference of 5°C in z_c will account for a cooking-value difference in the range of 10 to 15%, with the remaining problem of its particular interpretation.

In addition, the cooking value concept, like quality retention, has been strongly linked to a *first order kinetic* and *closed system*.

4.2.3. Quality Retention

A better way to examine the impact on quality of a given process – with the specified constraints – is the evaluation of the target attribute retention. Starting, again, from equation (4.7) and assuming first-order kinetics for the attribute deterioration, we can obtain an equation for surface retention:

$$\% Surface - retention = 100e^{-\frac{\ln 10}{D_r}\int_0^t 10^{\frac{T_S - T_r}{z_c}}dt}, \tag{4.34}$$

and relating the surface retention (equation (4.34)), with the cooking value (equation (4.33)), we obtain:

$$\% Surface - retention = 100e^{-\frac{\ln 10}{D_r}C_r}. \tag{4.35}$$

The main difference between equation (4.34) and the equation for cooking value is that the surface retention is a direct calculation of the process-impact over the food product's surface. To do the calculations for surface retention it is necessary to know, not only the value z_c, but also the D_r value.

In addition, in the case of retention, it is also possible to derive an equation for the average retention. The volume-average quality retention value is given by:

$$\% Average = \frac{100}{V}\int_0^V e^{-\frac{\ln 10}{D_r}\int_0^t 10^{\frac{T - T_r}{z}}dt}dV. \tag{4.36}$$

The main drawback of equation (4.36) is the requirement of information. It is imperative to have temperature data for the whole container for the whole process.

4.3. Pasteurization Value

For products with a pH less than 4.5, so-called acid products, food poisoning organisms of the type *Clostridium botulinum* do not germinate; consequently it is only necessary to inactivate molds and yeasts. This can be done at much lower temperatures, with the result that the F_0-values are very low, since the lethal rate at a temperature of 80°C is 7.76×10^{-5} min^{-1}. A more practical unit for quantifying the lethal effect of this type of process is the pasteurization unit P (Shapton 1966; Shapton *et al.* 1971) given by

$$P_{65}^{10} = \int_0^t 10^{(T-65)10}dt \tag{4.37}$$

where the reference temperature is 65°C.

For the pasteurization of beer, the *PU*, i.e. pasteurization unit, has been suggested (Ball & Olson 1957) and defined by

$$PU = \exp[2.303(T - 140)/18], \tag{4.38}$$

which is equivalent to P_{140}^{18} in Fahrenheit units. This has been developed into a standard pasteurization unit. 1 PU is equivalent to a process of 1 min at 140°F (60°C) (Portno 1968), with $z = 7°C$ for beer spoilage organisms. Using these criteria, Fricker (1984) suggested that a satisfactory process for stabilizing beer was 5.6 PU.

Horn *et al.* (1997) have used the concept of PUs for pasteurizing beer, adopting a reference temperature of 60°C and z value of 6.94. The experimental results showed that the average PU was of the order of 20 min. The authors also discussed staling of the beer and the development of a quality unit based on chemiluminescence analysis.

For the milk industry, Kessler (1981) proposed a P^*-value, based on a reference temperature of 72°C and z-value of 8°C with equation (4.37), where a satisfactory milk process was $P^* = 1$.

4.4. Minimally Processed Foods

The current trend in the food industry is to reduce the thermal process used for sterilizing food products and hence improve the overall quality. For this purpose much research work has been carried out on using processes that reduce the heating time. Some of these processes are discussed in this section.

4.4.1. Acidified Products

Apart from those products which fall naturally into this class, acidification, where organoleptically acceptable, is a method of reducing the heat process. The extent to which this is possible depends on the actual pH (see Section 4.5). It is essential with acidification processes that the acid penetrates the product completely; otherwise, there would be a danger of understerilization in the regions where the acid had not penetrated.

A hot-fill technique, which is used with acid fruit products, heats the product to or near to boiling, fills the preheated container, seals it rapidly, inverts the container to heat the lid, then uses air to cool it. The heat is sufficient to inactivate spoilage organisms likely to grow under the acid conditions. The pasteurization value given to the product depends on the most heat-resistant micro-organism or spoilage enzyme present. Sandoval *et al.* (1994) have produced a mathematical model for the process that involves calculating the temperature-time profile of the product, i.e., tomato purée, during the cooling process in air, and the effect of the external heat transfer coefficient. Several workers have also studied the process, including Nanjunaswamy *et al.* (1973) with mango products, Nath and Ranganna (1983) with guava pulp, and Silva and Silva (1997) with cupuaça (*Theobroma*

TABLE 4.5. Heat Treatment Recommendations for some pasteurized acid products.

Product	pH	^zF_T-value min
Tomato	3.9–4.6	$^{8.8}F_{93} > 20$
Green olive	3.6	$^{20}F_{62.5} > 15$
Pickles 1–2% acetic acid. no sugar:		
Onion, cucumber & capers	<3.7	$^7F_{87} > 5$
Pickles 1–2% acetic acid with sugar:		
Leek & carrot	3.7–3.9	$^7F_{87} > 20$–25

Data from Perez (2002) Centro Técnico Nacional de Conservas Vegetales, Spain.

grandiflorum) purées. See Table 4.5 for heat treatment recommendations for some pasteurized acid products.

4.4.2. Pasteurized/Chilled Products

Pasteurized foods, so-called *sous vide* products, are stored and retailed at low temperatures. Generally they have had a $6D$ process, which is equivalent to a thermal treatment of 12 min at 70°C. While this is adequate for inactivating the human pathogen *Listeria monocytogenes*, it is far from adequate for the inactivation of *Clostridium botulinum* spores. It is therefore necessary to hold the product at a chilled temperature to prevent the growth of botulinum spores. At 5°C type E spores regenerate in 42 h and type B spores in 30.3 h. Thus these products have a very limited shelf-life in the distribution chain. Considerable advantage has been taken, during the last few years, of microwave heating to achieve a low thermal process with recipe products packaged in plastic trays. The standards to be achieved in controlling this type of product have been extensively discussed (Betts & Gaze 1993; Chilled Foods Association 1998). It is necessary with this type of product to exercise every care in their manufacture, distribution, storage and reuse. Good manufacturing practices for pasteurized foods have been developed (CCFRA 1992).

4.4.3. Electrical Methods of Heating

Ohmic and microwave types of heating have been widely developed during the last few years. Jun and Sastry (2005) have used pulsed ohmic heating for the sterilization and reheating of food packaged in flexible pouches, e.g., chicken noodle soup and black beans. A 2-D thermal-electric model was developed for optimizing the design and layout of electrodes to ensure uniform heating using a CFD package FLUENT v. 6.1 software.

For packaged foods the only practical method that has been used to any large extent is microwave heating (see Chapter 2). This can only be applied at the present time in batch processes or continuous tunnels at ambient pressures; consequently, sterilization temperatures cannot be achieved. Because of the advantages of microwave heating for achieving low-impact thermal processes, there is considerable interest in extending microwave heating to sterilization processes (George

1994). Provided that a suitable method of control can be established to ensure that uniform heating can be reproducibly attained, then the process may be used to produce ambient shelf-stable products in plastic containers (Mullin 1995).

4.4.4. Other Processes

There are other methods of reducing the required thermal process, which involve using either water activity control or special additives such as antimicrobials, nitrites and sorbates. The bacteriocin nisin has been used to control clostridium spore outgrowth in a range of low-acid canned foods, including carrot purée, sausages, mushrooms, legumes, soups, hams and pasta products (De Vuyst & Vandamme 1994).

4.5. Process Achievement Standards

4.5.1. Sterilization

The foregoing sections have indicated the necessary criteria for determining F-, C- and P-values. While the details of determining these are discussed in other chapters, it is necessary to decide whether the value obtained for a process is adequate for the required purpose. In this section the standards used in commercial practice will be discussed.

A simple method of determining the efficacy of a process is to relate the F-value to the change in the microbial contamination using the equation

$$F = D_{ref} \log(N_0/N)$$
$$= D_{ref} \log 10^n,$$

(4.38)

where D_{ref} is the decimal reduction time at the reference temperature for the target organism, N_0 is the initial number of organisms, N the final number and n is the number of log-cycles for the reduction in microbial population. The equation has been modified by Kyzlink (1990) to take into account the effect of nonlinear survivor curves:

$$F = D_{ref} \log[(N_0/N) + t_x],$$

(4.39)

where t_x is usually of the order of 1 min.

Gillespy (1962) used the principle to establish minimum processes for low-acid canned foods on the basis of a maximum value for $D_{121.1}$ for *Clostridium botulinum* of 0.3 min. A process should aim at a microbial survival rate of one spore in 10^{12}, corresponding to 12 decimal reduction times, often referred to as a $12D$ process; this gives a minimum F-value of 3.6 min. A process that achieves this standard is often referred to as a "botulinum cook." For a critique of the system, the reader is referred to Kilsby (1985).

More recently Teixeira (2002) has discussed the origins of the 12-D concept, outlining the original work of Esty and Meyer (1922) and concluding that Stumbo

TABLE 4.6. Relationship between pH and minimum target F-values in minutes.

pH	7.0	6.0	5.7	5.5	5.2	5.0	4.6
F_{min}	3.0	3.0	2.6	2.3	1.9	1.6	1.2

(1948) was the first to recognize the logarithmic nature of TDT curves. The latter showed the D-value derived from a plot of the logarithm of survivors versus time.

Pflug and Odlaugh (1978) reviewed the literature on D-values for *C. botulinum* and concluded that a $D_{121.1}$-value of 0.25 min is adequate. Current US practice (Pflug 1987) and UK practice (Department of Health 1994) recommend $F_{min} = 3$ min.

Pflug *et al.* (1985) have suggested that for foods with pH in the range 4.6–6.0 and processing temperatures in the range 110–118°C the minimum F-value should be as given in Table 4.6. It is recommended that extreme care should be taken in using this concept; adequate experimental work should be carried out with individual products to ensure that the minimum processes are entirely adequate. It is also important to ensure that specified processes will deal with the microbial loads encountered in practice.

The F-values achieved in actual commercial processes are always significantly higher than the minimum value, due to safety margins and the cooking requirements of some products. Some values are tabulated in Chapter 6.

4.5.2. Cooking

There are no standards for cooking in the same sense as there is a minimum F-value. While it is possible to determine C_s-values for acceptable sterilization processes (see Tucker & Holdsworth 1991), each commercial process is set to a quality control standard established as a brand image. Consequently, different manufacturers use different processes. There is a need to be able to quantify organoleptic attributes, and this may be done by determining the C_s-value. Some reference values are given in Chapter 7.

References

Ball, C. O., & Olson, F. C. W. (1957). *Sterilization in food technology – Theory, practice and calculations*. New York: McGraw-Hill.

Betts, G. D., & Gaze, J. E. (1993). Growth and heat resistance of microorganisms in 'sous vide' products heated at low temperatures (54–65°C). In *Proceedings first European symposium: "Sous Vide" cooking* (pp. 27–39). Leuven, Belgium: ALMA University.

Bigelow, W. D., Bohart, G. S., Richardson, A. C., & Ball, C. O. (1920). Heat penetration in processing canned foods. *Bulletin No. 16L*. Washington, DC: Research Laboratory, National Canners' Association.

Brennan, J. G., Butters, J. R., Cowell, N. D., & Lilly, A. E. V. (1976). *Food processing operations* (2nd Edition). London: Applied Science Publishers.

CCFRA (1992). Food pasteurization treatments. *Technical Memorandum No. 27*. Chippinng Campden, Glos. UK: Campden & Chorleywood Food Research Association.

Cheftel, H., & Thomas, G. (1963). Principles and methods for establishing thermal processes for canned foods. *Bulletin No. 14*. Paris: J. J. Carnaud Forges de Base-Indre. [English translation: Israel Program for Scientific Translations, Jerusalem, 1965.]

Chilled Foods Association (1998). *Guidelines for good hygienic practice in the manufacture, distribution and retail sale of chilled foods* (4th Edition). London: Chilled Foods Association.

De Vuyst, L., & Vandamme, E. J. (1994). Nisin, a lantibiotic produced by *Lactococcus lactis* subsp. *lactis*: properties, biosynthesis, fermentation and applications. In L. De Vuyst & E. J. Vandamm (Eds.), *Bacteriocins of lactic acid bacteria* (pp. 151–221). Glasgow: Blackie Academic and Professional.

Deindoerfer, F. H., & Humphrey, P. H. (1959a). Analytical method for calculating heat sterilization times. *Appl. Microbiol., 7*, 256–264.

Deindoerfer, F. H., & Humphrey, P. H. (1959b). Principles of the design of continuous sterilizers. *Appl. Microbiol., 7*, 264–70.

Department of Health (1994). *Guidelines for the safe production of heat preserved foods*. London: HMSO.

Esty, J. R., & Meyer, K. P. (1922). The heat resistance of spores of *B. botulinus* and allied anaerobes. *J. Infect. Diseases, 31*, 650–663.

Fricker, R. (1984). The flash pasteurization of beer. *J. Inst. Brewing, 90*, 146–152.

George, R. M. (1994). The use of metallic packaging for improving heating uniformity during microwave sterilization. *Proceedings symposium – Food process engineering. Vol. 1* (pp. 10–15) Bath, UK: University of Bath.

Gillespy, T. G. (1951). Estimation of sterilizing values of processes as applied to canned foods. I. Packs heating by conduction. *J. Sci. Food Agric., 2*(3), 107–125.

Gillespy, T. G. (1962). The principles of heat sterilization. In J. Hawthorn & J. M. Leitch (Eds.), *Recent advances in food science Vol. 2 – Processing*. London: Butterworths.

Hersom, A. C., & Hulland, E. D. (1980). *Canned foods. Thermal processing and microbiology*. Edinburgh: Churchill-Livingstone.

Hicks, E. W. (1951). On the evaluation of canning processes. *Food Technol., 5*(4), 134–142.

Holdsworth, S. D. (1992). *Aseptic processing and packaging of food products*. London: Elsevier Applied Science.

Holdsworth, S. D., & Overington, W. J. G. (1975). Calculation of the sterilizing value of food canning processes. *Technical Bulletin No. 28*. Chipping Campden, Glos., UK: Campden & Chorleywood Food Research Association.

Horn, C. S., Franke, M., Blakemore, F. B., & Stannek, W. (1997). Modelling and simulation of pasteurization and staling effects during tunnel pasteurization of bottled beer. *Food and Bioproducts Processing, Trans. I ChemE., 75*C1, 23–33.

Jen, Y., Manson, J. E., Stumbo, C. R., & Zahradnik, J. W. (1971). A procedure for estimating sterilization of and quality factor degradation in thermally processed foods. *J. Food Sci., 36*(4), 692–698.

Jun, S., & Sastry, S. (2005). Modeling and optimisation of ohmic heating of foods inside a flexible package. *J. Food Proc. Eng., 28*(4), 417–436.

Kessler, H. G. (1981). *Food engineering and dairy technology*. Chapter 6. Freising, Germany: Verlag A. Kessler.

Kilsby, D. C. (1985). The 12-*D* concept for low-acid canned foods, is it necessary, desirable or attainable? In *Proceedings workshop on The destruction of bacterial spores*. Brussels: European Research Office of the US Army.

Kyzlink, V. (1990). *Principles of Food Preservation*. Amsterdam: Elsevier.

Lewis, M. J. (1987). *Physical properties and food processing systems.* Chichester: Ellis Horwood.

Lopez, A. (1987). *A complete course in canning, Vol. 2. Packaging, aseptic processing and ingredients.* Baltimore, MD: The Canning Trade Inc.

Maesmans, G., Hendrickx, M., De Cordt, S., & Toback, P. (1995). Theoretical consideration of the general validity of the equivalent point method in thermal process evaluation. *J. Food Eng., 24*, 225–248.

Mansfield, T. (1962). High temperature/short time sterilization. *Proc. 1st Int. Congress Food Sci. Technol., 4*, 311–316.

Manson, J. E., & Zahradnik, J. W. (1967). Computer thermal process determination for conduction heated foods. *Food Technol., 21*(9), 1206–1208.

Mullin, J. (1995). Microwave processing. In G. W. Gould (Ed.), *New methods of food preservation* (pp. 112–134). Glasgow: Blackie Scientific and Professional.

Nanjunaswamy, A. M., Saroja, S., & Ranganna, S. (1973). Determination for the thermal process for canned mango products. *Ind. Food Packer, 27*(6), 5–13.

Nath, N., & Ranganna, S. (1983). Heat transfer characteristics and process requirements of hot-filled guava pulp. *J. Food Technol. 18*, 301–316.

Newman, M., & Holdsworth, S. D. (1989). Methods of thermal process calculation for food particles. *Technical Memorandum No.* 321 (Revised). Chipping Campden, Glos., UK: Campden & Chorleywood Food Research Association.

Nunes, R. V., & Swartzel, K. R. (1990). Modeling thermal processes using the equivalent point method. *J. Food Eng., 11*, 103–111.

Patashnik, M. (1953). A simplified procedure for thermal process evaluation. *Food Technol., 7*(1), 1–5.

Perez, J. (2002). Canned food process validation in Spain. In Tucker, G. S. (Ed.), *Second International symposium on thermal processing – Thermal processing: validation challenges.* (Session 2:4). Chipping Campden UK: Campden & Chorleywood Food Research Association.

Pflug, I. J. (1987). *Textbook for an introductory course in the microbiology and engineering of sterilization processes* (6th Edition). Minneapolis, MN: Environmental Sterilization Laboratory.

Pflug, I. J., & Odlaugh, T. E. (1978). A review of z and F values used to ensure the safety of low-acid canned foods. *Food Technol., 32*, 63–70.

Pflug, I. J., Odlaugh, T. E., & Christensen, R. (1985). Computing minimum public health sterilizing values for food with pH values between 4.6–6.0. *J. Food Protection, 48*(10), 848–850.

Portno, A. D. (1968). Pasteurization and sterilization of beer – a review. *J. Inst. Brewing, 74*, 291–300.

Reichert, J. E. (1977). The C-value as an aid to process optimization. *Z. Lebensm.-Technol. u.-Verfahr., 54*(1), 1–7 [in German].

Sandoval, A. J., Barreiro, J. A., & Mendoza, S. (1994). Prediction of hot-fill-air-cool sterilization processes for tomato paste in jars. *J. Food Eng. 23*(1), 33–50.

Shapton, D. A. (1966). Evaluating pasteurization processes. *Process. Biochem., 1*, 121–124.

Shapton, D. A., Lovelock, D. W., & Laurita-Longo, R. (1971). The evaluation of sterilization and pasteurization processes for temperature measurements in degrees Celsius (°C). *J. Appl. Bact., 34*(2), 491–500.

Silva, F. F. M., & Silva, C. L. M. (1997). Quality optimization of hot-filled pasteurized fruit purées. *J. Food Eng. 32*(4), 351–364.

Skinner, R. H., & Urwin, S. (1971). Automation of the canning process calculations. *Process Biochem.*, *6*(11), 35–38.

Sleeth, R. B. (Ed.) (1978). *Introduction to the fundamentals of thermal processing.* Chicago: Institute of Food Technologists.

Stumbo, C. R. (1948). Bacteriological considerations relating to process evaluation. *Food Technol.*, *2*(2), 115–132.

Stumbo, C. R. (1949). Further considerations relating to evaluation of thermal processes for foods. *Food Technol.*, *3*(4), 126–131.

Stumbo, C. R. (1953). New procedures for evaluating thermal processes for foods in cylindrical containers. *Food Technol.*, *7*(8), 309–315.

Stumbo, C. R. (1973). *Thermobacteriology in food processing* (2nd Edition). New York: Academic Press.

Stumbo, C. R., & Longley, R. E. (1966). New parameters for process calculation. *Food Technol.*, *20*, 341–347.

Teixeira, A. A. (2002). Origins of the 12-D process. In Tucker, G. S. (Ed.), *Second International Symposium, Thermal processing – Thermal processing: validation challenges.* (Session 1:2). Chipping Campden UK: Campden & Chorleywood Food Research Association.

Tucker, G., & Holdsworth, S. D. (1991). Mathematical modelling of sterilization and cooking for heat preserved foods – application of a new heat transfer model. *Food and Bioproducts Processing, Trans. IChemE*, *69*, C1, 5–12.

5
Heat Penetration in Packaged Foods

5.1. Introduction

This chapter is concerned with the measurement and interpretation of heat penetration curves, obtained from temperature measurements inside the food in the container or package. It is the most important part of determining an adequate process for sterilizing or pasteurizing food products. However, because of the complex nature of food products, the number of different packages and the variety of methods of processing them, the interpretation of the data is often difficult. All thermally processed packaged foods require a validated process schedule which is traceable to heat penetration measurements. The methods of doing this and the treatment of the experimental data are outlined in Section 5.2

An important protocol for doing heat penetration studies has been established by the Institute for Thermal Processing Specialists (IFTPS 1995). Similarly May (1997) has produced an industrial laboratory guideline for performing heat penetration tests, which includes guidance on the procedures and for the interpretation of the results. It also highlights the critical control points for a canning operation.

Table 5.1 gives an outline of the factors which affect the rate of heat penetration in canned foods; these are discussed in greater depth in Section 5.5.

5.1.1. Heat Transfer and Product Characteristics

The way in which heat is transferred in the food inside the container or package depends on the nature of the food product and the style of packing the components. Some of the earliest published work on heat penetration into cans of food was carried out by Bigelow *et al.* (1920) at the National Canners' Association in Washington. The now-classical document, *Bulletin 16-L*, revised many times, is well worth study even after more than 85 years. Equally worth reading is the work by Magoon and Culpper (1921), which deals with heat penetration into canned fruits and vegetables, and all the factors affecting it. Both documents show the remarkable perspicacity of these pioneers, and while there have been many changes in technique and products, and subsequent mathematical development of the subject, these documents form the foundation of this subject.

Table 5.2 gives an idealized representation of some of the types of canned foods; it should be taken as guidance only since the formulation of packs can vary

TABLE 5.1. Factors affecting heat penetration and process severity.

Factors	Comment
(a) Process-related	
1. Retort temperature and profile	The higher the temperature the faster the heating rate.
	Depends on sterilizer type: in static retorts and hydrostatic cookers the temperature rises slowly and there is a lag in heating the cans; in rotary cookers the heating starts instantaneously.
2. Process time	The longer the process time the greater the heat penetration and the nearer to processing temperature the pack contents become.
3. Heat transfer medium	The external heat transfer coefficient, h, governs the temperature at the surface of the container. Steam has a very high h, but h for water or steam–air mixtures depends on the velocity and geometry factors.
4. Container agitation	Agitation and rotation improve the internal heat transfer, depending on the amount of headspace, and degree of agitation and rotation.
(b) Product-related	
5. Consistency	The composition, consistency and rheological behavior control the rate of heat penetration. For products of a thin nature or in thin covering liquid, convection heating occurs, while for thicker products heat transfer is mainly by conduction. Some products (broken-heating) show both types of heating.
6. Initial temperature	The higher the initial temperature of the contents the shorter the processing. The process is very sensitive to the initial temperature, especially for conduction-heating products, which often do not reach process temperature by the end of the process.
7. Initial spore load	The severity of the process depends on the initial spore load; good factory hygiene keeps this low.
8. Thermophysical properties	The thermal diffusivity is the controlling factor. This approximates to that of water for most products, but is lower for those which are oil-based and those of low water activity.
9. Acidity pH	The severity of the process depends on the pH of the product. Products with pH > 4.5 require the severest processes, whilst products with pH < 4.5 can be pasteurized; this includes acidified products.
10. Additives	Certain additives, e.g. nisin, nitrite, salt, and sugar, reduce the process time.
(c) Packaging-related factors	
11. Container materials	These include: tin plate, aluminum, glass, and plastic and laminated materials. The thermal conductivity and thickness of the material determine the rate of heat penetration. The lower the conductivity and the thicker the material, the slower the heating.
12. Container shape	The external surface area and thickness of the container determine the heat penetration rate. The most rapidly heating packages have the highest surface area and thinnest profile.

TABLE 5.2. Heat transfer in canned foods.

Rapid convection	Convection	Convection f Conduction broken-heating	Conduction $\alpha \approx \alpha_{water}$	Conduction $\alpha < \alpha_{water}$
Type 1:	Type 2:	Types 3 and 5:	Type 4:	Type 6:
Juices	Liquid	Cream soups (starch gel)	Cream style corn	High fat/oil
Broths	Spinach/brine	Noodle soups	Thick purées	Meat and marine packs
Milk	Cabbage/brine	Tomato juice	Solid pack apples/ vegetables	High sugar products
Type 2:	Low-starch purée		Jams	Low-moisture puddings
Fruits/syrup	Vegetable soups		Vegetables/meat in sauce	
Vegetables/brine	Sliced vegeta- bles/brine		Marine/sauce	
Meat and marine	Macadoine		Beans in tomato sauce	
Products/brine	Type 7: Vapour Vacuum-packed vegetables Sweetcorn Whole kernel corn		Spaghetti Rice Pet foods	

The various types described are in Section 5.1.1.

considerably. The main factors are the packing style of the components in the cans and the rheological properties of the foods. The basis for the classification was developed by workers at the American Can Co. (Jackson 1940; Jackson & Olson 1940). The table identifies seven types:

1. Thin liquid products which heat extremely rapidly due to internal convection, e.g. fruit juices, beverages, milk and thin soups. The thicker the consistency becomes, the slower the heating rate.
2. Liquid products containing solid food. The liquid portion heats rapidly by convection and transfers the heat into the particulate by conduction. For products such as peas in brine and strawberries in syrup, the process is determined by heat penetration into the covering liquid. The rate of heat penetration is determined to a large extent by the solid–liquid ratio and packing style. For larger products, such as whole potatoes or celery hearts in brine, heat penetration into the center of the product is necessary to ensure an adequate process and sufficient cooking.
3. Solid products in thick covering liquids, such as beans in tomato sauce and thick soups. Depending on the formulation, heating is first by convection and then, after thickening by starch gelation, by a conductive mechanism. These

products show a heat penetration curve with two different rates of heating, known as a broken-heating curve. The reverse process also occurs (see type 5).

4. Thick products which heat by conduction, and whose thermal diffusivities are about the same as water, there being insufficient or no covering liquid. This covers a wide range of products in thick sauces and gravies and solid pack products.

5. Products which start by heating with a conductive mechanism and then, because of thinning due to structure and rheological changes, heat by convection, e.g. some thickened puddings and some tomato juices. Again this produces a broken-heating type of heat penetration curve.

6. Thick products which heat by conduction but have thermal diffusivities less than water, i.e. they have a high fat or sugar content, e.g. fish in oil.

7. Vacuum-packed products. These contain very little water, sufficient to produce enough steam inside the can to heat the product, e.g. corn on the cob, whole-grain corn and some vegetables.

5.2. Experimental Determination

5.2.1. Temperature Monitoring

The measurement of the temperature at a selected point in the product in the container is paramount to process determination. Temperature sensors are usually thermocouples of various types.

The Ecklund system of thermocouples (Ecklund Custom Thermocouples, Cape Coral, FL, USA) is widely used since each thermocouple is designed to place the sensor element at the point most appropriate to the container size. A standard size is 1.6 mm in diameter, made of copper-constantan (copper/45% nickel alloy) type T. For very accurate work thin wires are used to reduce the conduction errors.

There are several types of system available: (a) moulded thermocouples made of rigid Bakelite insulation, which are available to suit specific can sizes; (b) stainless-steel needle thermocouples, which are relatively thin and rigid; (c) flexible thermocouples which can be used for measuring the heat penetration at the center of particulate materials; and (d) custom-made thermocouples, rigid rod-in-tube type, which can be made to any suitable length. The system allows for the thermocouple to be inserted prior to filling, using a non-projecting mounting with plug-in facilities for the leads.

The thermocouples can be attached to any suitable data-logging system or PC, and special programs can be introduced to give the heat penetration parameters, f_h and j, as well as F_0-value.

In Europe, while many laboratories use the Ecklund system, an alternative system, the Ellab heat penetration system, is widely used (Ellab A/S, Copenhagen, Denmark). This has type T thermocouples and has a dedicated system for f_n and j parameters. The system uses self-threading receptacles which are placed in an opening in the filled and sealed container. A packing gland with a compression fitting allows for the location of the thermocouple sensor at the appropriate point (Eisner 1976).

For pilot-scale rotary cookers special slip-ring devices, commutators and rotary transformers can be supplied by the manufacturers for conveying the signal from the container to the data-logging device.

The historical development of the subject, which has moved rapidly with the development of electronic devices and made many previous devices out of date, has been dealt with by Charlett (1955), Holdsworth (1974, 1983, 1985), May and Cossey (1989), and May and Withers (1993).

For many years the major problem was to measure temperatures inside cans being processed in continuous cookers, e.g. rotary sterilizers and hydrostatic cookers. Various devices including radiotelemetry were used, with limited success. However, more recently a device with encapsulated data-logging facilities, the Ball Datatrace unit, has become available (Datatrace Division, Mesa Medical Inc., Wheat Ridge, CO, USA). This can be fitted inside the can for internal heat penetration or can go through the cooker measuring the environmental temperature. The unit consists of a completely self-contained miniature computer/temperature sensor. The Micropack version has a cylindrical body 1.38 in in diameter and length varying from 2.16 to 6.16 in, with a probe 1–5 in long. The temperature is a thermistor, with range 10–150°C. The computer can store a maximum of 1000 readings and record at frequencies between 1 per second and 1 per day. The performance of this unit has been thoroughly examined and shown to be reliable for temperature measurements in canned foods (May & Cossey 1989; May 1991, 1992).

George and Richardson (2001) have described a non-invasive temperature sensor, which uses a passive electronic inductor-capacitor particle (<5 mm diameter) and an external detection coil. This will withstand temperatures up to 130°C and has a reproducibility within ± 1°C.

Marra and Romano (2003) have made a study of the effect of the size and position of a radio- transmitting temperature-measuring device using cans of differing sizes. This involved the development of a sophisticated mathematical model, using finite element technique. The results showed that the presence of the measuring device influenced the measured-temperature profile, according to the relative dimensions of the can, especially with the smaller cans sizes as would be expected. A basal position for the sensor had least effect on the temperature profile at the point of slowest heating.

More recently Shaw (2004) has reviewed the use of data loggers for the validation of thermal processes. This includes new units, e.g., Datatrace Microack III (20 mm in length) and Ellab's Tracksence Pro Mini Logger (16 mm in length by 20 mm wide). Both these systems only give time-lapsed data. TechniCAL RTemp wireless data collection systems overcome the problem to some extent by providing real time data. Problems of getting the signal out of the thick-walled retort may be overcome by internal aerials.

For retort pouches special methods of sealing the wires into the package and for locating the thermocouple sensor have been devised (Bhowmik & Tandon

1987; Spinak & Wiley 1982). These use a folded strip welded to the pouch sides which straightens up across the pouch when filled, so that the thermocouple sensor is correctly centerd. Alternatively, Peterson and Adams (1983) used machined PTFE blocks to support the thermocouple; Thorpe and Atherton (1972) recommended the use of a nylon cross spacer across the pouch to locate the thermocouple.

5.2.2. Thermocouple Errors

One of the main errors associated with thermocouples is the conduction of heat along the thermocouple wires. While this is of relatively little importance in convection heating packs, it can be very important in conduction heating packs. Conduction errors tend to be greatest during the initial heating and cooling periods. Errors are unimportant in the early stages of heating before the temperature has any significant lethal effect; however, they become significant during the rest of the processing. It is also important to note that the thermal conductivity of the copper wire is 17 times that of the constantan; consequently the main conduction errors occur with the copper wire.

Zang et al. (2002). have studied the effect of thermocouple and receptacle type on the f_h-value in a chicken/gravy pack showing conduction eating characteristics, in 202×204 & 211×300 cans. Three types of thermocouple were used, viz., flexible, with stainless steel receptacles, and with Delrin plastic receptacles, the latter two having stainless steel needle thermocouples. The highest j_h-value of 1.48 was observed for the plastic receptacles, and the lowest f_h for the stainless steel receptacle. The highest f_h of 22.4 min was observed for the flexible thermometers.

Ecklund (1956) investigated the thermocouples which he had designed, and compared them with a standard thermocouple, assuming that the latter was free from error. He found that the error was mainly associated with the lag factor j and devised a table of correction factors to be applied when using his thermocouples.

Cowell et al. (1959) investigated the conduction errors in simple thermocouples, using both theoretical and experimental results. They found that there was negligible error with very thin wires of thickness 40 s.w.g. (0.122 mm) diameter compared with thicker wires 27–20 s.w.g. (0.42–0.91 mm). They also confirmed Ecklund's (1956) work concerning the effect on the j-value; however, they found that the rate of heating factor f was also affected when thicker wires were used. For accurate work, especially with the cooling phase, they recommended that wires of low thermal conductivity were used, rather than copper. Hosteler and Dutson (1977) showed a similar dependence on wire size during the cooking of meat samples.

Beverloo and Weldring (1969) examined the effect of thermocouple construction materials and mode of assembly and found that the effect of errors on the F_0-value was very low. The overestimation of the temperature during heating was compensated for during the cooling.

Packer and Gamlen (1974) considered the heat transfer lag from the surroundings to the sensor element in convection heating packs. They devised a mathematical analysis to determine the magnitude of the error.

A finite-element model was developed by Kanellopoulos and Povey (1991) and the results compared with those from a type T thermocouple with wires of diameter 0.56 mm (24 s.w.g.). They considered that the temperature during the heating phase could be overestimated by 2°C and that the cooling corrections for the j-value were not always an adequate compensation for the cooling phase.

The important aspect of this work is the recommendation to use thin thermocouple wires of low thermal conductivity for accurate heat penetration work involving conduction heating and cooling products. For conventional type T thermocouples it is necessary to consider the effect of any errors involved, and to consider both heating and cooling phases when determining the achieved lethality.

5.2.3. Thermocouple Calibration

An important aspect of temperature measurement is the calibration of the thermocouples or other temperature measuring devices. It is essential for the purpose of process determination that the thermocouples are regularly checked against a traceable standard. This requires a constant temperature source, e.g. a molten salt or boiling water, and a thermometer calibrated to a national standard. Cossey and Richardson (1991) have described the use of a constant temperature block for the calibration of master temperature indicators, and Dobie (1993) discussed the methods of calibration. The ASTM guidelines (ASTM 1988) should be consulted for exact procedures.

5.2.4. Thermocouple Location: Slowest Heating Point

The main aim of process determination is to locate the thermocouple sensor at the point of slowest heating or critical point, in order to obtain the temperature history: Given that the critical point has been found, then all other points must have reached a higher temperature and received a greater lethal process. Consequently, if the process is safe at the critical point, then it will be safe at all other points in the container. The problem is how to locate this point. The usual method is to put a number of different thermocouples in the can at different points and thereby determine, by trial and error, the position of the slowest heating point. For small-sized cans of conduction-heating food the critical point will be near to the center of the food mass, but for larger cans this may not be so, because the center point may go on heating (known as overshooting – Kopelman et al. 1982) after the cooling has started, until in fact the cooling is felt at the center. The geometrical center will always be the slowest heating point during heating but not necessarily during cooling. This tends to produce a circular or triangular cross-sectioned torus symmetrical about the vertical axis, of lowest temperature (Hurwicz & Tischer 1955). This was found to apply to the processing of meat in 300 × 308 cans at processing temperatures of between 107 and 157°C. Flambert and Deltour (1972) found that the position of the critical point depended upon the h/d ratio for the particular can. Under the conditions of their work they showed that for h/d ratios less than 0.3 and greater than 1.9 and for h/d about 0.95, the critical point was at the geometrical center. For lower values, $0.3 < h/d < 0.95$, the critical point was constituted by two points located symmetrically along the vertical axis with respect to the central plane. For higher values, $0.95 < h/d < 1.9$, the critical area is ring-shaped and lies within the central plane. Naveh *et al.* (1983a, b) used a finite-element model to investigate the overshooting phenomenon. They showed that the extent of overshooting was proportional to the value of g, the temperature difference between the critical point and the retort temperature at the end of heating, and decreased as the external heat-transfer coefficient increased. Silva and Korczak (1994) similarly showed that the packaging dimensions and the heating rates affected the position of the least-lethality point. For packs with $0.1 < h/r < 0.9$ the point was located along the central vertical axis, and for $0.9 < h/r < 4.0$ it was located along the radius.

May (2004) has described the location of the slowest heating point in a variety of differing situations, e.g., convection in thin liquids, mixed convection-conduction liquids showing broken heating curves, conduction packs and packs going through a hydrostatic cooker. Wiese and Wiese (1992) have studied the heat penetration characteristics of a range of vegetable products, e.g., beans in tomato sauce, which show broken heating curves.

A similar study was reported by Sumer and Esin (1992) who examined the effect of the arrangement of cans on the slowest heating point for a can filled with water, peas in brine and 28% tomato paste. For the water the point was 10–15% of the height of the can from the bottom, for the peas 20–25% and around the geometric center for the tomato paste. A finite element model was developed for the distribution of temperature in the stacking patterns.

Ghani *et al.* (2002) have made a study of the destruction of *Bacillus stearothermophilus* in pouches containing a beef-vegetable soup. Simulations using PHOENICS packages were used to determine the relative concentration profiles of the spores and also the temperature profiles. It was shown that the slowest heating zone covered the whole cross section of the pouch at the early stages of heating after which it migrates and located itself about 30% pouch height.

For convection packs Datta and Teixeira (1988) showed that the point of slowest heating moves during the processing; there are in fact a cluster of points located on either side of the center line low down in the can. The volume in which these points were located was doughnut-shaped. They suggested that the thermocouple should be placed 15% of the height up from the base of the can. This is lower than the UK recommendation of 20% (CCFRA 1977).

It should be possible to demonstrate this effect by using a colored dye in a silicone rubber cylinder, which is decolorized by heating. There must be many dyes which have been examined as food colors and found to be unstable to heat. Alternatively there are chemicals that develop color on heating, e.g. leuco-anthocyanins, as witnessed by the pinking of pears and some other fruits when processed in A10 cans. Doubtless there are many other chemicals which could be used to demonstrate this effect.

For practical purposes the UK recommendations are that thermocouples are located as follows:

1. For conduction-heating packs, cylindrical and other shapes of can, the thermocouple should be arranged at the geometrical center of the food mass.
2. For convection-heating foods in cylindrical cans, the point of slowest heating is about one-fifth the height of the can up from the base. The exact point should be determined using multiple thermocouples for each can size and commodity.
3. For products heated in rotary retorts, the slowest heating point is in the geometric center of the food. The temperature should be fairly uniform in rotated and agitated packs, and therefore the actual position of the thermocouple is less critical.
4. For products which show a broken-heating curve, a similar position to that for convection heating products is used if the break occurs towards the end of the heating. For these cases the point of slowest heating should be determined experimentally. It will move upwards during the process when conduction heating starts.
5. For mixed products, which heat by both convection and conduction simultaneously, e.g. potatoes in brine, it is necessary to ensure that the center of the potatoes receive a satisfactory process and the thermocouple should therefore be located in the center of the potato with the largest dimensions.
6. Care should be taken to insert the thermocouple into the side of the can, if possible, since thermocouples inserted into the ends tend to change their position if the ends flex during the processing.

5.2.5. Model Systems

For the purposes of studying heat transfer in containers, obtaining thermocouple correction factors and demonstrating the various modes of heat transfer, simulants have been used. Food products are usually too variable for detailed experimental work, and the use of simulants removes this variability.

The most common of the simulants is known as bentonite (Jackson 1940; Jackson & Olson 1940). In one method of preparation the powder is mixed intimately with water to form a suspension, which is subsequently heated to boiling, cooled and allowed to stand for 24 h. Some workers prefer to use the cold suspension directly for their experiments. Robertson and Miller (1984) made a 10 kg batch of a suspension of calcium bentonite in water by adding it slowly to the vortex created in the water (c. 65°C) by a Silverson mixer. The suspension was mixed for 15 min to facilitate hydration and to prevent any clumping. The suspension was stored for 48 h to allow for full rehydration. A 1% suspension is used for simulating convection-heating packs and suspension of 5% or more for conduction-heating packs. Broken-heating packs can be demonstrated using a concentration of about 3.5%. In practice, it requires a good deal of experimental skill to obtain packs which behave uniformly. The thermal diffusivity of bentonite suspensions is very sensitive to the exact composition, and the suspension deteriorates with repeated use. Tong and Lenz (1993). Have determined the dielectric properties of bentonite pastes at different temperatures.

For studying convection, a wide range of materials have been used: water, solutions of ethylene glycol, sugar solution, oils with thermal properties independent of temperature and various solutions of known viscosity.

For conduction heating a range of suspensions, including agar, and starch-based materials are available. More recently, silicone elastomers (Dow Corning Ltd., Barry, South Glamorgan) have been used, e.g. Sylgard 170 and 184; although these are expensive materials, they give very reproducible results when used for simulating conduction-heating packs. The thermal diffusivity is, however, somewhat lower than that of water and water-based food products: $1.02 \times 10^{-7} \mathrm{m}^2 \mathrm{\,s}^{-1}$ compared with 1.4–$1.6 \times 10^{-7} \mathrm{m}^2 \mathrm{\,s}^{-1}$ (Bown *et al.* 1985).

The thermal diffusivities for some simulants are given in Table 2.2.

5.3. Graphical Analysis of Heat Penetration Data

5.3.1. The Linear Plot

The simplest method of plotting the results of a heat penetration experiment is to use linear graph paper. Figure 5.1 shows a typical heating and cooling curve,

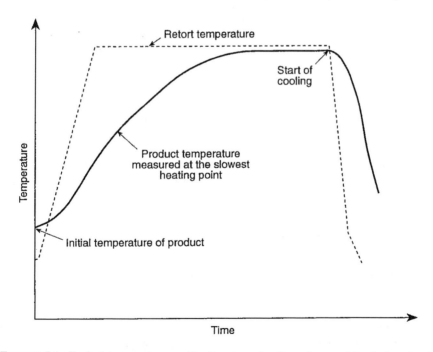

FIGURE 5.1. Typical temperature profile for steam heating of canned foods in a batch retort.

with the retort temperature profile and the corresponding temperature measured at the critical point. This type of plot is useful for observing the come-up of the retort and for converting the temperatures to a lethality curve for determining the F_0-value.

5.3.2. The Semi-Logarithmic Plot

If the temperature-time results from a heat penetration experiment are plotted on multi-cycle logarithmic-ordinary graph paper, it is possible to construct a straight line through the points. The data for conduction-heating products usually have a lag period before linear heating is achieved, represented by a curved portion to the graph. The extent of this depends on the external heat-transfer coefficient and tends to increase in curvilinearity with increasingly slower heating media or thicker packaging material. The data for convection-heating packs often follow a slightly sigmoidal pattern, and are usually less uniform than those for conduction-heating products.

Three methods of representing the results are in common use. The first is $\log(T_R - T)$ v. time t (Figure 5.2). When the logarithm of the temperature difference between the retort temperature and the critical point temperature, known as the temperature deficit, $(T_R - T)$, is plotted against the time, then a plot is obtained with curvilinear and linear portions with negative slope. When the deficit is greater than one then three-cycle logarithmic paper is satisfactory; however, when the deficit is less than one four cycles will be required. This method is mainly used in theoretical work.

The second method is $\log[(T_R - T)/(T_R - T_0)]$ v. time t (Figure 5.3), where $(T_R - T)/(T_R - T_0)$ is the reduced temperature and T_0 the initial temperature of the food in the can; in this case a graph similar to Figure 5.2 is obtained. This method is mainly used in theoretical work.

The third method is $\log T$ v. t (Figure 5.4). If the logarithmic scale is arranged to run in the opposite direction to the previous two (inverted scale), and the top line of the graph is numbered $1°$ below the retort heating temperature, the next line is numbered $2°$ below, and so forth, then a linear plot with positive slope is obtained. Using this method the lines are numbered with the actual temperatures and no conversion is necessary.

The cooling curves may be represented in a similar manner.

5.3.3. Analysis of Heat Penetration Graphs

A straight line can be obtained from the log temperature–time or semi-log plots by drawing the asymptote to the curve (Olson & Jackson 1942; Ball & Olson 1957). The equation of this line is then fully defined by its intercept and slope.

The intercept is obtained by extrapolating the asymptotic line to the axis and defining a temperature T_A such that

$$T_R - T_A = j(T_R - T_0) \tag{5.1}$$

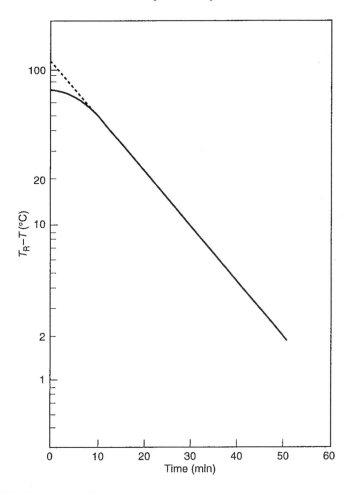

FIGURE 5.2. Temperature deficit plot.

or

$$j = \frac{T_R - T_A}{T_R - T_0},$$
(5.2)

where j is known as the lag factor, since it measures the lag in establishing a uniform heating rate; it is also designated j_h to represent the heating part of the process, and j_c is the corresponding value for cooling. Part of the lag is due to the slow come-up of the retort, and this is accounted for by determining a new zero time for the process. Ball and Olson (1957) used 58% of the come-up time as making a useful contribution to the process and this is widely accepted. It corresponds to adding 42% of the come-up time to process time at retort temperature. The j-value for this process is found by using the intercept of the

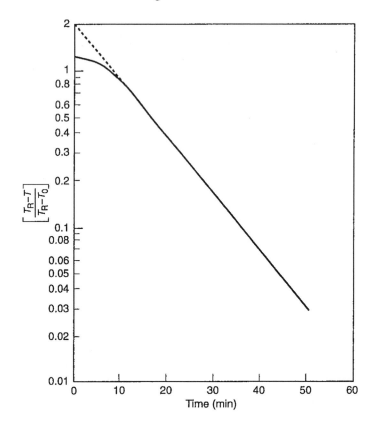

FIGURE 5.3. Reduced temperature plot.

line with the new zero axis, T_I (see Figure 5.5). For this purpose

$$j = \frac{T_R - T_I}{T_R - T_0}. \qquad (5.3)$$

The slope of the line is given by the tangent of the angle between the line and the t-axis, although it is best obtained as the time for the curve to traverse one logarithmic cycle. This time is known as the f-value and is the time to reduce a given temperature to one-tenth of its value. The f-value is the reciprocal of the slope and cotangent of the angle between the line and the time axis. For the heating phase it is designated f_h, and for the cooling phase f_c.

Combining these two heat penetration factors in the straight-line equation $y = mx + c$ gives the equation

$$\log u = -t/f + \log j, \qquad (5.4)$$

where u is the reduced temperature, $(T_R - T)/(T_R - T_0)$, the slope is $-1/f$ and the intercept $\log j$. This equation may also be re-expressed as

$$t = f \log(j/u) \qquad (5.5)$$

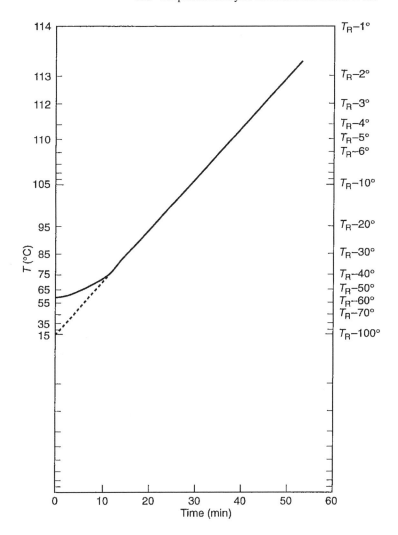

FIGURE 5.4. Inverted scale plot for $T_R = 115°C$ and $T_0 = 62°C$.

or

$$u = j \cdot 10^{-t/f} = je^{-2.303t/f}. \tag{5.6}$$

The value of j ranges from approximately 1 to 2, although both smaller and larger values are found in practice. The value, as can be shown theoretically, depends on the position at which the temperature is measured, the shape of the container or body being heated and cooled, and the initial temperature distribution.

The value of f is expressed in minutes, and depends on the thermal properties, in particular the thermal diffusivity, and the dimensions of the object being heated. The faster the rate of heat penetration, the steeper the slope and the smaller the

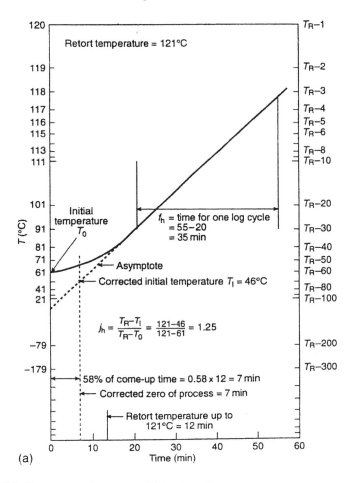

FIGURE 5.5. Heat penetration curves: (a) heating phase.

f-value. For most conduction-heating packs f_h is greater than 20 min, and for convection-heating packs values of between 4 and 11 min for the common sizes of cylindrical container are obtained. Table 5.3 shows some typical values of f_h for static steam-heating processes.

The values for f_c tend to be larger because of the slower heat transfer with water compared to steam. An empirical guideline is that $f_c = 1.3 f_h$, which is confirmed from the work of Hurwicz and Tischer (1956) for the processing of beef. These workers also found a greater variance in the values of f_c compared with f_h.

For broken-heating curves (see Figure 5.6) there are two f_h-values, usually designated f_1 and f_2, and a break point, T_p, at time t_p. The retort temperature tends to influence the value of f_2 and the break point parameters t_p and T_p. The closing temperature also affects the break point. The can size affects both f_1 and f_2. For a 3.5% suspension of bentonite in water in No. 2 cans the values of j varied

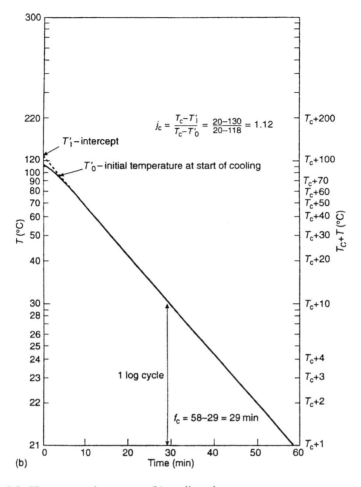

$$j_c = \frac{T_c - T'_i}{T_c - T'_0} = \frac{20 - 130}{20 - 118} = 1.12$$

FIGURE 5.5. Heat penetration curves: (b) cooling phase.

TABLE 5.3. Some typical f_h-values for products processed in static retorts using steam heating.

Can size	Conduction heating (min)	Convection heating (min)
211 × 2025oz	25	4.0
211 × 301 picnic	34	4.5
211 × 400 A1	39	5.0
300 × 2078T	34	4.5
300 × 408$\frac{3}{4}$ UT	47	4.5
301 × 40916Z	52	5.5
307 × 408 A2	62	6.0
401 × 411A2$\frac{1}{2}$	83	7.0
603 × 700 A10	198	11.0

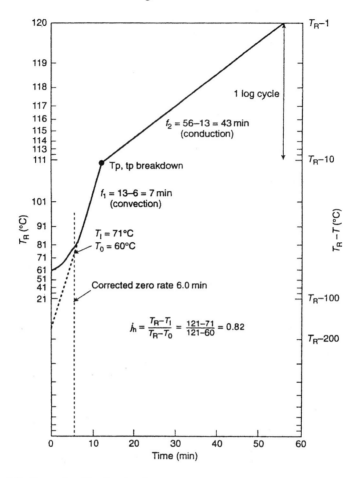

FIGURE 5.6. Example of broken heating curve.

from 0.99 to 1.72; f_1 from 4.87 to 6.84 min; f_2 from 44.6 to 62.0, and t_p from 6.7 to 8.3 min (Jackson 1940; Jackson & Olson 1940). Baked beans in tomato sauce exhibit a broken heating curve; initially the pack shows typical convection heating behavior and then as the tomato thickens by gelation of the starch leached from the beans, the pack shows increasing conduction behavior. For an A10 can processed at 121.1°C, and heated for 35 mins typical values for the heat penetration factors are $f_1 = 8.7$ min and $f_2 = 39.8$ min and $t_p = 10$ min. A typical broken heating curve for this type of product is shown in Figure 5.6. (Tucker 2002). The problems associated with this type of process are dealt with in Section 6.4.2.3.

  Giannoni-Succar and Hayakawa (1982) reported values $f_h = 29.37$ and $f_c = 34.45$ min for 8%w/w bentonite in can size 307 × 115, with corresponding j values of 2.13 and 1.56 respectively. The ratio f_c/f_h was 1.17, which is in general agreement with the quoted value of 1.3.

Bichier *et al.* (1995) found that for tomato sauce in 300×407 cans in a batch retort the $f_h = 44.4\text{–}45.2$ min, $f_c = 45.2$ min, $j_h = 1.69$ & $j_c = 1.69$, whereas for an agitated cook in a Steritort retort $f_h = 3.7\text{–}4.45$ min 7 $j_h = 1.44\text{–}2.05$ with the f_c & j_c very similar. The value for the stationary process agrees well with the value in Table 5.3.

5.4. Theoretical Analysis of Heat Penetration Curves

5.4.1. Conduction-Heating Packs

In Chapter 2 the general equation for determining the temperature distribution in a cylindrical container of conduction-heating food was presented (Olson & Jackson 1942; Ball & Olson 1957). From this series solution, a first-term approximation can be derived which applies to any point (r, z) in a cylinder of radius r and height $2b$:

$$u = A_{110} J_0(R_1 r/a) \sin\left(\frac{\pi(z+b)}{2b}\right) e^{-(R_1^2/a^2 + \pi^2/4b^2)\alpha t}, \qquad (5.7)$$

where $u = (T_R - T(t))/(T_R - T_0)$, A_{110} is a function of the initial temperature distribution (equal to 2.03970 for a uniform distribution), $J_m(x)$ is the mth-order Bessel function of x, R_n the nth positive root of $J_0(x)$ and α the thermal diffusivity. The first-term approximation is only valid for long processing times, and for the purposes of the cooling part it is necessary to examine whether this criterion applies.

Equation (5.7) is of the same form as the exponential heat penetration equation (5.6), viz.

$$u = j e^{-2.303 t/f}$$

from which

$$f = \frac{2.303}{\alpha\left[R_1^2/a^2 + \pi^2/4b^2\right]} \qquad (5.8)$$

$$j = A_{110} J_0(R_1 r/a) \sin\left(\frac{\pi(z+b)}{2b}\right) \qquad (5.9)$$

$$R_1 = 2.404826.$$

There are two important conclusions from this comparison: first, from equation (5.8) the f-value depends on the thermal diffusivity and the container dimensions; it is the same at all points inside the container, since it is independent of z and r. Second, from equation (5.9) the j-value depends on the initial temperature distribution and the position in the can; for conduction-heating packs the theoretical value of j at the center will be 2.03970.

The above analysis applies to containers heated by steam; if there is a finite heat-transfer coefficient on the surface of the container then the equations given in Chapter 2 should be used. This is discussed further in Section 5.5.6.

5.4.2. Convection-Heating Packs

Using the uniform internal temperature model, discussed in Chapter 2, the basic equation for the temperature in the fluid core is

$$u = 10^{-0.434Kt}, \tag{5.10}$$

where u is the reduced temperature $(T_R - T(t))/(T_R - T_0)$ and $K = UA/\rho V$, in which U is the overall heat-transfer coefficient, A is the internal surface area of the container, c is the specific heat and ρ is the density.

Comparing equation (5.10) with the heat penetration equation in the same form,

$$u = j \cdot 10^{-t/f}, \tag{5.11}$$

we obtain

$$f = 2.303/K = 2.303 c\rho V/UA \tag{5.12}$$

$$j = 1.$$

For many products the value of $c\rho$ is constant, so that f is directly proportional to the volume of the system and indirectly proportional to the internal area of the container and the heat-transfer coefficient. Although in theory $j = 1$, in practice a wide range of values, some less than 1, are observed. The reason put forward for this is that the U-value is initially high because of the high driving force ΔT, and with increasing time this decreases progressively to a constant value. The f_h-value will be small initially and will increase to a constant rate of heating. The initial low value will favour low values of j, which will be less than the theoretical value of 1.

Datta (1990) examined the case of natural convection in canned foods and concluded that there was no justification for the existence of an asymptotic straight-line temperature–time relationship, because of the complexity of the temperature field and the flow patterns. However, he concluded that over short ranges of processing time the errors would be sufficiently small to use the approximation for practical purposes. It should be remembered that the method is entirely empirical and care should be taken in the evaluation of convection heating.

Schmidt and Pflug (1966) studied the effect of solids (glass spheres of varying sizes) on the heat penetration factor f_h for jars of water/spheres. They found that the f_h-value decreased with increasing surface area-volume ratio for the spheres, but did not vary with sphere size. Jars of water alone had a higher f_h than filled jars.

Ikegami (1977) derived an empirical convection equation for heat transfer into a can of solids with a covering liquid:

$$u = \frac{jj' f_2}{f_2 - f} 10^{-t/f_2},\tag{5.13}$$

where f and j apply to the solid portion and f_2 and j' to the liquid portion. From various experimental results the j-values for the solid portion were approximately 2.0, the j'-values for the liquid portion were below 0.5. The formula was rigorously derived in Ikegami (1978).

Akterian (1996) developed a model for predicting heat penetration data for convection heat fruits in syrup and vegetables in brine packed in glass jars. This was based on the uniform internal temperature model (Section 2.3), and used the used concept of the coefficient of thermal inertia $E = f_h / \log_e 10$. The values for the f_h calculated from E values ranging from 7–25 min are much greater than those given in Table 5.3 for canned products because of the thickness of the walls of the glass bottles and the effect of the external heat-transfer coefficient. A table giving experimental data, including process data, E- & F-values and heat-transfer coefficients, for twenty products is useful for manufacturers using glass jars.

5.4.3. Computer Modeling

A computer model, NumeriCAL, has been developed by Manson (1992) for determining heating factors from heat penetration data. The program has the advantage of not using the Ball 42% come-up time but calculating this effect from the actual processing conditions. This means that pilot-plant data can be used more reliably. An extension of this model to determine the f_c and j_c factors for the cooling phase was proposed.

5.5. Factors Affecting Heat Penetration

5.5.1. Effect of Container Shape and Dimensions

5.5.1.1. f_h-value

The f_h for a Cylindrical Container of Conductive-Heating Product is Given by the Equation

$$f_h = 0.398/[\alpha(1/a^2 + 0.427/4b^2)],\tag{5.14}$$

and is obtained by substituting the values for R_1 and π^2 in equation (5.8). This equation can be used for determining the f_h-value for a container of different dimensions, with the same food product. Thus if the first container has a value f_1 and dimensions radius a and height $2b$ and the second container an unknown f_2 and dimensions c and $2d$, respectively, then the value of f_2 can be obtained by eliminating the thermal diffusivity α from the respective equations, giving

$$f_2 = \frac{1/a^2 + 0.427/4b^2}{1/c^2 + 0.427/4d^2} f_1. \tag{5.15}$$

The container dimensions for this type of calculation should be based on the mass of product. Thus the diameter is the nominal can diameter less $\frac{1}{8}$ in and the height is the nominal can height less $\frac{1}{8}$ in less the headspace of the container. For two-piece cans with only one double seam $\frac{1}{16}$ in is subtracted. An easy system, using can factors, for conversion to various can sizes is given by Ball and Olson (1957) in tabular form. This only applies to cans with seamed lids at the top and bottom; consequently, for modern two-piece cans with one seamed end and for many other types of package it is more appropriate to calculate the conversion directly using equation (5.15).

A similar equation can be obtained for a brick-shaped container or pouch (Olson & Jackson 1942),

$$f = 0.933/\left[\alpha(1/a^2 + 1/b^2 + 1/c^2)\right] \tag{5.16}$$

and if f_1 applies to a container of dimensions $2a$, $2b$, $2c$, and f_2 applies to one with dimensions $2p$, $2q$, $2r$, then f_2 is given by

$$f_2 = \frac{1/a^2 + 1/b^2 + 1/c^2}{1/p^2 + 1/q^2 + 1/r^2} f_1. \tag{5.17}$$

The container dimensions are determined in the same way as before.

If the object can be approximated to a slab then only one term is necessary:

$$f_2 = p^2 f_1/a^2. \tag{5.18}$$

For convection heating, a similar conversion can be carried out using the general equation

$$f = 2.303 \frac{c\rho V}{UA}, \tag{5.19}$$

from which the relation between the f-values for two cans of differing sizes with radii a_1 and a_2 and heights h_1 and h_2 can be obtained, by substituting for the volumes $(\pi a^2 l)$ and surface areas $(2\pi a^2 + 2\pi al)$ of the cans:

$$f_2 = \frac{V_2 A_1}{V_1 A_2} f_1 = \frac{a_2 + l_2}{a_2 l_2} \frac{a_1 l_1}{a_1 + l_1} f_1, \tag{5.20}$$

where a and l are, respectively, the radius and height of the container, suitably adjusted for the seam and headspace dimensions. Ball and Olson (1957) have again produced tabular data for easy conversion of can sizes. Kopelman et al. (1981) developed conversion factors for a wider range of processing conditions and other containers, by expanding the overall heat-transfer coefficient so that the wall thickness and the outside heat-transfer coefficients could be considered. This allowed conversions to made from metal to glass containers and heating from steam to water.

5.5.1.2. Lag Factor j

The Lag Factor j, as Previously Discussed, Depends on the Position in the Object. The j-Value for a Finite Cylinder is Derived from Equation (5.9)

$$j = 2.03970 J_0(R_1 r/a) \cos(\pi y/2b), \tag{5.20a}$$

where a is the radial position, r is the radius of the can, y is the axial position and $2b$ the height of the can. $J_0(x)$ is a zero-order Bessel function of x and R_1 is the first positive root of $J_0(x) = 0$. The j-value for a brick-shaped object of dimensions $2a$, $2b$, $2c$ is given by

$$j = 2.06410 \cos(\pi x/2a) \cos(\pi y/2b) \cos(\pi z/2c). \tag{5.21}$$

Olson and Jackson (1942) also showed that, contrary to the theory, the j-value was related to the diameter-length ratio for a finite cylinder, attaining a maximum value of 1.68 when the length of the can was equal to that of the diameter. This was explained as being due to the fact that the j-value was taken from the tangent to the curve rather than the asymptote. This was further investigated by Cowell and Evans (1961), who showed that the j-value derived from the tangent equation was related to the height–diameter ratio squared. For theoretical work on this subject of the difference between tangents and asymptotes, the reader is referred to the work of Hayakawa (1964), Alles and Cowell (1971), and Flambert and Deltour (1974).

5.5.2. Effect of Initial Temperature

From equation (5.9) the factor A_{110} contained expressions for the initial temperature distribution in the product being heated. This meant that the j-value was dependent upon this condition. Olson and Jackson (1942) developed an expression for j to deal with non-uniform initial temperature distributions,

$$j = 1.27 + 0.77 \frac{T_R - T_W}{T_R - T_0}, \tag{5.22}$$

where T_R is the retort temperature, and T_W is the temperature about 0.1 in (2.5 mm) from the wall just before heating begins. The equation was used to predict j-values for 603×700 cans of 5% bentonite in water, which had different initial temperature distributions. Moderate agreement was shown between the experimental and the practical results. The equation does show the magnitude of the effect, and it is worth considering when doing practical work.

The effect of variable initial temperature distributions will be seen with the cooling phase of large cans, when the critical point temperature is considerably different from the retort temperature.

The effect of varying the initial temperature has also been studied. Berry and Bush (1989) showed that for whole-kernel corn packed in 603×700 cans the j_h-value for $T_0 = 25°C$ was 1.36, and for $T_0 = 70°C$ it was 1.66. With smaller-sized cans the effect was less noticeable. The f_h-values were relatively unaffected.

For products displaying broken-heating curves the effect of initial temperature differences was more pronounced on both heat penetration factors and the break-point time. The j-values increased and the f-values decreased with increasing initial and retort temperatures.

5.5.3. Effect of Position Inside The Container

It has already been seen from equations (5.8) and (5.9) that the j-factor is dependent on position inside the container, whereas f_h is independent but depends on the container dimensions. The fact that the j-value varies is a very important property and was used by a number of workers to determine the total sterilizing value for the whole container, rather than just at the critical point. The method consisted of constructing a series of iso-j shells in the container and determining the corresponding lethalities.

Ball and Olson (1957) estimated the errors in the j-value associated with thermocouple positioning. For a 307×409 container of conduction-heating product, when the temperature sensor is located on the container axis, a displacement of 0.205 in (5.21 mm) from the center towards either end will produce an error of 1% in the j-value. If the sensor is located in a plane mid-way between the ends of the can, then a displacement of 0.138 in (3.5 mm) will produce a similar error. It was concluded that the error in the j-value was slight when the thermocouple sensor was off-center.

5.5.4. Effect of Headspace

Evans and Board (1954) made a theoretical and practical study of the effect of headspace on f- and j-values. The results showed that the f-value is greater when the headspace is taken into account and the j-value slightly less, for both heating and cooling phases. The influence of trapped air on the heating characteristics of a model food packaged in a semi-rigid plastic container has been studied by Ramaswamy and Grabowski (1996).

5.5.5. Effect of Variation of Physical Properties with Temperature

Using physical property data for "immobilized" water at 20 and 120°C, Evans (1958) showed that if the value of the thermal diffusivity α at 20°C was used, the j-value was unaffected, but the f_h-value was decreased. However, when the correct data were used at the appropriate temperature, f_h was 7.5% higher. It is, therefore, important to consider the effect of temperature on the value of α, since process calculations are very sensitive to small changes in α.

5.5.6. Effect of External Heat-Transfer Coefficients

The effect of using processing media other than saturated steam is to impose a heat transfer restriction on the process. This is measured by the magnitude of the external heat-transfer coefficient, h.

5.5.6.1. j-value

For a Finite Cylinder, with an External Heat-Transfer Coefficient, the j-Value at the Center is Given by

$$j = \frac{2 \sin M_1}{M_1 + \sin M_1 \cos M_1} \frac{2}{\left(1 + R_1^2/Bi^2\right) R_1 J_1(R_1)}, \tag{5.23}$$

where M_1 is the first positive root of

$$Bi \cot M_n = M_n,$$

R_1 is the first positive root of

$$J_0(R_n) = (1/Bi) R_n J_1(R_n),$$

and J_0 and J_1 are Bessel functions of zero and first order, respectively; Bi, the Biot number, is ah/k, where a is the radius, h the heat-transfer coefficient and k the thermal conductivity. Tabulated values are given by Ball and Olson (1957) for a finite cylinder, and Alles and Cowell (1971) for a rectangular container. Castaigne and Lacroix (1986) give tables of Bi against j-values for the center, surface and at the mean temperature for infinite slabs, infinite cylinders and spheres. The value of j for a finite cylinder was obtained from

$$j = j_{slab} \times j_{cylinder}.$$

For a brick-shaped package

$$j = j_{thickness} \times j_{length} \times j_{width}.$$

For a slab

$$j = \frac{2 \sin M_n}{M_n + \sin M_n \cos M_n}. \tag{5.24}$$

If this is designated $2A_n$, then the j-value for a three-dimensional object will be given by

$$j = 2A_n \cdot 2A_m \cdot 2A_p.$$

The first root of M_n is $M_1 = 1.5708 (= \pi/2)$, and $A_1 = 1/M_1$, since $\sin(\pi/2) = 1$ and $\cos(\pi/2) = 0$. Therefore,

$$j = 8 \times 1.5708^{-3} = 2.06410.$$

The j-value decreases with decreasing values of the heat-transfer coefficient, i.e. greater resistance.

5.5.6.2. f_h-value

The Equation for the f-Value Corresponding to Equation (5.23) is

$$f = \frac{2.303}{\left(M_1^2 + R_1^2\right)\alpha}. \tag{5.25}$$

Both M_1 and R_1 increase with increasing h, and consequently the value of f decreases with increasing surface conductance.

Tabulate values are given by Ball and Olson (1957) for a finite cylinder and Alles and Cowell (1971) for a rectangular container. Castaigne and Lacroix (1986) give tables of Bi against $f\alpha/R^2$ for infinite slabs, infinite cylinders and spheres. The values for a finite cylinder and a brick-shaped package are obtained as follows. For a finite cylinder:

$$1/f = 1/f_{slab} + 1/f_{cylinder}.$$

For a brick:

$$1/f = 1/f_{thickness} + 1/f_{length} + 1/f_{width}$$

and

$$f = \frac{2.303}{\alpha M_1^2[1/t^2 + 1/l^2 + 1/w^2]}, \tag{5.26}$$

which becomes, when M_1 is substituted,

$$f = \frac{0.9334}{\alpha(1/t^2 + 1/l^2 + 1/w^2)}. \tag{5.27}$$

5.5.7. Effect of Container Material and Thickness

Lu et al. (1991) compared heat penetration rates into metal and plastic cans (with metal lids) for a range of conduction-heating products. While the j-values were found to be about 10% lower, the f_h-values were correspondingly higher, reflecting the slower heat penetration through the walls of the plastic container. The location of the slowest heating point in the plastic can depend on the orientation of the lid. When the headspace was in contact with the lid the heating rate was slower, i.e. f_h was greater, than when the cans were inverted and the product was in contact with the lid. The overall effect of using plastic cans was to produce much lower lethalities, and in some cases sub-lethal processes.

Kebede et al. (1996) studied heat transfer into high-barrier plastic trays containing an 8% bentonite suspension. The containers had a net volume of 460ml and had dimensions $35 \times 97 \times 142$ mm and were made from (i) polypropylene/ethylene vinyl alcohol/polypropylene and (ii) crystallized polyethylene terephthalate. The f_h values for the two types of tray were similar: 31.1 and 32.6 min respectively.

5.5.8. Effect of Can Rotation

In general, the effect of can rotation is to improve the internal heat transfer and reduce the f_h-value. This subject is discussed when rotary sterilization is considered in Chapter 8.

5.5.9. Statistical Aspects of Heat Penetration Data

The literature values for the variation of f_h-values show that the standard deviation was between 7% and 25% of the average f_h-value (Hicks 1961; Herndon 1971; Lund 1978; Patino & Heil 1985). For canned spaghetti in tomato sauce, histograms of the distribution of f_h showed that it was skewed to the right (inverse gamma distribution), whereas the distributions of j_h and j_c were skewed to the left and the f_c distribution was symmetrical (Hawakawa et al. 1988).

5.5.10. Extrapolation of Heat Penetration Data

The safety of canned foods depends upon adequate process determination by means of heat penetration experiments. The extrapolation of heat penetration data determined at one retort temperature and for a particular initial temperature to other temperatures should be undertaken with extreme care.

For products with straight-line semi-logarithmic heating curves, the problems are less than with broken-heating type curves (Berry & Bush 1987, 1989). For rapidly heating products in small cans the process times measured at different retort temperatures were identical to extrapolated figures. However, for products of intermediate viscosity, or for larger can sizes, the effect of the retort temperature becomes more significant and less confidence can be placed in extrapolation.

The extrapolation to other initial temperatures depends more on the product, but in general it is less reliable than retort temperature extrapolations.

For broken-heating curves extrapolation to higher retort temperatures is acceptable, but not the reverse. The extrapolation from higher initial temperatures to lower gives conservative results, but not the reverse.

5.6. Simulation of Thermal Processing of Non-Symmetric and Irregular-Shaped Foods Vacuum Packed in Retort Pouches: A Numerical Example

The following simulation analysis has been summarized from the research work done by Morales-Blancas et al. (2005). The complete problem statement for a homogeneous, isotropic non-symmetric irregular-shaped domain, with thermo physical properties independent of temperature and subject to constant boundary conditions is supported by the following equations:

Governing equation:

$$\frac{\partial}{\partial x}\left(k_x \frac{\partial T}{\partial x}\right) + \frac{\partial}{\partial y}\left(k_y \frac{\partial T}{\partial y}\right) + \frac{\partial}{\partial z}\left(k_z \frac{\partial T}{\partial z}\right) = \rho C_p \frac{\partial T}{\partial t} \tag{5.28}$$

Where for homogeneous and isotropic material:

$$\frac{\partial^2 T}{\partial x^2} + \frac{\partial^2 T}{\partial y^2} + \frac{\partial^2 T}{\partial z^2} = \frac{1}{\alpha}\frac{\partial T}{\partial t} \tag{5.29}$$

Initial conditions at heating

$$T(x, y, z, 0) = T_0; \forall x, y, z; t = 0 \qquad (5.30)$$

Initial conditions at cooling

$$T(x, y, z, t_g) = T_g(x, y, z, t_g); \forall x, y, z; t = t_g \qquad (5.31)$$

For the cooling stage, initial temperature of each point of the domain (x,y,z) is temperature $T_g(x, y, z, t_g)$ reached at the end of heating stage (time t_g).

Boundary conditions at heating

$$k \frac{\partial T(x, y, z, t)}{\partial n} = U_1 \left(T_{a_1} - T_S \right), \quad t \leq t_g \qquad (5.32)$$

Boundary conditions at cooling

$$k \frac{\partial T(x, y, z, t)}{\partial n} = U_2 \left(T_{a_2} - T_S \right), \quad t > t_g \qquad (5.33)$$

The global heat transfer coefficient (U) will be used, which involves all resistances from the external to the internal conditions.

$$\frac{1}{U} = \left[\frac{1}{h_i} + \frac{\delta_p}{k_p} + \frac{1}{h_e} \right] \qquad (5.34)$$

Where

U: Global heat-transfer coefficient
$h_{i,e}$: Local heat-transfer coefficient (i: internal and e: external)
δ_p: Thickness of the packaging material.
k_p: Thermal conductivity of the packaging material

Considering negligible the internal heat resistance, we obtain (see Figure 5.7):

$$\frac{1}{U} = \left[\frac{\delta_p}{k_p} + \frac{1}{h_e} \right] \qquad (5.35)$$

Where equations (5.34) and (5.35) have been derived from steady state conditions but also can be utilized in unsteady state conditions if the conductance of the packaging material is negligible in relation to the conductance of the food product. According to Zuritz and Sastry (1986) equations (5.34) and (5.35) are valid if:

$$\delta \ll \frac{l(\rho C_p)_{food}}{2(\rho C_p)_{package}} \qquad (5.36)$$

5.6.1. Reverse Engineering by 3-D Digitizing

Reconstruction of the three-dimensional surface and shape of the food products was carried out by using a precision mechanical 3-D digitizer, accuracy of stylus tip 0.64 mm (Microscribe-3D^{TM}, Immersion Co., San José, CA) and a NURBS (Non-Uniform Rational B-spline) modeling software (Rhinoceros®,

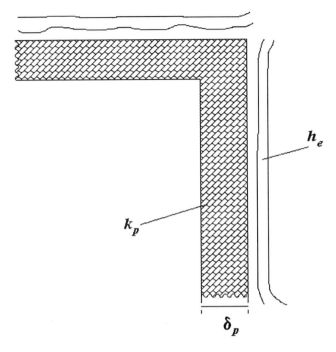

FIGURE 5.7. Heat transfer resistances from ambient to food product.

Robert McNeel & Associates, Seattle, WA). This technique recovers exact surface shape by capturing 3-D points and building 3-D surfaces directly from physical models (Figure 5.8).

5.6.2. Simulation of Heat Conduction Processes

In order to reach an appropriate numerical convergence, simulations were carried out considering an average mesh size ranging from 2.0 to 3.0 mm (see Figure 5.9), and a time step size ranging from 0.5 to 1 s.

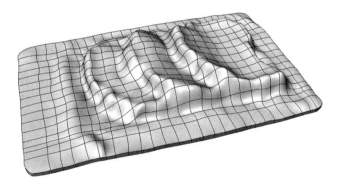

FIGURE 5.8. Reconstruction of the three-dimensional surface and shape.

FIGURE 5.9. Meshing of the 3D digitalized retort pouch.

5.6.3. Finite Element Analysis

The heat conduction model was solved using a Finite Element Analysis and Simulation software package (ALGOR®, ALGOR Inc., Pittsburgh, PA). FEA Models were obtained by solid meshing of the 3-D digitized irregular-shaped geometry models. Bricks/tetrahedra elements were used in order to get the highest quality meshing and the lowest number of elements.

5.6.4. Experimental Validation

Results showed sufficiently good agreement between predicted and measured temperature profiles for the slowest heating/cooling point of each product (Figure 5.10). The agreement between the experimental and predicted temperature

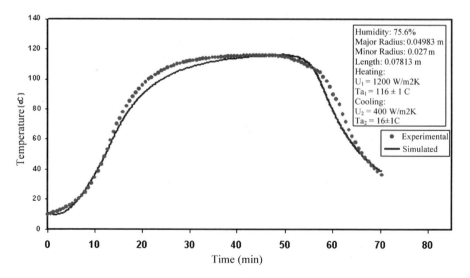

FIGURE 5.10. Validation example: Jack Mackerel product.

profiles was evaluated by using the %RMSE (Percentage of Root Mean Square Error). In general, %RMSE calculated values were less than 8.0%.

The proposed approach based on reverse engineering principles and CAD/CAE tools can be used to simulate and optimize thermal processes of odd-shaped foods.

References

Akterian, S. G. (1996). Studying and controlling thermal sterilization of convection- heated canned foods using functions of sensitivity. *J. Food Eng.*, *29*(3/4), 125–143.

Alles, L. A. C., & Cowell, N. D. (1971). Heat penetration into rectangular cans of food. *Lebensm.- Wiss. u.-Technol.*, *4*(2), 50–54.

ASTM (1988). *Standard Guide for the use in the establishment of thermal processes for foods packaged in flexible containers* (pp. 995–1006). Philadelphia, PA: American Society for Testing Materials.

Ball, C. O., & Olson, F. C. W. (1957). *Sterilization in food technology– Theory, practice and calculations*. New York: McGraw-Hill.

Berry, M. R. Jr, & Bush, R. C. (1987). Establishing thermal processes for products with broken heating curves from data taken at other initial and retort temperatures. *J. Food Sci.*, *52*, 558–561.

Berry, M. R. Jr, & Bush, R. C. (1989). Establishing thermal processes for products with straight-line heating curves from data taken at other retort and initial temperatures. *J. Food Sci.*, *54*(4), 1040–1042, 1046.

Beverloo, W. A., & Weldring, J. A. G. (1969). Temperature measurements in cans and the consequences of errors for the process calculation. *Lebensm.- Wiss. u. -Technol.*, *2*, 9–14.

Bhowmik, S. R., & Tandon, S. (1987). A method for thermal process evaluation of conduction heated foods in retortable pouches. *J. Food Eng.*, *52*(1), 202–209.

Bichier, J. G., Teixeira, A. A., Balaban, M. O., & Heyliger, T. L. (1995). Thermal process simulation of canned foods under mechanical agitation. *J. Food Process Eng. 18*(1), 17–40.

Bigelow, W. D., Bohart, G. S., Richardson, A. C., & Ball, C. O. (1920). Heat penetration in processing canned foods *Bulletin No. 16-L*, 1st Edition. Washington, DC: National Canners' Association,.

Bown, G., Nesaratnum, R., & Peralta Rodriguez, R. D. (1985). Computer Modelling for the control of sterilization processes. *Technical Memorandum No. 442*. Chipping Campden, Glos. UK: Campden & Chorleywood Food Research Association.

Castaigne, F., & Lacroix, C. (1986). Rapid heat treatment estimation tables for homogeneous products of simple geometrical shapes. *Sci. des Aliments*, *6*, 119–133 [in French].

CCFRA (1977). Guidelines for the establishment of scheduled heat processes for low acid canned foods. *Technical Manual No. 3*. Chipping Campden, Glos., UK: Campden & Chorleywood Food Research Association.

Charlett, S. M. (1955). Heat penetration into canned foods. *Food Manufacture*, *30*(6), 229–232, 245.

Cossey, R., & Richardson, P. S. (1991). The use of constant temperature blocks for the calibration of master temperature indicators. *Technical Memorandum No. 613*, Chipping Campden, Glos.,UK: Campden & Chorleywood Food Research Association.

Cowell, N. D., & Evans, H. L. (1961). Studies in the canning process. IV. Lag factors and slopes of tangents to heat penetration curves for canned foods heated by conduction. *Food Technol.*, *15*, 407–412.

Cowell, N. D., Evans, H. L., Hicks, E. W., & Mellor, J. D. (1959). Conduction errors in thermocouples used for heat penetration measurements in foods which heat by conduction. *Food Technol.*, *13*, 425–429.

Datta, A. K. (1990). On the theoretical basis of the asymptotic semi-logarithmic heat penetration curves used in food processing. *J. Food Eng.*, *13*, 177–190.

Datta, A. K., & Teixeira, A. A. (1988). Numerically predicted transient temperature and velocity profiles during natural convection heating of canned liquid foods. *J. Food Sci.*, *53*(1), 191–195.

Dobie, P. A. P. (1993). Calibration: DIY or ISO? *Food Manufacture*, *68*(4), 28–29.

Ecklund, O. F. (1956). Correction factors for heat penetration thermocouples. *Food Technol.*, *10*(1), 43–44.

Eisner, M. (1976). Ellab-temperature measuring system. *Ind. Obst-u. Gemüseverwert.*, *61*(13), 339–341.

Evans, H. L. (1958). Studies in canning processes II. The effects of the variation with temperature of the thermal properties of foods. *Food Technol.*, *12*(6), 276–280.

Evans, H. L., & Board, P. W. (1954). Studies in canning processes. I. Effect of headspace on heat penetration in products heating by conduction. *Food Technol.*, *8*(5), 258–262.

Flambert, C. M. F., & Deltour, J. (1974). Lag factor j and logarithmic slope f on the symmetry axes of cylindrical cans. *Lebensm.-Wiss. u. -Technol.*, *7*(6), 348–352 [in French].

George, R. M., & Richardson, P. S. (2001). Non-invasive temperature measurement in thermal processing – An innovative approach. In J. Welti-Chanes, G. V, Barbosa-Cánovas, & J. M. Aguilera (Eds.), *Eighth international conference: Engineering and food ICEF8, Vol.1* (pp. 656–659). Lancaster, PA: Technomic Pub. Co.

Ghani, A. G., Farid, M. M., & Chen, X. D. (2002). Theoretical and experimental investigation of the thermal inactivation of *Bacillus stearothermophilus* in food pouches. *J. Food Eng.*, *51*(3), 221–228.

Giannoni-Succar, E. B., & Hayakawa, K-I. (1982). Correction factor of deviant thermal processes applied to packaged heat conduction. *J.Food Sci.* *47*(2), 642–646.

Hayakawa, K. (1964). Development of formulas for calculating the theoretical temperature history and sterilizing value in a cylindrical can of thermally conductive food during heat processing. PhD thesis. New Jersey: Rutgers University of New Brunswick.

Hayakawa, K., Massaguer, P. de, & Trout, R. J. (1988). Statistical variability of thermal process lethality in conduction heating food—computerized simulation. *J. Food Sci.*, *53*(6), 1887–1893.

Herndon, D. H. (1971). Population distribution of heat rise curves as a significant variable in heat sterilization process calculations. *J. Food Sci.*, *36*, 299–305.

Hicks, E. W. (1951). Uncertainties in canning process calculations. *Food Res.*, *26*, 218–223.

Holdsworth, S. D. (1974). Instrumentation developments for heat processing. *Food Manufacture*, *49*(11), 35–38, 58.

Holdsworth, S. D. (1983). Developments in the control of sterilizing retorts. *Process Biochem.*, *18*(5), 24–28.

Holdsworth, S. D. (1985). Developments in the control of sterilizing retorts for packaging food. *Packaging*, *56*, 13–17.

Hosteler, R. L., & Dutson, T. R. (1977). Effect of thermocouple wire size on the cooking times of meat samples. *J. Food Sci.*, *42*(3), 845–846.

Hurwicz, H., & Tischer, R. G. (1955). Heat processing of beef. V. Temperature distribution during processing of beef at high retort temperatures. *Food Res.*, *20*, 399–414.

Hurwicz, H., & Tischer, R. G. (1956). Heat processing of beef. VI. Thermal diffusivity and "slopes" of heating and cooling curves for the high-temperature process. *Food Res., 21,* 147–155.

IFTPS. (1995). *Protocol for carrying out heat penetration studies.* Fairfax, VA: Institute for Thermal Processing Studies.

Ikegami, Y. (1977). Heat penetration in canned foods containing solids and liquid. *Canner's J., 56*(7), 548–552 [in Japanese].

Ikegami, Y. (1978). Heat transfer and food surface temperature change with time. *Canner's J., 57*(7), 593–596 [in Japanese].

Jackson, J. M. (1940). Mechanisms of heat transfer in canned foods during thermal processing. In *Proceedings of the 1st food conference of food technologists* (pp. 39–49).

Jackson, J. M., & Olson, F. C. W. (1940). Thermal processing of canned foods in tin containers. IV. Studies of the mechanisms of heat transfer within the container. *Food Res., 5*(4), 409–420.

Kanellopoulos, G., & Povey, M. J. W. (1991). A finite element model for conduction errors in thermocouples during thermal sterilization of conduction-heating foods. *Int. J. Food Sci. Technol., 26,* 409–421.

Kebede, E., Mannheim, C. H., & Miltz, J. (1996). Heat penetration and quality preservation during thermal treatment in plastic trays and metal cans. *J. Food Eng., 30*(1/2), 109–115.

Kopelman, I. J., Pflug, I. J., & Naveh, D. (1981). On the conversion factors in thermal processes. *J. Food Technol., 16,* 229–238.

Kopelman, I. J., Pflug, I. J., & Naveh, D. (1982). On the overshooting in thermal processing. *J. Food Technol., 17,* 441–450.

Lu, Q., Mulvaney, S. J., & Hsieh, F. (1991). Thermal processes for metal cans compared with retortable plastic containers. *J. Food Sci., 56*(3), 835–837.

Lund, D. B. (1978). Statistical analysis of thermal process calculations. *Food Technol., 32,* 76–78, 83.

Magoon, C. A., & Culpepper, C. W. (1921). *A Study of the factors affecting temperature changes in the container during the canning of fruits and vegetables.* United States Department of Agriculture Bulletin No. 956. Washington, DC: Government Printing Office.

Manson, J. E. (1992). A new approach to determining product heating factors from heat penetration data. *TechKNOWLEDGE Bulletin 201,* Metairie, LA., TechniCAL, Inc.

Marra, F., Romano, V. (2003). A mathematical model to study the influence of wireless temperature sensor during the assessment of canned food sterilization. *J. Food Eng 59*(2/3), 245–252.

May, N. S. (1991). Evaluation of the Datatrace Temp® Micropack system. Part II. Use inside food containers in static steam sterilization process. *Technical Memorandum No. 620.* Chipping Campden, Glos.,UK: Campden & Chorleywood Food Research Association.

May, N. S. (1992). Evaluation of the Datatrace Temp® Micropack system. Part III. External mounting for use in batch static and hydrostatic/hydrolock processes. *Technical Memorandum No. 661.* Chipping Campden, Glos., UK: Campden & Chorleywood Food Research Association.

May, N. S., & Cossey, R. (1989). Review of data acquisition units for in-container temperature measurements and initial evaluation of the Ball Datatrace® system – Part I. *Technical Memorandum No. 567.* Chipping Campden, Glos., UK: Campden & Chorleywood Food Research Association.

May, N. S., & Withers, P. M. (1993). Data acquisition units for thermal process evaluation. In *Food technology international Europe* (pp. 97–99). London: Sterling Publishers.

May, N. (Ed.), (1997). *Guidelines for performing heat penetration trials for establishing thermal processes in batch retort systems. Guidelines No. 16.* Chipping Campden, Glos: Campden & Chorleywood Research Association.

May, N. (2004). Cold point determination in containers. In G. S. Tucker (Ed.), *Third international symposium thermal processing – Process and package innovation for convenience foods* (Session 3:1). Chipping Campden UK: Campden & Chorleywood Food Research Association.

Morales-Blancas, E., Pérez, D., Rodríguez, C., and Simpson, R. (2005). Simulation of thermal processing of non-symmetric and irregular-shaped foods vacuum packed in retortable pouches. Presentation at IFT Meeting, New Orleans, LO, USA.

Naveh, D., Kopelman, I. J., Zechman, L., & Pflug, I. J. (1983a). Transient cooling of conduction heating products during sterilization: temperature histories. *J. Food Process. Preserv., 7,* 259–273.

Naveh, D., Pflug, I. J., & Kopelman, I. J. (1983b). Transient cooling of conduction heating products during sterilization: sterilization values. *J. Food Process. Preserv., 7,* 275–286.

Noronha, J., Van Loey, A., Hendrickx, M., & Tobback, P. (1996a). Simultaneous optimization of surface quality during the sterilization of packed foods using constant and variable retort temperature profiles. *J. Food Eng., 30*(3/4), 283–297.

Noronha, J., Van Loey, A., Hendrickx, M., & Tobback, P. (1996b). An empirical equation for the description of optimum variable retort temperature profiles that maximize surface quality retention in thermally processed foods. *J. Food Process. Preserv., 20*(3), 251–264.

Olson, F. C. W., & Jackson, J. M. (1942). Heating curves. Theory and practical application. *Ind. Eng. Chem., 34,* 337–341.

Packer, G. J. K., & Gamlen, J. L. R. (1974). Calculation of temperature measurement errors in thermocouples in convection heating cans. *J. Food Sci., 39,* 739–743.

Patino, H., & Heil, J. R. (1985). A statistical approach to error analysis in thermal process calculations. *J. Food Sci., 50,* 1110–1114.

Peterson, W. R., & Adams, J. P. (1983). Water velocity effect on heat penetration parameters during institutional size retort pouch processing. *J. Food Sci., 48,* 457–459, 464.

Ramaswamy, H. S., & Grabowski, S. (1996). Influence of trapped air on the heating behaviour of a model food packaged in semi-rigid plastic containers during thermal processing. *Lebens. -Wiss. u. -Technol., 29,* (1/2), 82–93.

Robertson, G. L., & Miller, S. L. (1984). Uncertainties associated with the estimation of the F_0 values in cans which heat by conduction. *J. Food Technol. (UK), 19,* 623–30.

Schmidt, E. D., & Pflug, I. J. (1966). A study of the variables that affect heat penetration rates in glass containers. *Quarterly Bulletin Michigan State Univ., 48*(3), 397–410.

Shaw G. H. (2004). The use of data loggers to validate thermal process. In P. Richardson (Ed.), *Improving the thermal processing of food* (pp. 353–364). Cambridge: Woodhead Publishing.

Silva, C. L. M., & Korczak, K. (1994). Critical evaluation of restrictions used to optimize sterilization processing conditions. Poster paper presented at the *4th Conference on Food Process Engineering*. UK: University of Bath.

Smout, C., Avila, I., Van Loey, A. M. L., Hendrickx, M. E. G., & Silva, C. (2000a). Influence of rotational speed on the statistical variability of heat penetration parameters and on the non-uniformity of lethality in retort processing. *J. Food Eng., 45*(2), 93–102.

Smout, C., Van Loey, A. M. L., & Hendrickx, M. E. G. (2000b). Non-uniformity of lethality in retort processes based on heat distribution and heat penetration data. *J. Food Eng.*, *45*(2), 103–110.

Spinak, S. H., & Wiley, R. C. (1982). Comparisons of the general and Ball formula methods for retort pouch process calculations. *J. Food Sci.*, *56*(1), 27–30.

Sumer, A., & Esin, A. (1992). Effect of can arrangement on temperature distribution in still retort sterilization. *J. Food Eng. 15*(4), 245–259

Thorpe, R. H., & Atherton, D. A. (1972). Sterilized foods in flexible packages. *Technical Bulletin No. 21*. Chipping Campden, Glos., UK: Campden & Chorleywood Food Research Association.

Tong, C. H., & Lenz, R. R. (1993). Dielectric properties of bentonite pastes as a function of temperature. *J. Food Proc.Preserv.*, *17*(2), 139–145.

Tucker, G. S. (2002). Capabilities and pitfalls of computer-based prediction software. In Tucker, G. S. (Ed.), *Second international symposium on thermal processing – Thermal processing: validation challenges* (Session 4:1). Chipping Campden UK: Campden & Chorleywood Food Research Association.

Tucker, G. S. (2004). Improving rotary thermal processing. In P. Richardson (Ed.), *Improving the thermal processing of foods* (pp. 124–137). Cambridge: Woodhead Publishing.

Wiese, K. L., & Wiese, K. T. (1992). A comparison of numerical techniques to calculate broken-line heating factors of a thermal process. *J. Food Proc. Preserv.*, *16*(5), 301–312.

Zang, Z., Weddig, L., & Economides, S. (2002). The effect of thermocouple and receptacle type on observed heating characteristics of conduction-heating foods packaged in small containers. *J. Food Proc. Eng.*, *25*(4), 323–335.

Zuritz, C. A., & Sastry, S. K. (1986). Effect of packaging materials on temperature fluctuations in frozen foods: mathematical model and experimental studies. *J. Food Sci.*, *51*, 1050–1056.

6
Process Evaluation Techniques

6.1. Determination of F-Values: Process Safety

Process evaluation, also called process determination, is the science of determining the F-value for a given process time or the process time required for a given F-value. If the heat penetration curve has been determined, then the f_h- and j-values will be known and the appropriate methods can be used. The methods can also be used to determine the effect of altering the can size or the temperatures on the F-value. The importance of these methods for maintaining the safety of sterilized canned and packaged products cannot be overstated.

The methods of determining F-values fall into two classes: calculation methods and microbiological methods. In this section the calculation methods will be discussed (see Chapter 4, Section 4.1.4 for the applicability and extension of equations 6.1 and 6.2).

The main equation to be solved is the basic integral equation for F_0

$$F_0 = \int_0^t 10^{(T - T_{\text{ref}})/z} \, dt, \tag{6.1}$$

or its Arrhenius kinetic form as discussed in Section 4.1.6. Equation (6.1) is often used in the exponential form

$$F_0 = \int_0^t e^{2.303(T - T_{\text{ref}})z} \, dt. \tag{6.2}$$

The solution of this equation requires knowledge of the time–temperature history of the product at a specified point, usually the point of slowest heating. This may be obtained from the heat transfer equations given in Chapter 2, either by substitution in equation (6.1) or by calculating the temperature history and using one of the methods outlined below. The practical methods of determining the temperature history under processing conditions are discussed in Chapter 5.

6.2. The General Method

The simplest of all methods, known as the general method, is universally used in experimental work, precisely because of its simplicity. It was devised by

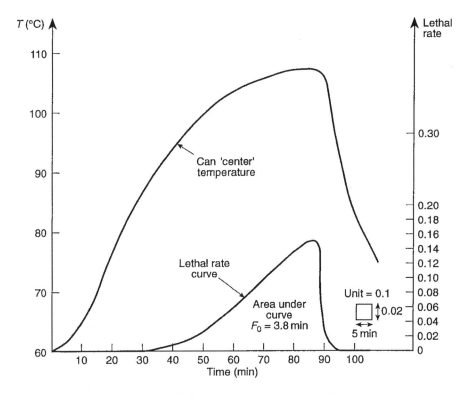

FIGURE 6.1. Graphical method of determining F_0-value from lethal rate curve.

Bigelow *et al.* (1920), and involves graphical or numerical integration of equation (6.1) when the temperature distribution is known, either from heat transfer equations or experimental data.

6.2.1. Graphical Methods

The method that prevailed for many years was to convert the time–temperature history graph into a lethal rate graph, using appropriate data, and then to determine the area under the curve (see Figure 6.1). The area can be determined by counting squares or using a planimeter. While this mechanical method would no longer be used for routine purposes, it is a good method for training courses. It illustrates clearly the relative microbial killing power for different parts of the total process and, in particular, how the heating-up part of the curve contributes relatively little to the total lethality of the process until the retort temperature is approached, often in the last few minutes before cooling begins. The simple method was made more complex by the development of lethal rate paper, in which the ordinate axis is the lethal rate $L = 10^{(T-T_{ref})/z}$ and the scale marked with the appropriate temperature (Schultz & Olson 1940). Other workers extended the use of the method for other z-values and extended processing temperatures (Cass 1947; Patashnik 1953; Hayakawa 1968, 1973; Leonhardt 1978).

6.2.2. Numerical Methods

There are various methods of determining the area F under the lethality curve by calculation rather than construction. Among them are the trapezoidal rule and Simpson's rule, which are used for finding the area of irregular geometric figures. Both make use of parallel chords, whose lengths are designated $y_0, y_1, y_2, \cdots, y_n$, spaced equally at a distance t (measured in minutes). Simpson's rule, generally the more accurate, requires an even number of areas, corresponding to an odd number of points, and is given by

$$F = \frac{1}{3}t(y_0 + 4y_1 + 2y_2 + 4y_3 + \cdots + 2y_{n-2} + 4y_{n-1} + y_n), \qquad (6.3)$$

while the trapezoidal rule is expressed as

$$F = t\left(\frac{1}{2}(y_0 + y_n) + y_1 + y_2 + \cdots + y_{n-1}\right). \qquad (6.4)$$

Patashnik (1953) used the latter equation, making the time interval $t = 1$ min, and selecting y_0 and y_n so that they were zero, to develop a very useful tabular method. In this method the temperatures are listed at equal time intervals, usually 1 min, the corresponding lethal rates are listed in an adjoining column (see Table 4.1), and the F-value obtained by simple addition. For equal time intervals, t, other than 1 min the general formula is given by the equation

$$F = \Delta t \sum_{1}^{n}(L_1 + L_2 + L_3 + \cdots + L_n). \qquad (6.5)$$

An example of the use of this method is given in Table 6.1. It is interesting to note that with beans in tomato sauce in a conduction-heating pack, processed at 120.7°C, it takes about 39 min to obtain a lethal rate of 0.001 min, and the heating phase only contributes 2.5 min to the total F_0-value of 4.22 min (about 60%) after 1 h 12 min, whereas the cooling phase contributes 1.65 min (about 40%) in 15 min. It is usual to take 0.001 min rather than 0.0001 min for the initiation of lethal rate build-up and decline. This fulfils the L_1 and L_n zero requirements with this method. The time increment t is equal to 1 min in this case.

Thus, $L = 4.24$ min and $F_0 = tL = 1 \times 4.24$ min.

Hayakawa (1968) developed a method using the Gaussian integration formula, which involved a template that could be used with the graphical method. The procedure is now available as a simple computer program with built-in statistical accuracy (Vinters et al. 1975; Pflug & Odlaugh 1978; Tung & Garland 1978), and can be used directly in conjunction with heat penetration equipment. Eszes and Rajkó (2004) have given a useful discussion of five different types of integral methods for obtaining the lethality of a given process from heat penetration data.

TABLE 6.1. Example of the use of Patashnik's method for process evaluation. Heat penetration data for UT can (73×115) of beans in tomato sauce processed under static conditions with saturated steam.

Time (min)	Retort temperature (°C)	Center temperature (°C)	Lethal rate (min)	L (min)
0	44.3	26.8	—	—
1	92.1	26.8	—	—
2	103.1	26.8	—	—
3	108.3	26.9	—	—
4	110.3	27.2	—	—
5	112.2	27.6	–	—
6	116.6	28.1	—	—
7	119.9	28.9	—	—
8	120.6	29.9	–	—
9	120.7	31.1	—	—
10	120.7	32.6	–	—
11	120.7	34.3	—	—
12	120.7	36.2	—	—
13	120.7	38.3	—	—
14	120.7	40.5	—	—
15	120.7	42.8	–	—
16	120.7	45.2	—	—
17	120.7	47.7	—	—
18	120.7	50.3	—	—
19	120.7	52.8	—	—
20	120.7	55.3	—	—
21	120.7	57.8	—	—
22	120.7	60.2	—	—
23	120.7	62.6	—	—
24	120.7	64.9	—	—
25	120.7	67.4	—	—
26	120.7	69.5	—	–
27	120.7	71.6	—	–
28	120.7	73.7	—	—
29	120.7	75.7	—	—
30	120.7	77.8	—	—
31	120.7	79.5	—	—
32	120.7	81.4	0.0001	0.0001
33	120.7	83.1	0.0001	0.0002
34	120.7	84.7	0.0002	0.0004
35	120.7	86.5	0.0003	0.0007
36	120.7	87.9	0.0004	0.0011
37	120.7	89.4	0.0007	0.0018
38	120.7	90.9	0.0009	0.0027
39	120.7	92.1	0.0012	0.0039
40	120.7	93.5	0.0017	0.0056
41	120.7	94.7	0.0023	0.0079
42	120.7	96.1	0.0032	0.0111
43	120.7	97.3	0.0042	0.0153
44	120.7	98.4	0.0053	0.0206
45	120.7	99.5	0.0070	0.0276
46	120.7	100.4	0.0090	0.0366

(*cont.*)

TABLE 6.1. (Continued)

Time (min)	Retort temperature (°C)	Center temperature (°C)	Lethal rate (min)	L (min)
47	120.7	101.3	0.0104	0.0470
48	120.7	102.3	0.0130	0.0600
49	120.7	103.2	0.0162	0.0762
50	120.7	104.0	0.0194	0.0956
51	120.7	104.8	0.0234	0.1190
52	120.7	105.6	0.0281	0.1471
53	120.7	106.4	0.0339	0.1810
54	120.7	107.1	0.0400	0.2210
55	120.7	107.7	0.0457	0.2667
56	120.7	108.3	0.0534	0.3201
57	120.7	108.9	0.0602	0.3803
58	120.7	109.6	0.0708	0.4511
59	120.7	110.1	0.0790	0.5301
60	120.7	110.6	0.0890	0.6191
61	120.7	111.1	0.1000	0.7191
62	120.7	111.6	0.1120	0.8311
63	120.7	112.1	0.1258	0.9569
64	120.7	112.5	0.1380	1.0949
65	120.7	112.9	0.1513	1.2462
66	120.7	113.4	0.1698	1.4160
67	120.7	113.7	0.1820	1.5980
68	120.7	114.1	0.1995	1.7975
69	120.7	114.5	0.2187	2.0162
70	120.7	114.8	0.2340	2.2502
71	120.7	115.1	0.2511	2.5013
72 (Cool)	112.9	115.2	0.2570	2.7583
73	86.3	114.9	0.2398	2.9981
74	59.6	114.6	0.2238	3.2219
75	42.9	114.3	0.2089	3.4308
76	36.1	114.2	0.2041	3.6349
77	31.9	114.0	0.1949	3.8298
78	28.0	113.1	0.1584	3.9882
79	25.7	111.5	0.1100	4.0982
80	25.4	109.0	0.0616	4.1598
81	20.6	106.8	0.0371	4.1969
82	14.4	103.8	0.0190	4.2159
83	13.2	101.9	0.0120	4.2279
84	12.0	100.1	0.0079	4.2358
85	11.9	96.7	0.0036	4.2394
86	11.8	94.2	0.0021	4.2415
87	11.8	92.7	0.0014	4.2429
88	11.9	90.5	0.0008	4.2437
89	11.8	88.7	0.0005	4.2442
90	11.7	86.2	0.0003	4.2445

Thus, $L = 4.24$ min and $F_0 = tL = 1 \times 4.24$ min.

6.2.3. *An Extension of General Method: Revisited General Method (RGM)*

A detailed description will be given according to the work done by Simpson *et al.* (2003). According to the referred authors the main characteristics of the Revisited General Method (RGM) are (a) the developed procedure is capable of integrating lethality calculation by the General Method with principles of heat transfer theory, (b) it is able to evaluate processes at different conditions from those used in heat penetration tests (retort temperature, initial temperature, etc.), (c) it is able to take into account slow come-up and cool down phases, and (d) the procedure performs with at least the same ease of use and reliability as the Formula Method but with better accuracy.

6.2.3.1. Heat Transfer Concepts

Most mathematical models for the prediction of time-temperature histories in food products at a given point normally need to assume one of the basic modes of heat transfer. Two extreme cases have their own analytical solutions (Chapter 2): (a) perfect mixing of a liquid (forced convection) and (b) homogeneous solids (pure conduction). Most foods are an intermediate case, and these extreme solutions would give a guideline for the usefulness of temperature-time histories (profiles) developed here.

6.2.3.2. Heat Transfer Model for Perfect Mixing

For forced convection (agitated liquids), it is possible to assume that the temperature inside the can is uniformly distributed but time dependent. In the case of process temperature (*TRT*), three steps are encountered during the process, as follow: (a) Starting a slow come up, (b) holding constant temperature (*TRT*), and (c) cooling down (*Tw*), normally at a constant temperature. Two dimensionless temperature ratios can be deduced from the heat transfer model (Chapter 2, equations 2.89 and 2.94) for perfect mixing, one for slow come up (equation 6.6), and one for constant process temperature (equation 6.7):

$$\frac{T - (a + bt) + b\left(\frac{fh}{\ln 10}\right)}{IT - a + b\left(\frac{fh}{\ln 10}\right)} = \frac{T' - (a' + b't) + b'\left(\frac{fh}{\ln 10}\right)}{IT' - a' + b'\left(\frac{fh}{\ln 10}\right)} = constant \quad (6.6)$$

$$\frac{TRT - T_{C.P.}}{TRT - IT} = \exp\left[-\frac{UA}{MCp}t\right] = constant \ or$$

$$\frac{T_w - T_{C.P.}}{T_w - IT'} = \exp\left[-\frac{UA}{MCp}t\right] = constant \quad (6.7)$$

Where:

A: Area

a *and* b: Constants of linear equation describing retort temperature profile ($TRT(t) = a + bt$)

b': New slope of the linear equation describing retort temperature profile ($TRT'(t) = a' + b't$)

a': New constant of the linear equation describing retort temperature profile ($TRT'(t) = a' + b't$)

C_p: Heat capacity of food

f_h *and* f_c: Heating and cooling rate factors (related to slope of semi-log heat penetration curve)

M: Product mass

T: Temperature

$T_{C.P.}$: Temperature in the coldest point

IT: Initial temperature

IT': Initial temperature at the cooling step

TRT: Retort temperature

T_w: Cooling temperature

t: Time

6.2.3.3. Heat Transfer Model for Pure Conduction

Although solutions for different geometries are not necessarily straightforward, in general, for any geometry, the dimensionless temperature ratio for constant retort temperature (or constant cooling temperature) can be expressed as (Carslaw & Jaeger 1959):

$$\frac{TP - T_{C.P.}}{TP - IT} = f(IT\ distribution,\ geometry,\ thermal\ properties,\ time) = constant$$

(6.8)

Where TP could be either TRT or T_W.

For slow come-up time with conduction heating Gillespy (1953) and Hayakawa (1974) have developed methods to determine center temperature where the heating profile was time-dependent (e.g. linear or exponential). Gillespy (1953) developed an equation for a slab of material being heated with a linear temperature gradient valid during come-up time. Hayakawa (1974) developed a similar expression for finite cylinders. Expressions for conduction heating products of other geometries (e.g. parallelepiped) with a linear temperature gradient can be found in Carslaw and Jaeger (1959) and Luikov (1968). According to Carslaw and Jaeger (1959) and Luikov (1968) it is possible to find a dimensionless temperature ratio equation suitable for a linear heating profile during come-up time in conductive heating products.

6.2.3.4. Heat Transfer Model: A General Approach

Although the heat transfer mechanisms are rather dissimilar, both models (pure conduction and forced convection), within certain limitations, can be described by dimensionless temperature ratio (equations 6.6, 6.7 and 6.8).

Corollary 1: *The importance and relevance is that it is possible to transform the raw data from heat penetration tests and use the General Method, not only to directly evaluate the raw data, but also to evaluate processes at different conditions (retort temperatures, initial temperatures, longer or shorter process times) than those originally recorded.*

6.2.3.5. Methodology, Implementation and Utilization of RGM

1. *Thermal Process Evaluation* (calculates lethality). The method allows for the calculation of the F value for a set of data obtained experimentally or by simulation. Given that the data are not continuous, the integration procedure should be done numerically (Gauss, Simpson, trapezoidal, etc.) or alternatively fit the data by interpolation method (e.g. cubic spline) and integrate the lethality analytically.
2. *Thermal Time Adjustment* (calculates process time). To determine the processing time so that the F value obtained (F_p) is equal or greater than the required lethality (F_r), the F_p value has to first be determined with the original heat penetration data. This F_p value may be bigger, smaller, or equal to F_r. Therefore, three situations arise: $F_p > F_r$, $F_p = F_r$, and $F_p < F_r$. Situations such as $F_p > F_r$ and $F_p < F_r$ are of interest for further analysis, and they will be called *Case 1 ($F_p > F_r$)* and *Case 2 ($F_p < F_r$)*.

Case 1 are all those heat penetration tests in which final lethality is bigger than the required lethality, so the processing time has to be shortened, i.e., to find a new processing time shorter than the real processing time, so that $F_p \geq F_r$ and $F_p - F_r$ is a minimum. Within this case, three different situations may arise. Figures 6.2, 6.3, and 6.4 were computer generated to show and analyze the three different situations (see Table 6.2).

In the first situation (Figure 6.2), shortening the process time was straightforward because, with the new process time (for an F value equal to or bigger than F_r), the temperature inside the can is uniform and it can be assumed that the cooling temperature profile would be the same as the original. In the second and third situations, the problem is different and more complicated.

In Figure 6.3, the temperature at the coldest point (for the adjusted process) would be lower than the retort temperature, giving a non-uniform temperature distribution inside the can. In this situation, the temperature at the cold spot (referred to the heating part) during the cooling phase will have inertia. To evaluate process lethality (for the adjusted process) it is necessary to generate data for the cooling phase. In this case, the use of equation (6.8) will generate data assuming no inertia. Although this is not completely accurate the resulting error in predicted lethality coming from the cooling phase will be on the conservative (safe) side.

In Figure 6.4, similarly to Figure 6.3, the temperature at the cold spot for the process is lower than the retort temperature, as well as for the adjusted process. Although both curves have inertia, again the utilization of equation (6.8) will not

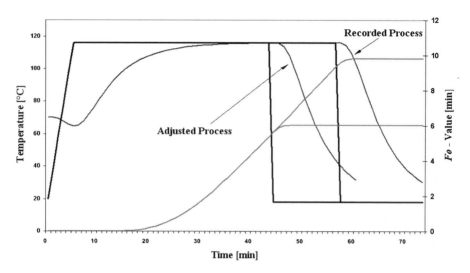

FIGURE 6.2. Simulated heat penetration data for analysis ($F_p > F_r$) for **Case 1** - Situation 1.

be accurate but will safely predict lethality from the cooling phase for the adjusted process. In this application, the use of equation (6.8) will predict inertia but less pronounced than in the real curve.

Case 2 are all those heat penetration tests where final lethality is lower than the required, so the processing time has to be extended, i.e., find a new processing

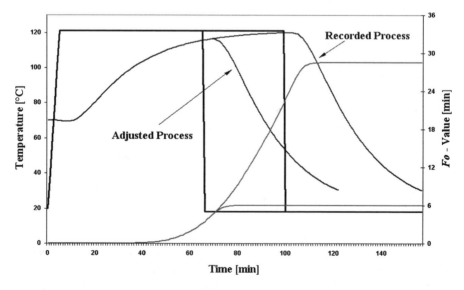

FIGURE 6.3. Simulated heat penetration data for analysis ($F_p > F_r$) for **Case 1** - Situation 2.

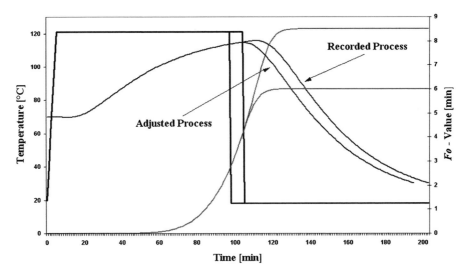

FIGURE 6.4. Simulated Heat Penetration Data for analysis ($F_p > F_r$) for **Case 1** - Situation 3.

time longer than the test processing time, so that $F_p \geq F_r$ and $F_p - F_r$ is a minimum. In this case, the cooling phase temperature profile must be displaced to the right in order to extend process time. Three situations need to be considered and analyzed that are presented in Figures 6.5, 6.6, and 6.7. In the first situation (Figure 6.5), the process adjustment is straightforward. The two assumptions needed are that the coldest temperature could be maintained at the same level and the new cooling temperature profile would be the same as the original one.

TABLE 6.2. Can dimensions and processing conditions for Figures 6.2–6.8.

Figure	Can size [mm]	Can (Common name)	TRT[a,b] [°C]	P_t [min]	F_0–Value[c] [min]
1	52 × 38	70 g tomato paste	116	50	9.8*
2	74 × 116	No. 1 tall	121	90	28.5*
3	99 × 141	Liter	121	95	8.5*
4	52 × 38	70 g tomato paste	116	28	3.1**
5	52 × 38	70 g tomato paste	116	13	0.2**
6	99 × 141	Liter	121	70	2.0**
7	74 × 116	No. 1 tall	130	40.1	6.0

[a] $TRT(t) = a + bt$, $(0 < t < CUT)$, where $a = 40$ [°C] and b can be evaluated, in each case, considering the known value of a and that $TRT(5)$ is the value reported as processing temperature per process.
[b] $TRT =$ Constant $(t > CUT)$
[c] Calculated with General Method.
*: $F_p > F_r$; **: $F_p < F_r$
$F_r = 6$ [min, CUT = 5 [min], Conduction heated product, $\alpha = 1.7 \times 10^{-7}$ [m/s^2], $T_w = 18$ [°C], IT = 70 [°C].

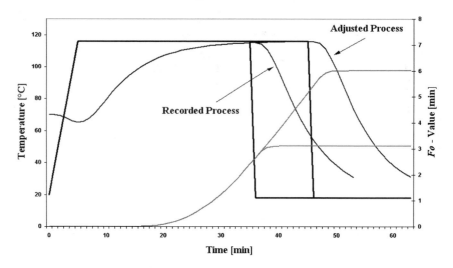

FIGURE 6.5. Simulated Heat Penetration Data for analysis ($F_p < F_r$) for **Case 2** - Situation 1.

In Figures 6.6 and 6.7, it was first necessary to generate more data for the heating process to be able to extend the process time utilizing the following equation:

$$t = f \cdot \log \left(j \frac{TRT - IT}{TRT - T} \right) \tag{6.9}$$

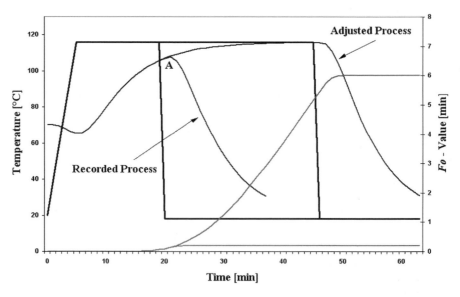

FIGURE 6.6. Simulated Heat Penetration Data for analysis ($F_p < F_r$) for **Case 2** - Situation 2.

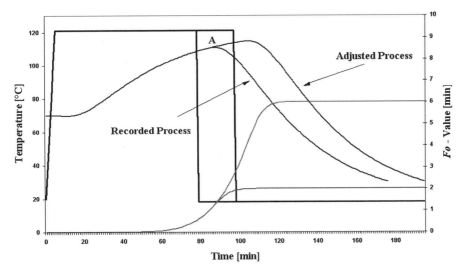

FIGURE 6.7. Simulated Heat Penetration Data for analysis ($F_p < F_r$) for **Case 2** - Situation 3.

Where:

$$j = \frac{TRT - T_A}{TRT - IT} \tag{6.10}$$

T_A: extrapolated initial can temperature obtained by linearizing entire heating curve of a can

j: dimensionless lag factor.

According to heat transfer theory, the recorded cooling temperature profile will have more pronounced inertia than the new cooling temperature profile, therefore, in these two situations (Figures 6.6 and 6.7), the use of equation (6.8) would generate not only inaccurate but also unsafe data. To avoid this problem, equation (6.8) was applied, considering only the data starting at point A as seen in Figures 6.6 and 6.7 (A pinpoints the maximum temperature recorded in the process). Since these data lack inertia, they could generate the new cooling temperature profile on the safe side.

Case 3 involves thermal process evaluation at conditions other than those experimentally recorded (or generated by simulation). Sometimes it is useful to obtain a process evaluation at conditions different from those used for the original heat penetration test, thereby avoiding or significantly reducing the number of new experiments. The new process conditions could be: initial food temperature, retort temperature, and/or cooling temperature. The new time-temperature data should be predicted using adequate mathematical models (if the type of food allows it), or using the dimensionless temperature ratio concept developed in this study, which is applicable to any kind of food. The dimensionless temperature ratio concept could be used for any kind of geometry. However, in real process situations, the

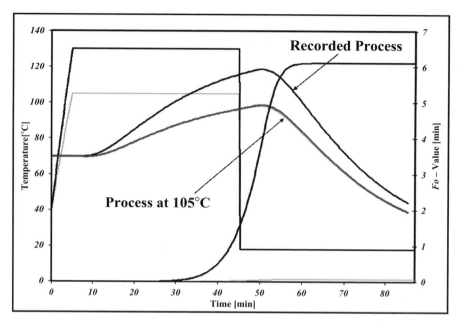

FIGURE 6.8. Simulated heat penetration data for analysis (changing processing conditions, *TRT*)

retort temperature is not always constant (e.g. come-up time), and will impair the theoretical validity of the concept derived for dimensionless temperature ratio as has been discussed in the literature (Shultz & Olson 1940). In the present work the retort temperature was divided into three parts as follows: (a) come-up time ($TRT(t) = a + bt$), (b) process temperature ($TRT = Constant$) and (c) cooling temperature ($T_w = Constant$).

Equation (6.6) was used to transform the original data to the newest processing conditions (TRT' and/or IT'), assuming a linear temperature profile during come-up time; equation (6.8) was used either for the constant retort temperature (TRT) or cooling water temperature (Tw) conditions, respectively. In the case of Figure 6.8 two aspects should be carefully considered when changing processing conditions (TRT' and/or IT'): (a) maintain the same come-up time, and (b) decide how to generate the new cooling temperature profile, specifically in the presence of different inertia. Firstly, although it is a limitation, come up time must be maintained. Secondly, the cool down temperature data transformation could lead to a new cooling temperature profile with less inertia (e.g. the new retort temperature is lower than the original). In this case, the transformed data would overestimate the F value, which could result in an unsafe process. In this situation (Very rare: See Figure 6.8 where an extreme situation is presented), a safe procedure should be to follow the recommendation explained in Case 2, Figures 6.6 and 6.7.

To manage the already transformed data (new processing conditions) it is necessary to follow the procedure explained in the section *Thermal time adjustment (calculates process time)* to adjust the process.

6.2.3.6. Critical Analysis and Future Refinements

The hypothesis behind the chosen strategy assumes that the extreme heat transfer models, perfect mixing and homogeneous heat conduction are able to represent any kind of food (mixed heat transfers mode). Although the heat transfer models are rather dissimilar, both can be fairly well described by the same mathematical function. If these two extreme models have heat transfer similarities, it is assumed that any kind of food will follow the same mathematical functionality as these two extremes.

The proposed methodology was rather accurate when used for thermal process adjustment. The processing time was always estimated (over-estimated) with an error less than 5% (in all cases under study). For the Formula Method, it was common to find errors of 10–20% or more. The Revisited General Method procedure safely predicted shorter process times than those predicted by the Formula Method. These shorter process times may have an important impact on product quality, energy consumption, plant production capacity, and adequate corrections for on-line control (process deviations).

Further testing with experimental data must be done on this developed procedure so as to check its performance and then make it available in a computer via user-friendly software. The developed procedure could be extended to pasteurization processes, UHT/HTST processes, etc. Future studies should consider the possibility of changing come-up time, non-linear temperature profile during come up time, and the capability to evaluate broken heating curves at different conditions than the originally recorded temperature (*TRT* and/or *IT*).

6.3. Analytical Methods

The analytical approach is instructive, first, because it demonstrates the problems encountered in this type of work, and second, because the results are useful in other methods of analysis. The method essentially substitutes the equation for the temperature distribution $T(t, r, x, y, z)$ obtained from the analytical heat transfer equations of Chapter 2 in the F_0 equation (6.1) to give

$$F_0 = \int_0^t 10^{(T(t,r,x,y,z) - T_{\text{ref}})/z} \, dt. \tag{6.11}$$

Some examples of different temperature profiles show how complex the situation can become.

For heating of canned foods, initially at a temperature T_0 and in a retort at temperature T_R, the general equation is

$$T = T_R - (T_R - T_0)F(t)$$

where $F(t)$ is the analytical solution for the temperature at the center point for the given geometrical shape (see Chapter 2). Substituting this in equation (6.11) gives

$$F_0 = \int_0^t 10^{[T_R-(T_R-T_0)F(t)-T_{\text{ref}}]/z} dt. \tag{6.12}$$

For heating and cooling, the equation for $F(t)$ is

$$F(t) = T_R - (T_R - T_0)F(t) + (T_C - T_R)[1 - F(t - t_h)], \tag{6.13}$$

where t is the total time for the whole process and t_h is the time for the heating part; when this is used in calculating the F_0, the total equation is the addition of the two parts for heating and cooling F_h and F_c respectively. This leads to the equation

$$F_0 = \int_0^{t_h} 10^{[T_R-(T_R-T_0)F(t)-T_{\text{ref}}]/z} dt$$

$$+ \int_{t_h}^t 10^{[T_R-(T_R-T_0)F(t)+(T_C-T_R)(1-F(t-t_h))-T_{\text{ref}}]/z} dt. \tag{6.14}$$

Hicks (1951) and Hurwicz and Tischer (1952) have discussed the solution of this equation for the case of cylindrical cans, using the first-term approximation. Whether the approximation is valid for the cooling period, where the value of t is small, is debatable.

6.3.1. Constant Temperature with Time

For $T_t = T$, at time t, e.g. during the holding time in a retort sequence or during passage through a heat exchanger, the equation reduces to

$$F_0 = 10^{(T-T_{\text{ref}})/z} t. \tag{6.15}$$

6.3.2. Linear Temperature Gradient

For $T = T_0 + at$, where a is the slope of the profile $(T_2 - T_1)/(t_2 - t_1)$, for temperatures T_1 and T_2 corresponding to times t_1 and t_2,

$$F_0 = (z/a \ln 10)[10^{(T_2-T_{\text{ref}})/z} - 10^{(T_1-T_{\text{ref}})/z}]. \tag{6.16}$$

6.3.3. Exponential Temperature Rise

This is a general case where the temperature is given by $T = Ae^{-Bt}$, in which A and B are the heat transfer parameters, for example j and f_h. This represents the first term of many of the series solutions to the heat transfer equation and will only apply in practice when the heating time is long and the second and subsequent terms in the series become negligible. The temperature in this case is the reduced temperature $(T - T_H)/(T_0 - T_H)$, where T_H is the heating or retort temperature and T_0 is the initial temperature, and the basic temperature distribution equation becomes

$$T = T_H - (T_0 - T_H)Ae^{-Bt}.$$

The solution is given by the equation

$$F_0 = \int_0^t 10^{(T_H - (T_H - T_0)Ae^{-Bt} - T_{ref})/z} dt, \qquad (6.17)$$

which becomes

$$F_0 = L \int_0^t 10^{-(T_H - T_0)Ae^{-Bt/z}} dt, \qquad (6.18)$$

where the lethal rate at temperature T_H is $L = 10^{(T_H - T_{ref})/z}$. If $A' = A \ln 10^{(T_H - T_0)/z}$, then the equation can be solved using the exponential integral

$$E_1(x) = \int_x^\infty e^{-x}/x \cdot dx \qquad (6.19)$$

and

$$F_0 = (L/B)[E_1(A'e^{-bt}) - E_1(A')]. \qquad (6.20)$$

Since the second term is generally negligible, the equation becomes

$$F_0 = (L/B)E_1(A'e^{-bt}). \qquad (6.21)$$

6.3.4. The Exponential Integral

The solution of many of the F_0 equations involves integrals of the type

$$\int_0^x e^{-e^{-x}} dx \quad \text{and} \quad \int_0^x 10^{-10^{-x}} dx,$$

which are types of exponential integral. Tables are available for solving such integrals (National Bureau of Standards 1940; Pagurova 1961). Newman and Holdsworth (1989) produced a computer program, based on the algorithm

$$E_1(x) = -\gamma - \ln(x) + \sum_{n=1}^\infty (-1)^{n+1} x^n/(n.n), \qquad (6.22)$$

where γ is Euler's constant (0.577 215 665).

The exponential integral $E_1(x)$, as defined in equation (6.19), was originally used by Ball (see Ball & Olson 1957) in the e^x form and by Gillespy (1951) in the 10^x form. Throughout the literature there are examples of incorrect exponential integrals being used (see Ball & Olson 1957; Flambert *et al.* 1977). Great care needs to be taken in defining and using this function.

6.4. Some Formula Methods

6.4.1. Introduction

It is easy today to dismiss the methods developed in the past as no longer relevant. Modern computers can now be used to calculate processes without any knowledge of the background for the calculations. However, it is instructive to see how the methods developed and the problems associated with calculating processes. For this purpose two methods have been selected for more detailed treatment: the Ball methods, because of their ingenuity and widespread use; and the Gillespy method, widely used in the UK canning industry. A number of other methods are mentioned, but in less detail.

The formula methods have their origin in the use of the simple F_0 integral equation (6.1) and substitution of an empirical model of the temperature distribution from heat penetration experiments. Each worker has used different representations for the modeling of the heating and cooling curves. The Ball methods (Ball 1923; Ball & Olson 1957) involved f_h and j parameters for straight-line heating and a hyperbolic function to model the onset of cooling. The method was further developed by a number of workers, in particular, Jakobsen (1954), Hayakawa (1970), and Stumbo (1973). Gillespy (1953), Hicks (1951), and Flambert and Deltour (1972a) used theoretically derived equations based on conduction heat transfer into cylindrical containers.

Among the reviews of the methods which are useful background reading are Holdsworth and Overington (1975), Merson *et al.* (1978) and, especially, Hayakawa (1977, 1978, 1982).

6.4.2. Ball's Methods

6.4.2.1. Development of the Methods

The pioneering work of C.O. Ball (1923) is the starting point for thermal process calculations, and it is useful and instructive to discuss the methods and their development in order to provide some background on the subject. The methods that he developed (Ball 1923, 1928) and subsequently modified (Ball & Olson 1957) are still very widely used in industry. Here we will use the accepted nomenclature, together with Fahrenheit units.

The method makes use of the straight-line heating section of the semi-logarithmic heat penetration curve discussed in Chapter 5. The following equation is used to model the temperature distribution during heating:

$$\log(T_1 - T) = (-t/f_h) + \log[j(T_1 - T_0)], \qquad (6.23)$$

where T is the required temperature distribution, T_1 is the processing (retort) temperature, T_0 is the initial temperature of the can contents, t is the time and f_h and j are the heat penetration parameters. The value of $T_1 - T$ at the end of heating is referred to as g, and $\log g$ is given by the equation

$$\log g = (-t_B/f_h) + \log[j(T_1 - T_0)], \tag{6.24}$$

where t_B is referred to as the process time. Substituting equation (6.24) in the lethality equation (6.1) and integrating (see Ball & Olson 1957) gives

$$F = \frac{f_h\, e^{g/z}}{t}[Ei(-80/z) - Ei(-g/z)], \tag{6.25}$$

where $-Ei(-x)$ is the exponential integral also denoted by $E_1(x)$ and defined in Section 6.3.4, and the interval of 80°F was chosen because it was safe to assume that any temperature more than 80° lower than the processing temperature has a negligible lethal rate compared with that of the whole process. Using another hyperbolic expression for the cooling part of the curve, Ball (1923) obtained his classical equation for the whole heating and cooling process in which the total lethality of the process is given by

$$A = (f_{he}/t)\Big\{ e^{g/z_e}[Ei(-80/z_e) - Ei(-g/z_e)]$$

$$+ [0.33172e^{-0.343m/z_e} + 0.5833(z_e/m)e^{0.300m/z_e}E]$$

$$+ e^{-m/z_e}\left[Ei(0.657m/z_e) - Ei\,\frac{(m+g-80)}{z_e} \right]\Big\}, \tag{6.26}$$

where

$$f_{he} = f_h/\ln 10 = f_h/2.303$$

$$z_e = z/2.303$$

$$E = \int_{u_2}^{u_1} e^{-u}[u^2 - (0.3m/z_e)^2]^{1/2}du,$$

with $u_1 = 0.643m/z_e$ and $u_2 = 0.3m/z_e$, and m is the difference in temperature between the maximum temperature attained in the product at the end of heating, T_g, and the cooling water temperature. Ball denoted by C the term in the curly brackets, and he derived tables of C as a function of g, m and z. Although the formula given in Ball and Olson (1957) has an incorrect negative sign, due to the use of $Ei(-x)$ instead of $-Ei(-x)$, the tables presented there are the same as the ones in the original publication (Ball 1923), based on the correct $Ei(x)$. For a discussion of this problem see Flambert et al. (1977), Larkin (1989), Merson et al. (1978), Steele and Board (1979), Steele et al. (1979), and Storofos (1991). Olson and Stevens (1939) derived a series of nomographs based on the above equation for process determination.

The method developed so far allows the determination of the F-value for a process, conversion for various can sizes, based on f_h and j, and new processes can be evaluated for differing temperatures T_1 and T_0. However, to permit the calculation of process time for a given F-value, the C-equation was modified as follows. Equation (6.26) can be represented more simply in the form

$$F = f L_1 \, e^{-2.303g/z} C \tag{6.27}$$

or

$$f L_1 / F = e^{+2.303g/z} / C;$$

if we then write $F_1 = 1/L_1$ and $U = F F_1$, then

$$f/U = e^{2.303g/z}/C. \tag{6.28}$$

F_1 is equivalent to the thermal death time at the retort temperature for an organism whose F-value is unity; U is equivalent to the thermal death time at 250°F for an organism whose thermal death time is F min. Extensive tables for f/U versus g are given in Ball and Olson (1957) for various z-values and differing values of $m + g$. For the purpose of illustrating the use of the method a table of f/U versus $\log g$ is given in Table 6.3 for $z = 18°F$. For tables in degrees Celsius and examples of application to determine F-values, see Hayakawa and Downes (1981).

Various modifications to the Ball method have been presented in the literature; these include the new formula method (Ball & Olson 1957) and Pflug's (1968) method, which uses Hick's (1958) tables.

The process time, t_p, for a commercial operation is measured from the time when the retort reaches processing temperature, T_1, to the time when the steam is turned off and the cooling water applied. However, there is a significant time for the retort to reach processing temperature, which makes a contribution to the total lethal effect; this is known as the "come-up time." or CUT, t_c. Ball (1923) determined a value of $0.42t_c$ for the contribution to the lethal effect, so that the effective process time, t_B, was given by

$$t_B = t_p + 0.42t_c. \tag{6.29}$$

The figure of 42% is generally regarded as a conservative estimate and is really only applicable to batch retorts with a linear heating profile. While the lethal effects of CUT at the product center of a container are small for most cans, with thin profile plastic packages the effect of CUT could be more significant. Spinak and Wiley (1982) that found that CUT effectiveness varied from 35–77% and Ramaswamy and Tung (1986) found that the effectiveness factor of 42% was very conservative for thin profile packages. Ramaswamy (1993) again using thin profile retort packages, and two retort temperature profiles, one linear and the other logarithmic, showed that the traditional 42% CUT was appropriate for the former, but for the latter the values were twice as large. Apart from package thickness, other factors had only a small influence on the CUT. For other types of retort, initial conditions and venting procedures, the reader should consult Alstrand and Benjamin (1949), Berry (1983), Ikegami (1974a, b), Succar and Hayakawa (1982), and Uno and Hayakawa (1981).

TABLE 6.3. Values of f_h/U for given $\log g$ ($z = 18°$F), based on Ball (1928).

Log g	.00	.01	.02	.03	.04	.05	.06	.07	.08	.09
−2.00	0.369	0.368	0.366	0.365	0.364	0.362	0.361	0.360	0.358	0.357
−1.90	0.383	0.382	0.380	0.379	0.377	0.376	0.375	0.373	0.372	0.370
−1.80	0.398	0.397	0.395	0.394	0.392	0.391	0.389	0.388	0.386	0.385
−1.70	0.415	0.413	0.412	0.410	0.408	0.407	0.405	0.403	0.402	0.400
−1.60	0.433	0.431	0.429	0.427	0.426	0.424	0.422	0.420	0.418	0.417
−1.50	0.452	0.450	0.448	0.446	0.444	0.442	0.441	0.439	0.437	0.435
−1.40	0.474	0.472	0.469	0.467	0.465	0.463	0.461	0.459	0.457	0.455
−1.30	0.498	0.495	0.493	0.490	0.488	0.485	0.483	0.481	0.478	0.476
−1.20	0.524	0.521	0.518	0.515	0.513	0.510	0.508	0.505	0.503	0.500
−1.10	0.552	0.549	0.546	0.543	0.541	0.538	0.535	0.532	0.529	0.526
−1.00	0.585	0.581	0.578	0.575	0.571	0.568	0.565	0.562	0.559	0.556
−0.90	0.623	0.618	0.614	0.609	0.604	0.599	0.599	0.595	0.592	0.588
−0.80	0.670	0.665	0.660	0.656	0.651	0.647	0.642	0.637	0.633	0.628
−0.70	0.717	0.712	0.708	0.703	0.698	0.693	0.689	0.684	0.679	0.674
−0.60	0.769	0.763	0.758	0.753	0.747	0.742	0.737	0.732	0.727	0.722
−0.50	0.827	0.820	0.814	0.808	0.802	0.797	0.791	0.785	0.780	0.774
−0.40	0.894	0.887	0.880	0.873	0.866	0.859	0.852	0.846	0.839	0.833
−0.30	0.974	0.966	0.957	0.949	0.940	0.932	0.924	0.917	0.909	0.901
−0.20	1.071	1.060	1.050	1.040	1.030	1.020	1.011	1.001	0.992	0.983
−0.10	1.187	1.175	1.162	1.150	1.138	1.126	1.115	1.103	1.092	1.081
−0.00		1.314	1.299	1.284	1.269	1.255	1.241	1.227	1.214	1.200
0.00	1.330	1.346	1.362	1.379	1.396	1.413	1.431	1.449	1.468	1.487
0.10	1.506	1.526	1.546	1.567	1.588	1.610	1.632	1.654	1.678	1.701
0.20	1.726	1.751	1.776	1.803	1.829	1.857	1.885	1.914	1.944	1.974
0.30	2.005	2.037	2.070	2.104	2.139	2.174	2.211	2.248	2.287	2.327
0.40	2.368	2.410	2.454	2.498	2.544	2.592	2.640	2.691	2.743	2.796
0.50	2.852	2.909	2.968	3.029	3.092	3.157	3.224	3.294	3.366	3.441
0.60	3.519	3.599	3.682	3.768	3.858	3.951	4.048	4.148	4.252	4.361
0.70	4.474	4.591	4.714	4.842	4.975	5.113	5.258	5.410	5.568	5.733
0.80	5.906	6.087	6.276	6.474	6.682	6.901	7.130	7.370	7.623	7.889
0.90	8.168	8.463	8.773	9.100	9.445	9.809	10.194	10.600	11.031	11.486
1.00	11.969	12.481	13.024	13.601	14.213	14.865	15.560	16.299	17.088	17.929
1.10	18.828	19.789	20.817	21.919	23.100	24.367	25.729	27.192	28.767	30.465
1.20	32.295	34.271	36.407	38.718	41.221	43.935	46.881	50.083	53.568	57.364
1.30	61.505	66.028	70.974	76.389	82.326	88.843	96.008	103.895	112.589	122.187
1.40	132.8	144.5	157.6	172.0	188.1	206.0	226.0	248.4	273.4	301.4
1.50	332.9	368.3	408.2	453.3	504.3	562.0	626.6	702.2	787.1	884.2

6.4.2.2. Application of the Method: Straight-line Heating

The examples here and in Section 6.4.2.4 show how the process lethality can be calculated for four different cases.

Example 1: *Calculation of F_0 Value for a Straight-Line Heating Profile.* To use this method, the values of f_h and j must either be known from heat penetration experiments or assumed for the given can size and contents. In this example we have $f_h = 50$ min and $j = 1.48$; in addition, we are given a retort temperature $T_1 = 245°$F, an initial temperature $T_0 = 162°$F, a process time $t_B = 85$ min, and we will take the microbial inactivation factor z to be $18°$F and $m + g = 180°$F.

The lethal rate L for the processing temperature T_1 is obtained from

$$L = 10^{(T-250)/z} = 10^{-5/18} = 0.53.$$

We then calculate the difference between the retort and initial temperatures

$$T_1 - T_0 = 83°\,F,$$

multiply this difference by j

$$j(T_1 - T_0) = 1.48 \times 83 = 122.84,$$

and take logs

$$\log[j(T_1 - T_0)] = 2.089.$$

We can calculate $\log g$ from

$$\log g = \log[j(T_1 - T_0)] - t_B/f_h$$
$$= 2.089 - 85/50 = 0.389,$$

which, using Table 6.3 and interpolation, gives

$$f_h/U = 2.32.$$

We can now calculate F_0 from equations (6.24) and (6.25) in the form

$$F_0 = (f_h \times L)/f_h/U$$
$$= (50 \times 0.53)/2.32 = 11.42 \,\text{min}.$$

Example 2: *Calculation of Process Time t_B for a Straight-Line Heating Profile.*
In this example we are given the required value of F_0 (10 min) and wish to calculate the processing time to achieve it. We now have $f_h = 52$ min and $j = 1.78$; in addition, we are given a retort temperature $T_1 = 245°\,F$, an initial temperature $T_0 = 100°\,F$, and we will take the microbial inactivation factor z to be $18°F$ and $m + g = 180°\,F$ as before.

The lethal rate L for the processing temperature T_1 is obtained, as in Example 1, from

$$L = 10^{(T-250)/z} = 10^{-5/18} = 0.53.$$

The difference between the retort and initial temperatures is

$$T_1 - T_0 = 145°\,F,$$

which we again multiply by j

$$j(T_1 - T_0) = 1.78 \times 145 = 258.1;$$

taking logs gives

$$\log[j(T_1 - T_0)] = 2.41.$$

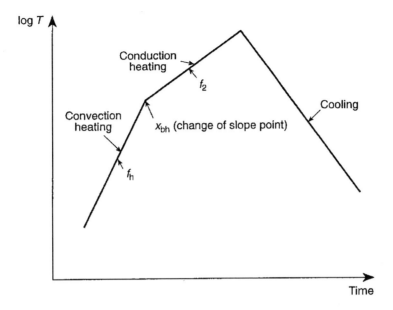

FIGURE 6.9. Semi-logarithmic curve representing broken heating conditions.

We now rearrange the equation for F_0 as used in Example 1, to give

$$f_h/U = (f_h \times L)/F_0$$
$$= (52 \times 0.53)/10 = 2.756;$$

using Table 6.3 and interpolation, we obtain

$$\log g = 0.483.$$

Finally, we calculate t_B from the expression

$$t_B = f_h \log\{[j(T_1 - T_0)] - \log g\}$$
$$= 52(2.41 - 0.483) = 100.2 \, \text{min}.$$

6.4.2.3. Products Exhibiting Broken-Heating Curves

Some packs heat quickly during the initial stages and then, because of the thickening of the constituents, the process slows down. The heating curve, plotted on a semi-logarithmic basis, for such products often shows two lines of differing slope, known as a broken-heating curve (see Figure 6.9). In order to calculate processes for these cases, it is necessary to know: the j-value; the f-value for the first part of the heating curve, f_h; the f-value for the second part of the heating curve, f_2; and the time at the intersection of the two lines, x_{bh}, taking into account the zero corrected for come-up time allowance. The method for process evaluation of the broken-heating curve pack given here is an extension of the Ball's method as developed by the American Can Co. (1950).

TABLE 6.4. Values of r_{bh} and log g for broken-heating curve calculations.

log g	r_{bh}	log g	r_{bh}
−1.0	0.947	1.05	0.710
−0.9	0.944	1.10	0.692
−0.8	0.941	1.15	0.675
−0.7	0.939	1.20	0.655
−0.6	0.934	1.25	0.635
−0.5	0.930	1.30	0.610
−0.4	0.927	1.35	0.590
−0.3	0.920	1.40	0.565
−0.2	0.915	1.45	0.540
−0.1	0.910	1.50	0.510
0.0	0.900	1.55	0.480
0.1	0.891	1.60	0.450
0.2	0.880	1.65	0.423
0.3	0.870	1.70	0.400
0.4	0.855	1.75	0.380
0.5	0.840	1.80	0.350
0.6	0.822	1.85	0.330
0.7	0.800	1.90	0.300
0.8	0.780	1.95	0.265
0.9	0.750	2.00	0.240
1.0	0.725		

The equations required for calculating the F_0-value are as follows:

$$\log g_{bh} = \log[j(T_1 - T_0)] - x_{bh}/f_h \tag{6.30}$$

$$\log g_{h2} = (f_h \log[j(T_1 - T_0)] + (f_2 - f_h) \log g_{bh} - t_B)/f_2 \tag{6.31}$$

$$F_0 = \frac{f_2 L}{(f_h/U_{h2})} - \frac{r_{bh} L(f_2 - f_h)}{f_h/U_{bh}}. \tag{6.32}$$

In equation (6.30) all the values required to calculate log g_{bh} are known from the heat penetration data, f_h, j and x_{bh}, along with the retort temperature, T_1, and the initial temperature of the can contents, T_0. The value of g_{bh} represents the number of degrees below retort temperature the can "center" point is at the intersection of the lines and, using the value f_h/U_{bh}, can be obtained from Table 6.3, for use in equation (6.32). The value of g_{h2} is calculated using equation (6.31) and represents the number of degrees the can "center" point is below the processing temperature at the end of the process. The value of f_h/U_{h2} is obtained from Table 6.3, for use in equation (6.32). The value of r_{bh} is obtained from Table 6.4.

The equation for log g_{h2} is based on the condition that $f_c = f_2$; if, however, $f_c = f_h$, then it is necessary to correct the value of log g_{h2} by adding the term $0.071(1 - f_c/f_2)$.

The equations required for calculating the process time, t_B, are as follows:

$$\log g_{bh} = \log[j(T_1 - T_0)] - x_{bh}/f_h \tag{6.30}$$

$$f_h/U_{h2} = f_2/[F_0/L + r_{bh}(f_2 - f_h)/(f_h/U_{bh})] \tag{6.33}$$

$$t_B = f_h \log[j(T_1 - T_0)] + (f_2 - f_h) \log g_{bh} - f_2 \log g_{h2}. \tag{6.34}$$

The value of $\log g_{bh}$ is calculated as before using equation (6.31), which then allows the values of f_h/U_{bh} and r_{bh} to be found from Tables 6.3 and 6.4. The value of $\log g_{h2}$ is then found from the value of f_h/U_{h2} calculated in equation (6.33). All the factors are then known for calculating t_B using equation (6.34).

If $f_c = f_2$, then the value of $\log g_{h2}$ is obtained directly from the value of f_h/U_{h2}; if $f_c = f_h$, then the correction factor discussed above, $0.071(1 - f_c/f_2)$, must be subtracted from the value of $\log g_{h2}$.

The examples in Section 6.4.2.4 illustrate the application of these equations.

The method has been computerized and FORTRAN programs published by Griffin *et al.* (1969). A method of establishing thermal processes for products with broken-heating curves from data taken at other retort temperatures and initial temperatures has been developed by Berry and Bush (1987).

6.4.2.4. Application to Broken-Heating Curves

Example 3: *Calculation of F_0 Value for a Broken-Heating Curve Process.* In this example we have $f_h = 14$ min and $j = 1.5$; other heat penetration factors are $f_2 = 68$ and $x_{bh} = 15$ min. In addition, we are given a retort temperature $T_1 = 240°$ F, an initial temperature $T_0 = 162°$ F, a process time $t_B = 52$ min; and we will again take the microbial inactivation factor z to be $18°$F and $m + g = 180°$ F.

We begin by calculating the difference between the retort and initial temperatures

$$T_1 - T_0 = 78° \mathrm{F},$$

and multiply this difference by j

$$j(T_1 - T_0) = 1.5 \times 78 = 117,$$

and take logs

$$\log[j(T_1 - T_0)] = 2.068.$$

We now calculate $\log g_{bh}$ from equation (6.30),

$$\log g_{bh} = \log[j(T_1 - T_0)] - x_{bh}/f_h$$

$$= 2.06889 - 15/14 = 0.996.$$

Based on this value of $\log g_{bh}$, we can use Table 6.3 and interpolation to derive an estimate for f_h/U_{bh} of 11.6; and we can use Table 6.4 and interpolation to estimate r_{bh} at 0.73.

We now calculate $\log g_{h2}$ using equation (6.31), adding a correction factor, to give

$$\log g_{h2} = \{f_h \log[j(T_1 - T_0)] + (f_2 - f_h) \log g_{bh} - t_B\}/f_2 + 0.071(1 - f_h/f_2)$$

$$= 0.487,$$

and use Table 6.3 to estimate f_h/U_{h2} at 2.78.

The lethal rate L for the processing temperature T_1 is obtained from

$$L = 10^{(T-250)/z} = 10^{-10/18} = 0.278.$$

This gives us the last piece of information we need. We can now calculate F_0 by substitution into equation (6.32):

$$F_0 = 5.86 \text{ min}.$$

Example 4: *Calculation of Process Time t_B for a Broken-Heating Curve Process.* In this example we are given $F_0 = 7$ min, and wish to calculate the processing time to achieve it. We now have $f_h = 12.5$ min and $f_2 = 70$; other factors are $j = 1.5$ and $x_{bh} = 15$ min as in Example 3. In addition, we are given a retort temperature $T_1 = 235°F$, an initial temperature $T_0 = 180°F$, and we will again take the microbial inactivation factor z to be $18°F$ and $m + g = 180°F$.

We begin by proceeding exactly as in Example 3. Thus,

$$T_1 - T_0 = 55°\text{F}$$

$$j(T_1 - T_0) = 1.5 \times 55 = 82.5$$

$$\log[j(T_1 - T_0)] = 1.92$$

$$\log g_{bh} = \log[j(T_1 - T_0)] - x_{bh}/f_h$$

$$= 1.92 - 15/12.5 = 0.72,$$

using the last value to estimate $f_h/U_{bh} = 4.714$ using Table 6.3 and $r_{bh} = 0.796$ using Table 6.3.

The lethal rate L for the processing temperature T_1 is obtained exactly as in Example 3 and gives the value, $L = 0.1468$. We now use equation (6.33) to calculate $f_h/U_{h2} = 1.843$. Then we calculate $\log g_{h2}$ using and interpolating from Table (6.3), substracting a correction factor, to give $\log g_{h2} = 0.245 - 0.0583 = 0.18668$. Finally, we substitute these values into equation (6.34) to give

$$t_B = 52.3 \text{ min}.$$

6.4.2.5. Comments on Ball's methods

Despite the widespread use of the Ball (1923) procedures, it is important to note the following points:

1. The method applies to a j_c-value of 1.41.
2. The curvilinear portion of the cooling curve stops at $t_c = 0.141 f_c$.
3. The method overestimates the F-value when $j_c < 1.41$ and, conversely, underestimates the F-value when $j_c > 1.41$.
4. The cooling phase treatment is less satisfactory than some of those developed by other workers.

These points should be taken into account, especially when processes with long come-up times are being considered, e.g. flexible pouches and rectangular-shaped plastic containers of food. In these cases it is important to use one of the general methods to determine the process lethality for the heating and cooling phases. The Ball method tends to underestimate the F_{total}-value.

6.4.3. Gillespy's Method

6.4.3.1. Standard Heating and Cooling Conditions

In his two classical papers, Gillespy (1951, 1953) published methods of determining process lethality for standard and complex heating profiles. The standard method (Gillespy 1951) has been in use for many years in the UK canning industry and is well established in industrial training programs. The simplified methodology with its computer algorithms did not appear until published by Board and Steel (1978).

The method is based on Ball's asymptotic approximation of the heating, which uses a dimensionless time t equal to the process time divided by the f_h-value:

$$T_R - T = j(T_R - T_0)10^{-t}. \tag{6.35}$$

The F-value for the process is obtained from $F = LU$, where L is the lethal rate at processing temperature and U is value of the process in terms of time at retort temperature. In Gillespy's notation the classical U/f is denoted by u and is obtained from the equation

$$u_H = (w - s) - \mu\gamma + \log\mu + \sum_{n=1}^{\infty} \frac{(-\mu)^{-(n-1)}10^{-n(w-s)}}{n \cdot n!} \tag{6.36}$$

where w is the processing time corrected for "come up" divided by f_h, and s is the time, in units for f, for the temperature on the asymptote to the heating curve to rise from T_A to T_S, where $T_S = T_R - z$, and $s = (T_R - T_A)/z = \log[(T_R - T_0)/z] + \log j$, $\mu = \log_{10} e = 0.4343$ and $\gamma = 0.5772$ (Euler's constant). Equation (6.36) is equivalent to the exponential integral form for heating given in equation (6.21). The term $w - s$, also written as υ, can be obtained from Table 6.5. For estimating the cooling contribution it was assumed that $f_h = f_c$. Board et al. (1960) showed that Gillespy's method often overestimated the contribution of the cooling phase. This was especially for the case of conduction-heating packs, which cooled slower than Gillespy anticipated. In the simplified version, the factor for estimating the cooling phase was replaced by a single constant 0.08, and Gillespy's equation became

$$\upsilon = B/f_h - \log_{10}[j(T_R - T_0)/z] + 0.08;$$

for a z-value of 18°F (i.e. $\log_{10} z = 1.255$), this equation is often written

$$\upsilon = B/f_h - \log_{10}[j(T_R - T_0)/z] + 1.335. \tag{6.36b}$$

TABLE 6.5. Values of u for v ranging from -0.2 to 3.49 (Gillespy's method).

v	0.00	0.02	0.04	0.06	0.08
-0.2	0.003	0.003	0.004	0.005	0.005
-0.1	0.006	0.008	0.009	0.010	0.012
0.0	0.014	0.016	0.019	0.021	0.024
0.1	0.027	0.030	0.034	0.038	0.042
0.2	0.047	0.051	0.057	0.062	0.068
0.3	0.074	0.080	0.087	0.094	0.102
0.4	0.110	0.118	0.126	0.135	0.144
0.5	0.154	0.164	0.174	0.184	0.195
0.6	0.206	0.218	0.229	0.241	0.253
0.7	0.266	0.279	0.292	0.305	0.318
0.8	0.332	0.346	0.360	0.375	0.390
0.9	0.404	0.419	0.435	0.450	0.466
1.0	0.482	0.498	0.514	0.530	0.546
1.1	0.563	0.580	0.597	0.614	0.631
1.2	0.648	0.665	0.683	0.700	0.718
1.3	0.736	0.754	0.772	0.790	0.808
1.4	0.826	0.844	0.863	0.881	0.900
1.5	0.918	0.937	0.955	0.974	0.993
1.6	1.012	1.031	1.050	1.069	1.088
1.7	1.126	1.126	1.145	1.164	1.184
1.8	1.203	1.222	1.241	1.261	1.280
1.9	1.300	1.319	1.339	1.358	1.378
2.0	1.397	1.417	1.436	1.456	1.475
2.1	1.495	1.515	1.534	1.554	1.574
2.2	1.593	1.613	1.633	1.653	1.672
2.3	1.692	1.712	1.732	1.751	1.771
2.4	1.791	1.811	1.831	1.851	1.870
2.5	1.890	1.910	1.930	1.950	1.970
2.6	1.990	2.009	2.029	2.049	2.069
2.7	2.089	2.109	2.129	2.149	2.169
2.8	2.189	2.209	2.229	2.248	2.268
2.9	2.288	2.308	2.328	2.348	2.368
3.0	2.388	2.408	2.428	2.448	2.468
3.1	2.488	2.508	2.528	2.548	2.568
3.2	2.588	2.608	2.628	2.648	2.668
3.3	2.688	2.708	2.728	2.748	2.768
3.4	2.788	2.807	2.827	2.847	2.867

The process time B can be estimated from

$$B = f_h\{v + \log_{10}[j(T_R - T_0)] + 1.335\}. \qquad (6.36c)$$

6.4.3.2. Examples of the Use of Gillespie's Simple Method

Example 5: *Determination of the F-Value for a process.* In this example we have an initial temperature $T_0 = 140°$ F and a process temperature $T_R = 240°$ F; the heat penetration factors are $f_h = 53$ min and $j = 1.30$; and the processing time corrected for come-up is $B = 84$ min.

We begin by calculating υ from equation (6.36b):

$$\upsilon = B/f_h - \log_{10}[j(T_R - T_0)/z] + 1.335$$
$$= 84/53 - \log(1.3 \times 100) + 1.335$$
$$= 0.805 \text{(heating and cooling)}.$$

Table 6.5 (and interpolation) converts this υ-value into a u-value of 0.332. F_0 is then calculated from

$$F_0 = uf_h L_{TR} = 0.332 \times 53 \times 10^{(240-250)/18} = 4.72 \text{ min}.$$

For heating only, the value of 1.335 in the equation for υ should be replaced by 1.255; then we would obtain $\upsilon = 0.725$ and $u = 0.283$, and hence $F_0 = 4.17$ min.

Note that for unknown j, a value of 1 should be assumed for a convection pack and 2 for a conduction pack; two-thirds of the come-up time of the retort from 212°F to the retort temperature should be counted as part of the process time B.

Note also that for z-values other than 18°F, it is necessary to substitute $\log z + 0.08$ as the final term in equation (6.36b).

Example 6: *Determination of Process Time B for a Given F_0-Value.* Using the same initial temperature $T_0 = 140°$F, process temperature $T_R = 240°$F, heat penetration factors $f_h = 53$ min and $j = 1.30$ as in Example 5, we now wish to find the processing time for an F_0-value of 5.

We must first determine u:

$$u = F_0/f_h L = 5/(53 \times 0.278) = 0.339.$$

Table 6.5 (and interpolation) converts this to a υ-value of 0.810. We can now use equation (6.36c) to calculate the process time:

$$B = 53(0.810 + 2.114 - 1.335) = 84.2 \text{ min}.$$

Note that if a j-value has been assumed, the come-up time correction must be deducted to obtain the operational time.

6.4.3.3. Some Comparisons with Other Methods

Board *et al.* (1960) made an extensive analysis of several methods of process determination, including the general method, Ball's formula method and Gillespy's method for a variety of products. Whereas all methods gave similar results for the heating phase, the cooling lethalities were considerably different. Ball's method gave figures which were much lower than the general method and significantly lower than Gillespy's, for example (see Table 6.6).

Gillespy's method generally agrees with the general method when cooling is by pure conduction. However, in products which cool more rapidly due to a combination of conductive cooling and convective cooling, Gillespy's method tends to overestimate the cooling contribution to the lethality. Using a computerized

TABLE 6.6. Comparison of F-values for cooling phase for static processing (from Board *et al.* 1960)

Product	Can size	F_c-Value for cooling phase		
		General method	Ball method	Gillespy method
Spaghetti bolognaise	301 × 411	1.47	0.63	2.29
Baked beans	603 × 700	1.50	1.20	4.02
Vegetable soup	301 × 411	0.323	0.106	0.305
Cream style corn	401 × 411	1.89	0.62	1.79
Meat loaf	401 × 411	1.46	0.51	1.54
Tomato pulp	401 × 411	1.11	0.60	1.63

version of Gillespy's method, Board and Steel (1978) showed that F-values for the total process were generally lower than those obtained by the general method. It was concluded that the mechanism of internal cooling in the can was mainly responsible for the discrepancies. Bown (1989) showed that Gillespy's (1951) method gave very close results to a numerical finite-difference model, which was developed for process sensitivity studies.

6.4.3.4. Complex Heating Profiles

Gillespy (1953) developed a method of process determination for cases where the heating profile was time-dependent, e.g. linear or exponential. Relatively little use has been made of this method, although it was used to determine processes in the heating and cooling legs of hydrostatic cookers at a time when it was believed that the temperature gradients in the legs were linear functions of time. It has subsequently been shown that the conditions are somewhat different. The method is applicable to packs which have a slow "come up," e.g. conduction-heating products in flexible pouches or plastic containers.

The method used the usual step rise in temperature and Duhamel's theorem to determine the response to a time-dependent heating profile. Using the new temperature distribution, the F-value was calculated at the point of slowest heating. The temperature at the center of the can being processed was given by

$$T = T_R - \Phi(t)(T_{\text{ref}} - T_0) \tag{6.37}$$

and the variation of T_R with time t by

$$T_R = T_0 - \Theta(t)(T_{\text{ref}} - T_0). \tag{6.38}$$

Using Duhamel's theorem, the temperature response to the new conditions is given by

$$T = T_R - (T_{\text{ref}} - T_0) \int \Phi(t - \lambda)\Theta'(\lambda)\,d\lambda. \tag{6.39}$$

In the Gillespy notation a jump in temperature is designated $J(t)$, and a break $B(t)$. The heating and cooling profile can then be formulated in terms of a series of jumps, breaks or holds. The method can be illustrated by reference to determining the fraction of the come-up time, which is equivalent in lethal effect to time at process temperature, for a slab of material being heated in a water retort with a slowly rising linear gradient temperature. The temperature history is as follows:

1. For $t < 0$, retort temperature $T_R = T_0$, the slab initial temperature.
2. At $t = 0$, the retort temperature rises instantaneously to T_W, the water temperature at the start, a jump of $(T_W - T_0)$.
3. For $0 < t < t_c$, the retort temperature rises linearly from T_W to processing temperature T_R, a break of $(T_R - T_W)/t_c$.
4. For $t > t_c$, the retort temperature is T_R, a break of $-(T_R - T_W)/t_c$.

The pack center temperature is then given, for $0 \le t \le t_c$, by

$$T = T_W + t(T_R - T_W)/t_c - J(t)(T_W - T_0) - B(t)(T_R - T_W), \quad (6.40)$$

$$T = T_R - J(t)(T_W - T_0) - B(t)(T_R - T_W)/t_c$$

$$+ B(t - t_c)(T_R - T_W)/t_c. \quad (6.41)$$

For a slab,

$$J(t) = \frac{4}{\pi}\left(e^{-\pi^2 x} - \frac{1}{3}e^{-9\pi^2 x} + \frac{1}{5}e^{-25\pi^2 x} - \cdots\right), \quad (6.42)$$

$$B(t) = \int_0^z J(z)\,dz$$

$$= \frac{4L^2}{\alpha\pi^3}\left(\frac{\pi^3}{32} - e^{-\pi^2 x} + \frac{1}{27}e^{-9\pi^2 x} - \frac{1}{125}e^{-25\pi^2 x} + \cdots\right) \quad (6.43)$$

Values of $J(t)$ and $B(t)$ can readily be evaluated, knowing the thermal diffusivity, α, and slab thickness, L, for various times t, where $x = t/L^2$. Thus the time–temperature history may be calculated and the F-value determined by a numerical technique (Patashnik 1953). To determine the process time necessary to achieve a desired F-value with no come-up time, i.e. at $t = 0$, note that the retort temperature rises instantaneously from T_0 to T_R. Thus if a particular process of 10 min come-up time $+20$ min at T_R gives an F_0-value of 4, and it is calculated that 27 min at the same T_R are required to deliver the same F_0-value without come-up time, then 7 min of come-up time is equivalent to time at the same T_R, i.e. 70%.

Hayakawa (1974) developed a method of estimating temperature histories for products undergoing time-variable processing temperatures. He made a comparative study of his method with Gillespy's without much success, and suggested that there might be errors in Gillespy's tables, without identifying them. He considered that Gillespy's method underestimates food temperatures in the heating phase when $j < 2.04$ and, conversely, overestimates temperatures when $j > 2.04$.

More recently, Noronha et al. (1995) have produced a semi-empirical method for determining the effect of time-variable boundary conditions on process values and compared this with other methods, using eight different processes. Most of the methods compared reasonably well, except for one or two anomalies.

6.4.4. Hayakawa's Method

Hayakawa (1970, 1971) developed an important method of process determination by producing a series of empirical equations to represent the various parts of the heating and cooling curves. The formulae were further modified (Hayakawa 1983c) and presented as follows:

1. Curvilinear portion of the heating curve:
   (a) for $0.001 \leq j \leq 0.4$,

$$T_1 - T = (T_1 - T_0)10^{-t/Bn},\qquad(6.44)$$

where

$$n = (t/f - \log j)(t_1/f)\qquad(6.45)$$

$$B = t_1(t_1/f - \log j)^{-n};\qquad(6.46)$$

   (b) for $0.4 \leq j < 1$,

$$T_1 - T = (T_1 - T_0)\cot[Bt + \pi/4]\qquad 0 \leq t \leq t_1\qquad(6.47)$$

where

$$B = \frac{1}{t_1}\left\{\arctan\left[\frac{\log(T_1 - T_0)}{\log[j(T_1 - T_0)] - t_1/f}\right] - \frac{\pi}{4}\right\}\qquad(6.48)$$

$$t_1 = 0.9f(1 - j);$$

   (c) for $j > 1$

$$T_1 - T = (T_1 - T_0)\cos Bt\qquad(6.49)$$

$$B = \frac{1}{t_1}\left\{\arccos\left[\frac{\log j(T_1 - T_0) - t_1/f}{\log(T_1 - T_0)}\right]\right\}.\qquad(6.50)$$

2. Linear portion of heating:

$$T_1 - T = j(T_1 - T_0)10^{-t/f}\quad\text{for } t_1 \leq t.\qquad(6.51)$$

3. Curvilinear portion of the cooling curve: Use $f = f_c$, $j = j_c$, $T_1 - T = T - T_W$, $T_1 - T_0 = T_g - T_W$ and $t = t_c$ in the heating equations.
4. Linear portion of the cooling curve: Use the same substitutions in the linear heating curve equation (6.51).

The method involves the use of universal tables of U/f versus g/K_s, which can be used for converting from one z-value to another, since $K_s = z_b/z_a$. The general procedure is similar to that of the Ball method but enables a much better evaluation of the cooling contribution. The method is considered one of the best available for covering both the heating and the cooling ranges. Some comparative data are given in Table 6.7.

TABLE 6.7. Comparison of lethalities calculated by the general method and formula methods.

Product	Size	General method	Ball (1923) method	Gillespy (1951) method	Hayakawa (1970) method	Ball and Olson (1957)	Pflug (1968) method	Griffin (1971) method	Jakobsen (1951) method	Reference
Spaghetti bolognaise	301 × 411	1.47	cooling only 0.63	2.29						Board et al. (1960)
Baked beans	603 × 700	1.50	1.20	4.02						
Vegetable soup	301 × 411	0.32	0.11	0.30						
Cream style corn	401 × 411	1.89	0.62	1.79						
Meat loaf	401 × 411	1.46	0.51	1.54						
Tomato pulp	401 × 411	1.11	0.60	1.63						
Chicken chop suey	rectangular	9.60	heating and cooling 7.00	6.68						Unpublished experimental data
Vegetable curry	rectangular	9.00	10.70	5.60						
Sweet and sour pork	rectangular	9.3	9.85	10.30						
Chicken chop suey	rectangular	7.00	heating only 4.80	4.90						Unpublished experimental data
Vegetable curry	rectangular	7.20	8.60	8.60						
Sweet and sour pork	rectangular	7.50	8.20	8.50						
Meat and cereal	3 kg	6.10	heating only —	4.30						Board and Steel (1978)

(Cont.)

TABLE 6.7. (Continued)

Product	Size	General method	Ball (1923) method	Gillespy (1951) method	Hayakawa (1970) method	Ball and Olson (1957)	Pflug (1968) method	Griffin (1971) method	Jakobsen (1951) method	Reference
Ham	1 kg	0.80	–	0.30						
Meat and cereal	0.3 kg	11.72	–	10.35						
Small abalone	0.5 kg	3.70	–	3.70						
Cream	0.2 kg	1.20	–	1.10						
			cooling only							
Potato soup	211 × 400	0.9	–	–	1.1	3.1	3.0			Hayakawa (1970)
Shrimp soup	211 × 400	1.5	–	–	1.4	4.4	3.9			
Vegetable/beef soup	211 × 400	0.5	–	–	0.7	4.1	3.7			
			heating and cooling							
Potato soup	211 × 400	10.5	–	–	10.7	13.6	11.7			Hayakawa (1970)
Shrimp soup	211 × 400	8.4	–	–	8.4	11.3	9.9			
Vegetable/beef soup	211 × 400	6.5	–	–	7.0	10.2	10.0			
			cooling only							
Theoretical (Hicks 1951)	11.6 × 10 cm	2.44	0.89	–	1.38	–	–	0.94	1.18	Skinner and Jowitt (1977)
Not specified TP1 (simple heating)		1.15	0.78	–	0.83	–	–	0.62	1.04	
Not specified TP2 (broken heating)		0.16	0.19	–	0.17	–	–	0.13	0.25	

6.4.5. Other Methods

Table 6.8 gives a wide range of formula methods which have been proposed; however, many of these have not so far been adopted by industry and are mainly of research and teaching interest. Table 6.6 gives some comparative data on different methods; however, very few of these comparisons cover a wide range of processing parameter. Each method should be judged on its own merits. Smith and Tung (1982) studied the percentage errors in process lethality for five formula methods, applied to conduction-heating packs. The degree of error in descending order was: Ball and Olson (1957) > Ball (1923) > Steele and Board (1979) > Hayakawa (1970) > Stumbo (1973). It was found that the error was a function of the H/D ratio of the can for all the methods, with a maximum value at $H/D = 1$. The error also increased with increasing values of g, the temperature difference between the centre of the can and the retort temperature at the end of cooking; thus greater errors were experienced with higher retorting temperatures ($140°C$) than with more conventional temperatures ($120°C$). Stumbo's method was found to be the one with least errors; however, it was found to be more sensitive to slight variations in f_h and j_c. Pham (1987) showed that his method was as good as the best of the above (Stumbo's method), and extended it to cover the cases where f_h and f_c were not the same (Pham 1990), with similar success. Since all the methods showed underestimates of the process lethality, the calculated process times would be larger than necessary and hence introduce a margin of safety. However, since the errors were dependent on the processing conditions, the extra safety margins would be variable.

The above applies to conduction-heating packs, where the value of g can be quite high at the end of a process. For convection-heating packs, where temperatures are more uniform, especially in agitated packs, process calculations do not involve such high errors.

Much of the earlier work was concerned with metallic cans; however, with the development of sterilizable plastic pouches and plastic trays with thin rectangular profiles, it is necessary to examine process calculation methods. This is partly because of the profile but also because of the processing methods, which involve the use of pressurized steam or hot water. These introduce further lags to the heat transfer in the system and result in relatively slow come-up times. Spinak and Wiley (1982) compared the general and the Ball formula methods and showed general satisfaction with the Ball method. Ghazala et al. (1991) examined a number of methods for estimating the lethality, including Ball (1923), Stumbo (1973), Steele et al. (1979), Pham (1987) and the authors' computerized model, including sensitivity tests for various processing parameters. Although the Pham and Stumbo methods had the lowest percentage of overall errors, the difference between these and the other methods was not particularly great— probably because the j_c-value for these packs was nearer to the assumed values in the models.

In conclusion, while many developments are taking place in computer technology that enable easy solutions to the problems, it is nonetheless important

TABLE 6.8. Some methods of process lethality determination.

Method	Reference	Comment
1. General methods	Bigelow et al. (1920)	Graphical integration of lethal rates obtained from heat penetration experiments
	Schultz and Olson (1940)	Lethal rate paper developed for $z = 18°F$
	Hayakawa (1973)	Lethal rate paper for any z-value
	Patashnik (1953)	Numerical integration – Simpson's rule applied directly is more accurate
	Hayakawa (1968)	Gaussian numerical integration – requires only a limited number of points
	Simpson et al. (2001, 2003, 2005)	A semi-analytical expression used to integrate process lethality, which was more versatile and quicker than the General method of integration
2. Formula methods (a) Empirical	Ball (1923)	First formula method published and the basis for most developments in process determination used f_h- and j-values for heating and cooling, and a hyperbola for representing the curvilinear parts of the heat penetration curve. Applicable for $j_c = 1.41$ and cooling time $t_c = 1.41 f_c$
	Ball and Olson (1957)	Simplified Ball's (1923) method using P_h and P_c factors for determining the process lethality. Applicable for $j_c = 1.41$ and cooling time $t_c = 1.41 f_c$ and for $f_h \neq f_c$. Broken-heating curve method developed
	Gillespy (1951)	Used Ball's (1923) asymptotic approximation for heating and developed an approximate method for cooling
	Jakobsen (1954)	Hyperbolic sine function used to represent temperature history; $j_c = 2.04$. This was modified by Shiga (1970)
	Hicks (1958)	Revision of P_h and P_c tables of Ball and Olson (1957)
	Herndon et al. (1968)	Computer-derived tables based on Ball (1923)
	Griffin et al. (1969)	Computer-derived tables for broken-heating curves
	Griffin et al. (1971)	Computer-derived tables for cooling curves
	Stumbo &Longley (1966)	Lethality parameters determined by hand-drawn curves and areas measured with a planimeter. Applicable to variable j_c and for $f_h \simeq f_c$
	Hayakawa (1970)	Curvilinear parts of heating and cooling curve represented by exponential cotangent and cosine functions. Applicable for $f_h \neq f_c$ and to broken-heating curves. One of the best methods in this group
	Skinner and Urwin (1971)	Computer program for Stumbo and Longley (1966) method, including algorithm for f/U and g
	Vinters et al. (1975)	Computer algorithm developed for Ball's (1923) tables

Shiga (1970, 1976)	B_h and B_c functions developed from Ball & Olson's (1957) P_h and P_c functions. Linear relation between F_h (heating lethality) and t_h^2 (heating process time)
Skinner and Jowitt (1977)	Review and comparison of Formula methods: Ball (1923); Jakobsen (1954); Griffin et al. (1969); Hayakawa (1970)
Hayakawa (1977, 1978, 1983a)	Reviews of formula method models
Downes and Hayakawa (1977)	°C tables developed
Hayakawa and Downes (1981)	New parameters for process determination based on °C
Tung and Garland (1978)	Computer method for process determination based on Stumbo (1973)
Steele and Board (1979)	Concept of dimensionless sterilizing ratios (S) developed which are independent of temperature scale used. $S = (T_{retort} - T_{centre})/z$. Uses cooling treatment developed by Griffin et al. (1971)
Kao et al. (1981)	Computer programs for process determination, which handle different z- and j_c-values. Applicable for $j_c = 0.4$ to 23 and z-values 10 to 26°F
Hayakawa (1983b)	Parametric tables of U/f against g, for $z = 10°C$, $T_0 = 70°C$ and $j = 1.0$. Uses notation of Hayakawa (1970)
Smith and Tung (1983)	Comparison of formula methods: Ball (1923); Stumbo and Longley (1966); Hayakawa (1970); Steele and Board (1979)
Spinak and Wiley (1983)	Comparison of General and Ball's (1923) formula method for retort pouches
Pham (1987)	Algebraic formulae developed for determining tables of U/f against g. Uses Smith and Tung (1982) methodology
Pham (1990)	Extension of Pham (1987) to $f_h \neq f_c$
Sablani and Shayya (2001)	Replace Stumbo's 57 tables by two neural network programs, viz., ANNG for the g-parameter and ANNF for the fh/U parameter
Larkin and Berry (1991)	Modified Ball's original method using a modified hyperbolic function for the cooling curve
Teixeira (1992)	Time–shift method

(Cont.)

Table 6.8. (Continued)

Method	Reference	Comment
	Weise and Weise (1992)	Comparison of techniques
	Chiheb et al. (1994)	A finite Nth-order linear model, with variable retort temperatures.
	Denys et al. (1996)	Semi-empirical model for broken-heating curves.
	Norohna et al. (1995)	Semi-empirical method for time-variable boundary conditions
(b) Theoretical temperature distributions	Riedel (1947)	Analytical method for dealing with complex heating and cooling profiles
	Gillespy (1953)	Extension of the Reidel method; deals with linear and exponential heating profiles. Uses Duhamel's theorem for dealing with time-variable boundary conditions
	Hicks (1951)	Analytical equations for heating and cooling. Heating phase results agree with, and cooling treatment is better than, Ball (1923)
	Teixeira et al. (1969)	Numerical solution of equations
	Flambert & Deltour (1972b)	Analytical equations for heating and cooling
	Lenz and Lund (1977)	Dimensionless treatment of microbial destruction kinetics and Arrhenius equation
	Hayakawa (1977)	Numerical integration of analytical equations
	Naveh et al. (1983a, b, c)	Numerical solution of heat transfer equation. Application to transient cooling of conduction heating products
	Bhowmik and Tandon (1987)	Finite-difference equations and numerical solution
	Ghazala et al. (1991)	Finite-difference equations, analysis of errors and application to thin packages
	Tucker (1990, 1991, 2002)	CTemp includes facilities for studying process deviations and can handle various shapes of package. Thermal diffusivities derived from f_h-values
	Teixeira et al. (1992)	Time-shift model
	Hayakawa et al. (1996)	Statistical distribution free equation applicable when cooling phase lethality is small

for the student to study the methods that have been developed to understand the background of process calculations. It is also important to check new methods for accuracy, sensitivity and applicability to the process required. With the development of newer types of processes, especially those with slow come-up times, it is essential to make sure that the asymptotic approximation, which underlies many of the methods of process calculation, is justified. Whichever method is used, it is always necessary to bear in mind the consequences of an inadequate process and the potential danger to the public. Food safety is paramount, and everyone applying these methods has a responsibility for recommending processes which are inherently safe. The ultimate responsibility rests with the correct operational procedures being applied, which is a major engineering problem.

6.5. Mass-Average Sterilizing Values

In Chapter 4 the concept of a total integrated or mass-average sterilization value, F_s, for the entire contents of a container was introduced. This was primarily to compensate for any problems which might arise if the critical point was not the point of slowest heating. This would mean that the point at which the temperature history was measured might not give the lowest F-value in the container. In Chapter 5 the evidence for this was reviewed, and it is generally concluded that for the domestic sizes of cans there is no problem, but for large cans the point of slowest heating could be displaced into a toroidal ring some distance from the axis of the can. This was because at the end of heating and the start of cooling the center temperature continues to rise for some time before the effect of the cooling water is felt. To deal with these circumstances the concept of integrating the lethal effects through the whole can volume was developed. It has been widely applied by Stumbo (1953), who introduced the method, after thoroughly reconsidering the bacteriological basis for process calculations. At the time this provoked a flurry of activity and discussion on the basis of process calculations and the need to introduce new methods (Ball 1949; Hicks 1952). Stumbo's methods are detailed in his excellent text (Stumbo 1973). His method involves the use of 57 tables, and these have been replaced by two tables using a neural network analysis (Sablani & Shayya 2001). Stumbo's method has also computerized for determining F_0-values by Germer *et al.* (1992).

An excellent review and critical analysis of the methods of process evaluation of processes has been presented by Storofos *et al.* (1997). Their work covers much of the material presented in this text, but in more detail.

A comparison of methods for treating broken heating curves has been presented by Wiese and Wiese (1992). Denys *et al.* (1996) have made a study of broken-heating using a 3–5% cereal starch suspension starch with end-over-rotation of 73×109.5 mm cans 81×172 mm glass jars in a Barriquand Steriflow process simulator. A semi-empirical method was developed for predicting the break-point in the heating curve. The method was also applicable to the study of process deviations.

Table 6.9. Some methods of calculation of mass-average sterilization values.

1. Analytical solutions:	
Gillespy (1951)	Uses first-term approximation for determining heating profile. Derives F_s-value in terms of F_c-value. See equation (4.19)
Stumbo (1953)	Uses Ball (1923) for lethality calculation. Derives F_s-value in terms of F_c-value, see equation (4.17)
Ball and Olson (1957)	Used P_h- and P_c-values to calculate local concentrations of heat-vulnerable components in each of 11 isothermal regions
Hayakawa (1969)	Dimensionless number method. Solution of heat distribution equations included all non-negligible terms
Jen et al. (1971)	Extension of Stumbo's (1953) method
Newman and Holdsworth (1989)	Revised Jen et al. (1971) figures and extended the concept to other geometrical shapes. Includes extensive computer programs in Basic
2. Numerical methods:	
Teixeira (1969)	Finite-difference equations. Applicable to variable heating and cooling medium temperatures
Flambert & Deltour (1972a)	Numerical solution. Location of slowest heating point discussed
Manson et al. (1970)	Application to brick-shaped containers
Manson et al. (1974)	Application to pear-shaped containers
Tucker and Holdsworth (1991)	Numerical model for brick-shaped containers, with finite surface heat transfer coefficients

Two outstanding papers (Gillespy 1951; Hicks 1951) produced two other methods of calculating F_s-values. Gillespy was always critical about the need for F_s-values in terms of commercial canning operations, since his method showed that F_c was always the minimum value in the whole pack. Holdsworth and Overington (1975) considered that this was due to the form of the equation for the volume integral, which did not permit Stumbo's conclusion. Other workers have also considered this approach, and Table 6.9 lists some of these.

With the development of interest in the thermal inactivation of heat-vulnerable components during the sterilization operation, these methods, using appropriate z-values, have been widely applied to optimizing processes. This is because the heat-vulnerable components are distributed throughout the volume of the food product and the concept of determining the effects of heating at the critical point only is inappropriate compared with microbial inactivation. This subject is developed more extensively in Chapter 7.

6.6. Some Factors Affecting F-Values

6.6.1. Introduction

The main factors that affect the magnitude of the F-value of a process and that are contained in the equations for determining F-values may be listed as follows:

1. Process Related Factors.
   (a) Retort temperature history, heating, holding and cooling temperatures and times: The accuracy of temperature measurements has a direct effect on

lethality values, and temperature-measuring devices need to be calibrated to a traceable standard.

(b) Heat transfer coefficients: For processes heated by steam and vigorously agitated in boiling water, these are usually so high that no effect is observed; however, with other methods of heating, which have much lower values, it is important to estimate the effective values as accurately as possible.

2. Product Related Factors.

   (a) Initial temperature of can contents: The uniformity of the initial filling temperature should be carefully controlled. The higher the initial temperature the shorter the required process time.

   (b) Pre-retorting delay temperature and time: This is related to the filling temperatures and results from malfunctioning of the process, and delays affect the initial temperature of the can contents.

   (c) Thermal diffusivity of product: Most models are very sensitive to changes in the value of this property.

   (d) z- and D-value of the target microbial species.

3. Container-Related Factors.

   (a) Container materials: Apart from tin-plate and other metallic containers, all other materials, e.g. plastics and glass, impede the heat transfer into the container.

   (b) Container shape: The most rapidly heating containers have the largest surface area and the thinnest cross-section.

   (c) Container thickness: The thicker the container wall, the slower the heating rate.

   (d) Headspace: This is of particular importance to agitated and rotary processes. The headspace and the rate of rotation need to be carefully controlled.

   (e) Container stacking: The position of the containers inside the retort and the type of stacking also affect the heat transfer to individual containers.

6.6.2. Statistical Variability of F-Values

From a consideration of all the factors that affect the F-value delivered to an individual can of food, it can be seen that there is likely to be a distribution in the F-values. Variations in heat penetration data were discussed in Chapter 5; however, here the effects of various parameters on lethality will be discussed.

One of the earliest F_0 distributions was published by Fagerson et al. (1951), who made a study of convection-heating products. It was found that the distribution was non-symmetrical rather than normal. Powers et al. (1962), for similarly heating products, found non-normal distributions, and these varied according to the product and the length of processing time. Positively skewed distributions were obtained for short processing times (less than 50 min) and negatively skewed distributions for longer times, using over 700 experimental values. Extreme-value analysis was used to determine, for a range of F_0-values, the proportion of F_0-values outside a certain confidence level. While most of variation was due to processing time, the filled weight of product also affected the F_c-values, but not the F_h- or F_0-values. The coefficient of variation (CV), the standard deviation divided by the mean, for the results varied in the range 16.3–57.4%. The extreme-value method of analysis was recommended for determining the safety margins for canned foods. A Monte Carlo procedure was used by Lenz and Lund (1977) to study the effect of thermal diffusivity and energy of activation for thermal inactivation of micro-organisms. An 8–15% CV was observed in process lethality, and with increase of heating time there was a reduction in CV. These workers also studied the effect of cooling water temperature from 5.5 to 10°C on variation of F-value but found no significant effect. Lund (1978) also reviewed the available literature and the uncertainties underlying process calculations. For home canning, Thompson *et al.* (1979) found that the CV values for nine different foods ranged from 43% to 82%, with all but one less than 60%. Pflug and Odlaugh (1978) and Thompson *et al.* (1979) showed that an error in retort temperature measurement of ±0.55°C could cause an error of 12% in the mean F_0-value. Naveh *et al.* (1982c) studied the effect of f_h on the process lethality and produced a graphical method of analysis. Robertson and Miller (1984) used a 90 min process at 121°C to study seven replicate cans of 55 bentonite over eight runs. The results showed the process lethality values for the heating phase, F_h, varied from 3.4 to 7.1 min or 13–26% around the mean value. Increasing the headspace in the cans significantly increased the f_h-value. Hayakawa *et al.* (1988) used a computerized method of determining the statistical variability of thermal process lethality, for a conduction-heating product, spaghetti in tomato sauce. Experimental CV values for two can sizes, 211 × 300 and 307 × 409, were respectively 67.0% and 28.5%, which agreed with the computer predictions. The F-value distributions were heavily skewed in favour of the higher values. Predictive equations were also developed and compared with regression equations (Wang & 1991; Hayakawa *et al.* 1994). These workers showed that variations in the process temperature had the greatest effect on the F-value, and that high-temperature processes required greater control of process parameters than more conventional processes. Bown (1989), using a computer simulation technique, studied the effect of five variables on lethality calculations. These were ranked in order of greatest influence as: thermal diffusivity; z-value; initial temperature; temperature of pre-retorting delay; and duration of pre-retorting delay. The results are interesting and shown in Figure 6.10. Akterian *et al.* (1990) also reported on a sensitivity analysis for a range of convection-heating packs. Other work on lethality distribution has been reported by Campbell *et al.* (1992a, b), Smout (2000a, b), and Vargo *et al.* (2000a, b) who have shown that the statistical distribution of lethalities approximated to a gamma-distribution skewed.

Turning to microbiological parameters variations in D- and z-values are among the most variable and difficult to quantify (Hicks 1961). While laboratory experiments can be used to estimate these quantities (Chapter 3), the circumstances are highly idealized compared with the commercial canning environment. For z-values Hicks (1961) reported CVs of 2–9% and found that the effect of errors in z-values was less when g was small. Patino and Heil (1985) estimated that

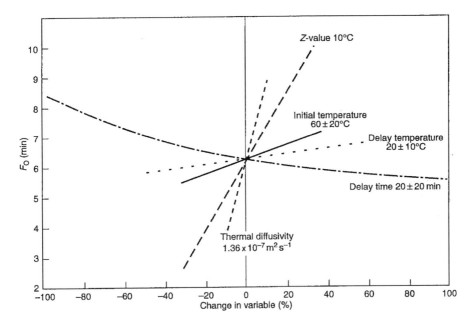

FIGURE 6.10. Sensitivity of F_0 to various factors (Bown 1989).

the standard deviation (SD) for D-values was 3–6%, and for z-values between 0.5 and 1.0°F. These were lower than the results obtained by Lenz and Lund (1977): SDs for D-values of 10% and for z-values of 2.0°F. Perkins *et al.* (1975) made a study of the influence of the z-value for *Clostridium botulinum* on the accuracy of process calculations and concluded that a more realistic estimate of the potential hazard in cans of product underprocessed by a few degrees or minutes may be made by assuming that any contaminating anaerobes may have a z-value of 14°F or 7.78°C. Pflug and Odlaugh (1978) reviewed the existing z- and F-value data and their variability, concluding that the minimum *botulinum* cook of $F_0 = 3$ min, using $z = 18$°F and $T_{ref} = 250$°F, represented a safe process for low-acid foods. Patino and Heil (1985) showed that the effect of z-value variation on F-values varied between $F/z = 0.12$ and 0.81. A useful method of converting F-values based on a z_1-value to one based on a z_2-value was presented by Pflug and Christensen (1980). The method involved the use of Ball's (1923) method and Stumbo's (1973) tabulated data, together with an f_h-value, assumed or known. The statistical modeling of lethality distribution in canned foods has been reviewed by Hall (1997).

The effect of container stacking patterns on the delivery of target F_0-values during the thermal processing of 2.7 kg cans (134 × 102 × 215.5 mm) of corned beef was studied by Warne and Moffit (1985). They compared the F-values obtained for specifically positioned cans in retort baskets with standard commercial packing and with tight packing, with an extra 21 cans per basket. For the former the F_0-values ranged from 9.8 to 11.7 min, but for the latter, where the cans were located in the center, the values range from 2.3 to 5.4 min.

6.7. Microbiological Methods

6.7.1. Introduction

The F_s-value was defined in equation (4.18),

$$F_s = D_r \log(n_0/n_t),$$

where D_r is the D-value for the organism under consideration at the reference temperature 121.1°C or 250°F, n_0 is the initial microbial spore loading and n_t is the spore loading after time t. This can be used to obtain an independent estimate of the total integrated F-value for the whole container.

A similar study was reported by Sumer and Esin (1992), who studied the effect of the arrangement of cans on the slowest heating point for a can filled with water, peas in brine and 28% tomato paste. For the water the point was 10–15% of the height of the can from the bottom, for the peas 20–25% and for the tomato paste the point was around the geometric center. A finite element model was developed for the distribution of temperature in the stacking patterns.

If the micro-organisms can be encapsulated in a bead, glass sphere or a synthetic particle, then this can be placed at the point of slowest heating and the F-value determined at that point. Thus provided that it is possible to determine the number of microbial species, usually spores, at the start and at the conclusion of the trial it should be possible to determine the safety of the process.

6.7.2. Inoculated Pack Method

The inoculated pack method, also known as the count reduction technique, is the simplest method and involves inoculating the food product with organisms of known heat resistance, preferably with a high D-value, 12–16 min at 115.5°C in phosphate buffer (Hersom & Hulland 1980). The extent of survival of the spores is determined by culture or incubation tests. The spores should be of appropriate heat resistance for the particular type of product being dealt with. For low-acid canned foods heat-resistant spores of PA 3679 are generally used, rather than spores of *Clostridium botulinum*, which require special facilities for handling them. The level of inoculum is usually between 10^3 and 10^5 spores per can. The heat resistance and amount of inoculum should be judged so that the process being considered should provide a low level of spoilage.

The method of inoculation depends on the nature of the product. If the product is liquid, then mixing will ensure uniform dispersion; if the product is a solid–liquid mixture then a pipette can be used to disperse the inoculum into the covering liquid; and in the case of more solid materials a syringe can be used to deliver the inoculum as near to the center of the food mass as possible.

The inoculated cans are then treated in a series of different processes, such that at the lowest level of sterilization a 100% spoilage should be achieved and at the highest level 0%. In the case of gas-producing spoilage it is possible to estimate the degree of spoilage from the number of blown cans.

6.7.3. Encapsulated Spore Method

The simplest form of this is to use a capillary tube filled with the inoculum. Alternatively, small bulbs of approximately 5 mm outside diameter that hold 0.05 ml can be used. These are filled by means of an accurate syringe. The advantage of using bulbs is that they can be located at the point of slowest heating or in particles of food or synthetic material. The time required for the bulb and contents to reach the temperature of the surroundings, up to 121.1°C, is about 15 s, and the contribution of the come-up time to the total process time is about 6 s at the operating temperature – thus 9 s is subtracted from the contact time to give the time of exposure (Hersom & Hulland 1980). The bulbs may be located at the appropriate point in the can by using a Tufnol rod 6 mm wide and 3 mm deep, cut to an appropriate length to fit the can diagonally from top to bottom. The stem of the capillary is inserted through a small hole in the rod and the bulb secured by a short collar of silicon elastomer. Each bulb contains approximately 10^6 spores of, for example, the highly heat resistant *Bacillus stearothermophilus* suspended in citrate buffer. The process lethality can then be determined from the initial number of spores and the number of survivors. The heat resistance can be obtained by using replicate bulbs, processed under the same conditions. For full details of these techniques the reader should consult standard texts, such as Townsend *et al.* (1954), Ball and Olson (1957), Stumbo (1973), Yawger (1978), and Pflug (1987).

6.7.4. Biological and Chemical Indicators

We now turn to the development of some other methods of determining process lethality, i.e. time–temperature indicators.

We have already discussed the use of encapsulated spores in relation to simple devices. The development of these for aseptic processing of particulates has been discussed by Holdsworth (1992). The use of the spores in spheres and tubes has been extended by the development of biological indicator units (BIUs) by Pflug *et al.* (1980a, b) and Pflug (1982). In this technique a sample tube is filled with spores, such that when a BIU rod is inserted into a can the spores are at the slowest point of heating. The tubes are made of plastic and a large batch is prepared, from which a representative sample is calibrated by being submitted to a known thermal process and an F-value determined from heat penetration. The technique was developed further by Rodriguez and Teixeira (1988) by using aluminium instead of plastic as the containment material. This improves the heat transfer and removes a lag involved in the use of plastic materials. If the spores do not have a z-value of 10°C, then the F-value may be determined at the z-value using the method of conversion previously described (Pflug & Christensen 1980). Two procedures have been suggested by Odlaugh *et al.* (1981) for converting the biological F-values into F_0 values: One involves converting F^z (biological) into F_0 (biological), $z = 10°C$, and comparing with F_0 (heat penetration); the other involves calculating F^z from heat penetration data and the BIU z-value and comparing with F^z (biological).

Pflug *et al.* (1980b) showed good agreement between F_0-values from heat penetration compared with F-values (BIU). In general the biological values tended to be slightly lower than the F_0 from heat penetration measurements. Many workers (see Table 6.10) have developed chemical indicators as time–temperature indicators. These are based on, for example, the inactivation of enzymes, colour destruction, vitamin degradation, sugar inversion and other chemical reactions. Hendrickx *et al.* (1992) has developed a time–temperature indicator based on the inactivation of the heat-stable fraction of peroxidase immobilized on porous glass beads in dodecane, which has a z-value of $10°C$. The results from using this unit were in excellent agreement with thermocouple results.

Rönner (2002) has described a commercial system, developed at SIK in Sweden, which uses heat resistant spores of *B. stearothermophilus* and *B. stearothermophilus* immobilized in an 8 mm diameter porous polymer bead, i.e., a modified polyacrylamide gel (PAG). These can be used for validating sterilization processes either in containers or in continuous sterilization processes. A similar system using ascospores has been developed for pasteurization processes.

The performance of a range of commercial indicators has been discussed by Lee *et al.* (1979). The calibration of a biomedical steam sterilizing indictor, Thermalog S, in F_0 units was carried out by Bunn and Sykes (1981). There is a continuous search for systems that can be used in practice. The commercial units should always be tested in order to find out what their limits of performance are and whether they are suitable for the desired purpose. Some indicators are relatively simple and only give an indication that the cans have received a process; others are more complex and are intended to give qualitative information.

The validity of the principles underlying the operation of BUIs and other types of time–temperature indicator have been discussed by Hendrickx *et al.* (1992), in particular the use of indicators with z-values other than $10°C$. It would appear that provided both the thermocouple and the BIU go through the same process, which they should if they are correctly located, then there should be no problem. This assumes there is no variation in the z-value with temperature. Elaborate mathematical analysis would appear to support a contrary view that the z-values for the indicators should be the same as the z-value of the target organism. While this may be ideal in theory, it is difficult in practice. Berry and Bradshaw (1986) also make the important point that the reaction kinetics should be of the same order and character. These workers considered that the thermal inactivation of *B. stearothermophilus* does not follow first-order kinetics over the entire heating period. It is, in fact, necessary to correct for the initial period when considering

Table 6.10. Some biological and chemical indicators used for determining sterilization values.

System	Comments	Reference
(a) General:		
Bioindicators	Review of principles	Myers and Chrai (1980)
Bioindicators	F-values for different z-values	Odlaugh et al. (1981)
Chemical markers	Kinetic theory and principles	Ross (1993)
Biological/chemical indicators	Review of applications	Van Loey et al. (1996); Hendrickx et al. (1994, 1995)
Biological/chemical indicators	Comprehensive list of potential indicators	Van Loey et al. (2004)
Multi-component indicators	Theoretical considerations	Maesmans et al. (1993)
Heat process indicators	Review of validation techniques	Tucker (2001b, 2004)
(b) Biological indicators:		
B. stearothermophilus	Contained in plastic tube	Pflug et al. (1980a, b)
B. stearothermophilus	In microporous beads 90% water	Rönner (2002)
B. subtilis	In microporous beads 90% water	Rönner (2002)
Bioindicators	Conversion of F-values for different z-values	Pflug (1982)
Biological indicators	Improved heat transfer using aluminium tubes	Rodriguez and Teixeira (1988)
(c) Chemical/Biochemical indictors:		
Anthocyanins	Chemical kinetics and application	Tanchev (1985)
Thiamin	Chemical kinetics and application	Berry (1989); Mulley et al. (1975)
Sucrose inversion	Chemical kinetics and application	Adams et al. (1984)
Sucrose inversion	Kinetics and application to F_0 determination	Siriwattanyaotin et al. (2006)
Methylmethionine sulfonium	Chemical kinetics and application	Berry (1989); Berry et al. (1990)
α–amylase	Kinetics and application to pasteurization	Van Loey (1996); Van Loey et al. (1997b, c)
α–amylase	Kinetics and application	Guiavarc'h. (2002); Guiavarc'h et al. (2002)
α–amylase, stabilized by polyols	Kinetics and application	De Cordt et al. (1994a)
α–amylase, reduced water content	Kinetics and application	Haentjens et al. (1998)
α–amylase, B. amyloliquefactens	Kinetics and pasteurization	Tucker (2001a)
α–amylase, B. lichenoformis	Kinetics and pasteurization	De Cordt et al. (1992); Maesmans et al. (1994)
α–amylase	Kinetics and pasteurization	Tucker (1999)
Marker M-1	Biochemical-indicator factors	Kim and Taub (1993)
Marker M-2	Biochemical indicator – kinetic factors	Lau et al. (2003)
Marker M-3	Biocemical indicator – kinetic factors	Lau et al. (2003)
Nitrophenyl glucoside hydrolysis	$z = 21.7°C$ & D = 10 min.	Adams and Langley (1998)

(cont.)

Table 6.10. (Continued)

System	Comments	Reference
Nitrophenyl glucoside hydrolysis	$z = 23°C$ applied to C_0-values	Williams and Adams (1997)
Protein-based indicators	Kinetics and application	De Cordt *et al.* (1994b)
Peroxidase (horseradish)	Kinetics and application to sterilization	Van Loey (1997a)
Lactulose (in milk)	Kinetics and application to milk processing	Andrews (1984); Andrews and Morant (1987)
(d) Commercial systems:		
Various	Comparative performance	Lee *et al.* (1979)
	Performance tests on indicator strip	Bunn and Sykes (1981)

the equivalent process for a BIU. Moreover, the time-lag before first-order kinetic behavior is established depends on the product factors. This means that it is necessary carefully to compare the kinetic characteristics of the marker organism and the analogue if satisfactory results are to be obtained. Chemical indicators have also been used to determine heat transfer coefficients between food particles and liquids (Maesmans *et al.* 1993, 1994; Weng *et al.* 1992; Van Loey *et al.* 1995).

Tucker (1999) has described a method of validating pasteurization processes using a *Bacillus amylolquifaciens* alpha-amilase time temperature indicator. This was fabricated by injecting a silicone particle with the indicator, and was used for yog-fruit (strawberries, pineapple, and apricot pieces) pasteurizations. The z value was $9.7 \pm 0.3°C$ and the pasteurizing temperatures from 70–90°C

Proteins and glucose, e.g., whey proteins, in food products will interact under heating conditions to produce a range of compounds that may be used to determine the degree of heating applied. Three compounds that are formed have been used as chemical markers, viz., **M-1**, 2,3-dihydro-3,5-dihydroxy-6-methyl-(4H-pyran-4-one), (Kim & Taub 1993; Kim *et al.* 1966a, b; Ramaswamy *et al.* 1996; Wang, *et al.* 2004), **M-2**, .4-hydroxy-5-methyl-3(2H)-furanone (Prakash *et al.* 1997; Lau 2000; Lau *et al.* 2003) & **M-3**, 5-hydroxymethylfurfural (Lau *et al.* 2003). For M-2 Lau *et al.* (2003) found that the kinetic parameters were $E_A = 64.2$–122.7 KJ/mol and $k_{121.1°C} = 0.94$–0.166 min^{-1}.

6.7.5. Conclusion

Microbiological methods are essentially very time-consuming and expensive. They do, however, fulfil a useful purpose for independently verifying processes. They are particularly useful for products which show considerable variation in f_h-values – e.g. spinach and leafy packs, as well as celery and artichoke hearts. The use of chemically based time–temperature indicators is important for practical control of canning operations. These should be carefully tested before using.

6.8. A Guide to Sterilization Values

Several definitive sources are available for the selection of suitable processes. The ones that are used world-wide are those produced by the National Food Processors' Association, USA, in particular *Bulletin 26-L* (NFPA 1982), which gives recommendations for thermal processes for low-acid canned foods in metal containers; and *Bulletin 30-L* (NFPA 1971), which gives recommendations for thermal processes for low-acid canned foods in glass containers. These are extremely valuable since they not only give recommended processes but also refer the processor to their laboratories for further advice on difficult products.

In the UK the publications of the Campden and Chorleywood Food Research Association are widely used and respected. The main ones are: *Technical Bulletin No. 4* (Atherton & Thorpe 1980), which gives a wide range of equivalent process times and temperatures for both sterilizing and cooking a range of fruit and vegetable products, for both rotary and static process; and *Technical Manual No. 6* (Atherton 1984), which deals with the processing of uncured meat products.

In France the main source of information is '*Barèmes de Sterilisation pour Aliments Appertisés*' (Institut-Appert 1979), which gives a time and temperature for given can sizes of specific products.

Some general principles are as follows. For low-acid products (pH 4.5) the minimum process is taken as $F_0 = 3$ min, but more severe processes equivalent to $F_0 = 6$ min or longer are usually applied to control spoilage organisms. There are a number of low-acid products that receive special treatment by virtue of statutory legislation, e.g. milk and dairy products. In the UK there are Statutory Instruments (SIs) that deal with pasteurization and sterilization of milk (SI 1033 of 1977), cream (SI 1509 of 1983) and milk-based drinks (SI 1508 of 1983). The process for sterilized milk is that it should be maintained at a temperature above $100°C$ for such a period so that it complies with the turbity test. Sterilized cream and milk-based drinks should be heated to a temperature not less than $104°C$ for not less than 45 min. For UHT sterilization, milk should be heated for not less than 1 s at a temperature not below $132.2°C$, and cream should be heated for not less than 2 s at a temperature not below $140°C$.

Another group of products that receive sub-minimal processes are canned cured meats, which contain salt or sodium nitrite as an inhibitor and some of which must be further protected by storage at refrigerated temperatures. These generally receive processes between $F_0 = 0.5$ and 1.5 min, the severity of the process being inversely proportional to the concentration of the inhibitor. These products require careful processing and, in particular, special control of the initial spore loadings.

For acid products, with pH below 4.5, the risk of growth and production of toxin by *C. botulinum* is highly unlikely. For products with pH 4.0–4.5 processes are designed to control the survival and growth of spore-forming organisms, e.g. *Bacillus coagulans, B. polymyxa, B. macerans*, and butyric anerobes, e.g. *C. butyricum* and *C. pasteurianum*. A process of $F_{121.1}^{10} = 0.7$ min is recommended by Hersom and Hulland (1980) for this purpose. Townsend *et al.* (1954) recommended a process of $F_{93.3}^{8.3} = 10$ min for products with pH in the range 4.3–4.5 and $F_{93.3}^{8.3} = 5$ min for pH 4.0–4.3. For products with lower pH, temperatures below 100°C are usually adequate; however, heat-resistant moulds, e.g. *Byssoclamys fulva* and *B. nivea*, may be present, which, if they survive and germinate to produce enzymes, can cause structural breakdown of fruit products, in particular strawberries.

Table 6.11 gives a guide to sterilization values which are used for some products. It is important to treat these as guidance only, since the specific conditions of the process, in particular the hygienic aspects, may be entirely different from the conditions being used in another factory. Whichever process is chosen, it is essential that the process is checked and verified under appropriate conditions, especially if there has been a change in container size, product specification or operating procedures. Stumbo *et al.* (1983) have produced an extensive and valuable guide for lethalities for both conduction and convection packs.

6.9. Computerized Process Calculations

APNS – (Apparent Position Numerical Solution), a semi-empirical method developed by Noronha *et al.* (1995), also allows the prediction of heat penetration profiles for products with broken-heating curves by changing the apparent thermal diffusivity at the break point. It is also applicable to packs showing a high degree of convection heating. The APNS method is based on the solution of the heat transfer equation for a sphere. In order to predict the heating and cooling characteristics of other geometrical bodies an apparent position is found within the sphere that represents the actual geometry. This involves the use of the actual experimental f_h & j_h-values; however the initial stages of cooling are not modeled very satisfactorily and Chen *et al.* (2005) have devised a method for improving this for cylindrical containers. This involves incorporating the actual temperature differences to the modified model, known as **MAPNS**.

CTemp (CCFRA Chipping Campden Glos.)
A commercial computer program *CTemp* has been developed for heat penetration and process evaluation studies by Tucker and Badley (1990) and Tucker *et al.* (1996). *CTemp* uses a finite-difference solution of the heat transfer model for a finite cylinder and rectangular parallelepiped geometries. It allows the prediction of heat penetration profiles for foods packed in a variety of containers, viz., cans, glass jars and plastic containers, using the appropriate surface heat transfer

Table 6.11. Some F_0-values for canned foods[a].

Product	Can size	Min F_0
Meat based:		
Curled meats and vegetables	Up to 16Z	8–12[b]
Ham 3.3% brine	—	0.3–0.5[c]
Ham 4.0% brine	—	0.1–0.2[c]
Ham 'sterile'	1 and 2 lb	3–4[b]
Luncheon meat 3–4% brine	—	1.0–1.5[c]
Luncheon meat 4–4.5% brine	—	1.0[c]
Luncheon meat 5–5.5% brine	—	0.5[c]
Meat in gravy	all	12–15[b]
Meat (sliced) in gravy	oval	10[b]
Meat loaf	A2	6[d]
Meat pies	tapered, flat	10[b]
Pet food	A2	12[d]
Pet food	Up to 16Z	15–18[b]
Pet food	A10	6[d]
Sausages 2.5% brine	—	1.5[c]
Sausages Vienna in brine	—	5[d]
Sausages in fat	Up to 1 lb	4–6[b]
Sausages Frankfurters	Up to 16Z	3–4[b]
Vegetables:		
Asparagus	all	2–4[d]
Beans in tomato sauce	all	4–6[b]
Carrots	all	3–4[b]
Celery	A2	3–4[b]
Chili con carne	—	6[d]
Corn, whole-kernel in brine	A2	9[d]
Corn, whole-kernel in brine	A10	15[d]
Corn, cream-style	A2	5–6[d]
Corn, cream-style	A10	2.3[d]
Green beans in brine	A2	3.5[d]
Green beans in brine	up to A2	4–6[b]
Green beans in brine	A10	6[d]
Green beans in brine	A2 to A10	6–8[b]
Mushrooms in brine	A1	8–10[b]
Peas, in brine	A2	7[d]
Peas, in brine	up to A2	6[b]
Peas, in brine	A2 to A10	6–8[b]
Peas, in brine	A10	10[d]
Fish:		
Mackerel in brine	301 × 411	3–4[d]
Herrings in tomato sauce	oval	6–8[b]
Poultry:		
Chicken, boned	—	6–8[d]
Chicken breast in jelly	up to 160 oz	6–10[b]
Poultry/game, whole in brine	$A2\frac{1}{2}$–A10	15–18[b]

(cont.)

Table 6.11. (Continued)

Product	Can size	Min F_0
Other:		
Baby foods	1/6	3–5[b]
Meat soups	Up to 16Z	10[b]
Tomato soup, non-cream	all	3[b]
Cream soups	A1 to 16Z	4–5[b]
Cream soups	Up to A10	6–10[b]
Milk puddings	Up to 16Z	4–10[b]
Cream	4 and 6 oz	3–4[b]
Cream	16Z	6[b]
Evaporated milk	up to 16 oz	5[b]

[a] Extensive tables of recommended F_0-values (American and European) are given in Eisner (1988).
[b] Collected UK data.
[c] Codex Alimentarius (1986).
[d] Ahlstrand and Eklund (1952).

coefficient and an apparent thermal diffusivity (based on the first term approximation of the solution of the heat transfer series). The *CTemp* program has been used to study the effects of process deviations in Reel and Spiral cookers (Dobie *et al.* 1994).

CAN-CALC© (University of Florida, Gainesville, FL),

This is a simulation software package that has been developed by Teixeira *et al.* (1999) and uses a similar concept to *CTemp,* which calculates the internal cold-spot temperature in response to any applied retort temperature profile. Various geometrical shaped containers are modeled using an apparent thermal diffusivity in the solution for the heat transfer equation for a spherical body.

Numerical™ On-Line The most recent version of this is, which is used in a process management system **LOG-TEC™-E,** with the FMC retort processing systems see p. 240 (Martens & Clifton 2002).

STERILIMATE

STERILIMATE is a software package for computer-aided design of sterilization processes (Kim *et al.* 1993). It is based on the early finite-difference programs developed to compute center temperatures, (Teixeira *et al.* 1969; Manson *et al.* 1970).

T-PRO

T-PRO is a Microsoft® Windows® based PC software designed to meet the unique information needs of the thermal process industry. Its predecessor, **TPRO for DOS,** assisted thermal processors beginning in 1986 with analysis and calculation of thermal process information. It has been used around the world to perform heat penetration calculations and produce graphic output.

The MS Windows version of **TPRO** attempts to keep the flexibility and functionality of the original **TPRO for DOS** while bringing a more up-to-date interface as well as new features to the user. The flexibility to change any parameter that may effect thermal process calculations is still part of the system, as well as flexible ways to do general method and Ball method calculations. All the functionality of the original **TPRO** is preserved (http://www.norbackley. com/tpro_frame.htm).

General Note:
Most of these systems have the propensity to calculate the achieved lethality and to respond to process deviations, and some can be used for retort control purposes (see Chapter 9).

References

Adams, J. B., & Langley, F. M. (1998). Nitrophenyl glucoside hydrolysis as a potential time-temperature indicator reaction. *Food Chem.*, *62*(1), 65–68.

Adams, J. P., Simunovic, J., & Smith, K. L. (1984). Temperature histories in a UHT indirect heat exchanger. *J. Food Sci.*, *49*, 273–277.

Akterian, S., Videv, K., & Das, H. (1990). Sensitivity of sterilization effect from process parameters. *J. Food Sci. Technol. (India)*, *27*(1), 45–49.

Alstrand, D. V., & Benjamin, H. A. (1949). Thermal processing of canned foods in tin containers 5. Effect of retorting procedures on sterilization values for canned foods. *Food Res.*, *14*, 253–257.

Alstrand, D. V., & Ecklund, O. F. (1952). The mechanics and interpretation of heat penetration tests in canned foods. *Food Technol.*, *6*(6), 185–189.

American Can Co. (1950). Calculation of the processes for canned foods. Maywood, IL: Research Division Staff Memorandum.

Andrews, G. R. (1984). Distinguishing pasteurized, UHT and sterilized milks by their lactulose content. *J. Soc. Dairy Technology*, *37*, 92–99.

Andrews, G. R., & Morant, S. V. (1987). Lactulose content and organoleptic assessment of ultra heat treated and sterilized milks. *J. Dairy Research*, *54*, 493–507.

Atherton, D. (1984). The heat processing of uncured canned meat products. *Technical Manual No. 6*. Chipping Campden, Glos., UK: Campden & Chorleywood Food Research Association.

Atherton, D., & Thorpe, R. H. (1980). The processing of canned fruits and vegetables. *Technical Bulletin No. 4*. Chipping Campden, Glos., UK: Campden & Chorleywood Food Research Association.

Ball, C. O. (1923). Thermal process time for canned foods. *Bulletin No. 37*. Washington, DC: National Research Council.

Ball, C. O. (1928). Mathematical solution of problems on thermal processing of canned food. *Univ. California Pub. in Public Health*, *1*(2), 145–245.

Ball C. O. (1949). Process evaluation. *Food Technol.*, *3*(4), 116–118.

Ball, C. O., & Olson, F. C. W. (1957). *Sterilization in food technology – Theory, practice and calculations*. New York: McGraw-Hill.

Berry, M. F. (1989). Chemical indexing method for determination of lethality in particulate foods. In J. V. Chambers (Ed.), *Proc. 1st int. congress on Aseptic processing Technologies* (pp. 128–156). USA: Purdue University.

Berry, M. F., Singh, R. K., & Nelson, P. F. (1990). Kinetics of methyl methionine sulfonium salt destruction in a model particulate system. *J. Food Sci.*, *55*(2), 502–505.

Berry, M. R. Jr (1983). Prediction of come-up time correction factors for batch-type agitating and still retorts and the influence on thermal process calculations. *J. Food Sci.*, *48*, 1293–1299.

Berry, M. R., & Bradshaw, J. G. (1986). Comparison of sterilization values from heat penetration and spore count reduction in agitating retorts. *J. Food Sci.*, *51*(2), 477–479, 493.

Berry, M. R., & Bush, R. C. (1987). Establishing thermal processes for products with broken-heating curves from data taken at other retort and initial temperatures. *J. Food Sci.*, *52*(4), 556–561.

Bhowmik, S. R., & Tandon, S. (1987). A method for thermal process evaluation of conduction heated foods in retortable pouches. *J. Food Sci.*, *52*, 201–209.

Bigelow, W. D., Bohart, G. S., Richardson, A. C., & Ball, C. O. (1920). Heat penetration in processing canned foods *Bulletin No. 16L*. Washington, DC: National Canners' Association.

Board, P. W., Cowell, N. D., & Hicks, E. W. (1960). Studies in canning processes. III. The cooling phase of processes for products heating by conduction. *Food Res.*, *25*, 449–459.

Board, P. W., & Steel, R. J. (1978). Calculating heat sterilization processes for canned foods. *Food Technol. Australia*, *20*(5), 169–173.

Bown, G. (1989). The influence of product and process variability on process control in the thermal sterilization of foods. PhD thesis. University of Leeds, UK.

Bunn, J. L., & Sykes, I. K. (1981). A chemical indicator for the rapid measurement of F_0-values. *J. Appl. Bact.*, *51*, 143–147.

Campbell, S., & Ramaswamy, H. S. (1992a). Heating rate, lethality and cold-spot location in air entrapped retort pouches during over-processing. *J. Food Sci.*, *57*(2), 485–490.

Campbell, S., & Ramaswamy, H. S. (1992b). Distribution of heat transfer rate and lethality in a single basket water cascade retort. *J. Food Proc. Eng.*, *15*(1), 31–48.

Carslaw, H. S., & Jaeger, J. C. (1959). Conduction of Heat in Solids. Oxford University Press. London.

Cass, G. V. (1947). A note on the use of Schultz and Olson lethal rate paper for calculation of thermal processes for canned food products in tin containers. *Food Reearch*, *12*, 24–26.

Chen, G., Corvalan, C, Campanella, O. H., & Haley, T. H. (2005). An improved method to estimate temperatures and lethality during the cooling stage of sterilized cylindrical cans. *Food and Bioproducts Processing, Trans IChemE*, *83*(C1), 36–42.

Chiheb, A., Debray, E., Le Jean, G., & Piar, G. (1994). Linear model for predicting transient temperature during sterilization of a food product. *J. Food Sci.*, *59*(2), 441–446,

Codex Alimentarius Commission (1986). ALINORM 86/16, Appendix VI. Rome: World Health Organization.

De Cordt, S., Hendrickx, M., Maesmans, G., & Tobback, P. (1992). Immobilized α-amylase from *Bacillus licheniformis*: a potential enzymic indicator for thermal processing. *Int. J. Food Sci. Technol.*, *27*, 661–673.

De Cordt, S., Avila, S., Hendrickx, M., & Tobback, P. (1994a). DSC and protein-based time-temperature indicators: case study of α-amylase stabilized by polyols and/or sugar. *Biotech. Bioeng.*, *44*(7), 859–865.

De Cordt, S., Maesmans, G., Hendrickx, M., & Tobback, P. (1994b). Feasibility of protein-based TTI development. In R. P. Singh & F. A. R. Oliveira. (Eds.), *Minimal processing of foods and process optimization* (pp. 52–64). Boca Raton, FL: CRC Press.

Denys, S., Noronha, J., Stoforos, N. G., Hendrickx, M., &Tobback, P. (1996). A semi-empirical approach to handle broken-line heating: determination of empirical parameters and evaluation of process deviations. *J. Food Proc. Preserv.*, *20*(4), 331–146.

Dobie, P. A. P., Tucker, G. S. & Williams, A. (1994). *The use of "CTemp" model to assess the effects of process deviations in reel and spiral cooker coolers.* Technical Memorandum No. 697. Chipping Campden, Glos. UK: Campden & Chorleywood Food Research Association.

Downes, T. W., & Hayakawa, K.-I. (1977). A procedure for estimating the retention of components of thermally conductive processed foods. *Lebensm.-Wiss. u. -Technol.*, *10*, 256–259.

Eisner, M. (1988). *Introduction to the technique of rotary sterilization.* Milwaukee, WI: private author's edition.

Eszes F., & Rajkó. R. (2004). Modelling heat penetration curves in thermal processes. In P. Richardson (Ed.), *Improving the thermal processing of food* (pp. 307–333). Cambridge, UK: Woodhead Publishing Ltd.

Fagerson, I. S., Esselen, W. B. Jr, & Licciardello, J. L. (1951). Heat transfer in commercial glass container during thermal processing. II. F_0 distribution in foods heating by convection. *Food Technol.*, *5*(6), 261–262.

Flambert, C. M. F., & Deltour, J. (1972a). Localization of the critical area in thermally-processed conduction heated canned food. *Lebensm.-Wiss. u. -Technol.*, *5*, 7–13.

Flambert, C. M. F., & Deltour, J. (1972b). Exact lethality calculation for sterilization process. I. Principles of the method. *Lebensm.-Wiss. u. -Technol.*, *5*, 72–73.

Flambert, C. M. F., Deltour, J., Dickerson, R. W., & Hayakawa, K. (1977). Lethal effect of food temperature on linear portion of heating or cooling curve. *J. Food Sci.*,*42*, 545–546.

Germer, S. P. M., Montes, H. B. E., & Vitali, A. De A. (1992). Mathematical modelling of data from [Stumbo's] tables for computerized F_0 [sterilization] values. [in Portuguese]. *Coletaena do Instituto de Tecnologia de Alimentos*, *22*(1), 73–82.

Ghazala, S., Ramaswamy, H. S., Smith, J., & Simpson, B. (1991). Thermal process time calculations for thin profile packages: comparison of formula methods. *J. Food Process Eng.*, *13*, 269–282.

Gillespy, T. G. (1951). Estimation of sterilizing values of processes as applied to canned foods. I. Packs heating by conduction. *J. Sci. Food Agric.*, *2*, 107–125.

Gillespy, T. G. (1953). Estimation of sterilizing values of processes as applied to canned foods. II. Packs heating by conduction: complex processing conditions and value of coming-up time of retort. *J. Sci. Food Agric.*, *4*, 553–565.

Griffin, R. C. Jr, Herndon, D H., & Ball, C. O. (1969). Use of computer-derived tables to calculate sterilizing processes for packaged foods. 2. Application to broken-line heating curves. *Food Technol.*, *23*(4), 519–524.

Griffin, R. C. Jr, Herndon, D. H., & Ball, C. O. (1971). Use of computer-derived tables to calculate sterilizing processes for packaged foods 3. Application to cooling curves. *Food Technol.*, *25*(2), 134–143.

Guiavarc'h, Y. (2002). Application of time-temperature integrators. In Tucker, G. S. (Ed.), *Second international symposium on thermal processing – Thermal processing: validation challenges* (Session 3:2.). Chipping Campden, Glos., UK: Campden & Chorleywood Food Research Association.

Guiavarc'h, Y., Deli, V., Van Loey, A. M., & Hendrickx, M. (2002). Development of and enzymic time temperature indicator for sterilization processes based on *Bacillus licheniformis* α-amylase at reduced water content. *J. Food Sci., 67*(1), 285–291.

Haentjens, T. H., Van Loey, A. M., Hendrickx, M. E., & Tobback, P. P. (1998). The use of alpha-amylase at reduced water content to develop time/temperature indicators for sterilization processes. *Lebensm-Wiss u-Technol., 31*, 467–472.

Hall, J. E. (1997). Statistical modelling of lethality distribution in canned foods. *R&D Report No. 50*. Chipping Campden, Glos., UK.: Campden & Chorleywood Food Research Association.

Hayakawa, K. (1968). A procedure for calculating the sterilizing value of a thermal process. *Food Technol., 22*, 905–907.

Hayakawa, K. (1969). New parameters for calculating mass average sterilizing values to estimate nutrients in thermally conductive food. *Can. Inst. Food Sci. Technol. J., 2*, 167–170.

Hayakawa, K. (1970). Experimental formulas for accurate estimation of transient temperature of food and their application to thermal process evaluation. *Food Technol., 24*(12), 1407–1418.

Hayakawa, K. (1971). Estimating food temperatures during various heating or cooling treatments. *J. Food Sci., 36*, 378–385.

Hayakawa, K. (1973). Modified lethal rate paper technique for thermal process evaluation. *Can. Inst. Food Technol. J., 6*(4), 295–297.

Hayakawa, K. (1974). Response charts for estimating temperatures in cylindrical cans of solid food subjected to time variable processing temperatures. *J. Food Sci., 39*(6), 1090–1098.

Hayakawa, K. (1977). Mathematical methods for estimating proper thermal processes and their computer implementation. *Adv. Food Res., 23*, 75–141.

Hayakawa, K. (1978) A critical review of mathematical procedures for determining proper heat sterilization processes. *Food Technol., 38*(3), 59–65.

Hayakawa, K. (1982). Advances in thermal process calculation procedures. *Food Process Engineering, AIChE Symp. Series, 78*(218), 42–48.

Hayakawa, K. (1983a). Mathematical procedures for estimating thermal processes applied to canned or other packaged food. In T. Motohivo & K. Hayakawa (Eds.), *Heat sterilization of food* (pp. 138–147). Tokyo: Koseicha-Koseikaku Co. Ltd.

Hayakawa, K. (1983b). New parametric values for thermal process estimation. In T. Motohivo & K. Hayakawa (Eds.), *Heat sterilization of food* (pp. 191–199). Tokyo: Koseicha-Koseikaku Co. Ltd.

Hayakawa, K. (1983c). Heat transfer in canned foods during heat sterilization processes. In T. Motohivo & K. Hayakawa (Eds.), *Heat sterilization of food* (pp. 109–120). Koseicha-Koseikaku Co. Ltd., Tokyo.

Hayakawa, K. & Downes, T. W. (1981). New parametric values for thermal process estimation by using temperatures and z values expressed in degree Celsius units. *Lebensm.-Wiss. u. -Technol., 14*, 60–64.

Hayakawa, K., Massaguer, P. de, & Trout, R. J. (1988). Statistical variability of thermal process lethality in conduction heated food – computerized simulation. *J. Food Sci., 53*(6), 1887–1893.

Hayakawa, K., Wang, J., & Massaguer, P. de (1994). Variability in thermal process lethality. In T. Yano, R. Matsuno & K. Nakamura (Eds.), *Developments in food engineering Part 2* (pp. 704–706). London: Blackie Academic & Professional.

Hayakawa, K.-I., Wang, J., & Massaguer, P. R. De (1996). Simplified predictive equations for variable thermal process lethality. *J. Food Proc. Eng.*, *19*(3), 289–300.

Hendrickx, M., Weng, Z., Maesmans, G., & Tobback, P. (1992). Validation of a time–temperature-integrator for thermal processing of foods under pasteurization conditions. *Int. J. Food Sci. Technol.*, *27*, 21–31.

Hendrickx, M., Maesmans, G., De Cordt, S., & Tobback, P. (1994). Evaluation of the integrated-time-temperature effect of thermal processes on foods: State of the art. In T. Yano, R. Matsuno, & K. Nakamura (Eds.), *Developments in food engineering - ICEF 6*, *Part 2* (pp. 692–694).

Hendrickx, M., Maesmans, G., De Cordt, S., Noronha, J., Van Loey, A., & Tobback, P. (1995). Evaluation of the integrated time-temperature effect in thermal processing of foods. *Critical Reviews in Food Science and Nutrition, 35*(3), 231–262.

Herndon, D. H., Griffin, R. C. Jr, & Ball, C. O. (1968). Use of computer-derived tables to calculate sterilizing processes for packaged foods. *Food Technol.*, *22*(4), 473–484.

Hersom, A. C., & Hulland, E. D. (1980). *Canned foods. Thermal processing and microbiology* (7th Edition). Edinburgh: Churchill-Livingstone.

Hicks, E. W. (1951). On the evaluation of canning processes. *Food Technol.*, *5*, 134–142.

Hicks, E. W. (1952). Some implications of recent theoretical work on canning processes. *Food Technol.*, *6*, 175–179.

Hicks, E. W. (1958). A revised table of the P_h function of Ball and Olson. *Food Res.*, *23*, 396–400.

Hicks, E. W. (1961). Uncertainties in canning process calculations. *J. Food Sci.*, *26*, 218–223.

Holdsworth, S. D. (1992). *Aseptic processing and packaging of food products*. London: Elsevier Applied Science.

Holdsworth, S. D., & Overington, W. J. G. (1975). Calculation of the sterilizing value of food canning processes. Literature review. *Technical Bulletin No. 28*. Chipping Campden, Glos., UK: Campden & Chorleywood Food Research Association.

Hurwicz, H., & Tischer, R. G. (1952). Heat processing of beef. I. A theoretical consideration of the distribution of temperature with time and in space during processing. *Food Res.*, *17*, 380–392.

Ikegami, Y. (1974a). Effect of various factors in the come-up time on processing of canned foods with steam. *Report Tokyo Inst. Food Technol. Serial No. 11*, 92–98 [in Japanese].

Ikegami, Y. (1974b). Effect of "come-up" on processing canned food with steam. *Canner's J.*, *53*(1), 79–84 [in Japanese].

Institut Appert (1979). *Barèmes de Sterilisation pour Aliments Appertisés*. Paris: Institut Appert.

Jakobsen, F. (1954). Note on process evaluation. *Food Res.*, *19*, 66–79.

Jen, Y., Manson, J. E., Stumbo, C. R., & Zhradnik, J. W. (1971). A procedure for estimating sterilization of and quality factor degradation in thermally processed food. *J. Food Sci.*, *36*(4), 692–698.

Kao, J., Naveh, D., Kopelman, I. J., & Pflug, I. J. (1981). Thermal process calculations for different z and j_c values using a hand-held calculator. *J. Food Sci.*, *47*, 193–197.

Kim, K. H., Teixeira, A. A., Bichier, J., & Tavares, J. (1993). STERILIMATE: software for designing and evaluating thermal sterilization processes. *American Society of Agricultural Engineers,* Paper No. 93–4051, Michigan, USA: St. Joseph.

Kim, H. J., & Taub, I. A. (1993). Intrinsic chemical markers for aseptic processing of particulate foods. *Food Technol.*, *47*, 91–98.

Kim, H.-J., Choi, Y.-M., Yang, T. C. S., Taub, I. A.,Tempest, P., Skudder, P., Tucker, G., & Parrott, D. L. (1996a).Validation of ohmic heating for quality enhancement of food products. *Food Technol., 50*, 253–262.

Kim, H.-J., Taub, I. A.,Choi, Y.-M., & Prakash, A. (1996b). Principles and applications of chemical markers of sterility in high-temperature-short-time processing of particulate foods. In T. C. Lee and H. J. Kim (Eds.), *Chemical markers for processed and stored foods* (pp. 54–69). Washington, DC: American Chemical Society.

Larkin, J. W. (1989). Use of a modified Ball's formula method to evaluate aseptic process-ing of foods containing particulates. *Food Technol., 43*(3), 66–72, 75.

Larkin, J. W., & Berry, M. R. (1991). Estimating cooling process lethality for different cooling 'j' values. *J. Food Sci., 56*(4), 1063–1067.

Lau, H. (2000). Microwave pasteurization and sterlization. Ph.D. Dissertation. Pullman, WA.:Washington State University.

Lau, H., Tang, J., Taub I. A., Yang, T. C. S., Edwards, C. G., & Mao, R. (2003). Kinetics of chemical marker formation in whey protein gels for studying high temperature short time microwave sterilization. *J Food Eng., 60*(4), 397–405.

Lee, C.-H., Montville, J., & Sinskey, A. J. (1979). Comparison of the efficacy of steam sterilization indicators. *Appl. Environ. Microbiol., 36*(6), 113–117.

Lenz, M. K., & Lund, D. B. (1977) The lethality-Fourier number method: experimental verification of a model for calculating temperature profiles and lethality in conduction-heating canned foods. *J. Food Sci., 42*(4), 989–996, 1001.

Leonhardt, G. F. (1978). A general lethal-rate paper for the graphical calculation of processing times. *J. Food Sci., 43*, 660.

Lund, D. B. (1978). Statistical analysis of thermal process calculations. *Food Technol., 32*(3), 76–78.

Luikov, A. V. (1968). Analytical Heat Diffusion Theory. Academic Press, Inc. New York.

Maesmans, G., Hendrickx, M., De Cordt, S., & Tobback, P. (1993). Theoretical considera-tions on design of multicomponent time temperature integrators in evaluation of thermal processes. *J. Food Process. Preserv., 17*, 369–389.

Maesmans, G., Hendrickx, M., De Cordt, S., & Tobback, P. (1994). Feasibility of the use of a time–temperature integrator and a mathematical model to determine fluid-to-particle heat transfer coefficients. *Food Res. Int., 27*, 39–51.

Maesmans, G., Hendrickx, M., De Cordt, S., Van Loey, A., & Noronha, J. (1994). Evalua-tion of process value distribution with time temperature indicators. *Food Res. Int., 27*(5), 413–424.

Manson, J. E., Zahradnik, J. W., & Stumbo, C. R. (1970). Evaluation of lethality and nutrient retentions of conduction-heating foods in rectangular containers. *Food Technol., 24*(11), 1297–1302.

Manson, J. E., Stumbo, C. R., & Zahradnik, J. W. (1974). Evaluation of thermal processes for conduction heating foods in pear-shaped containers. *J. Food Sci., 39*, 276–281.

Martens B. & Clifton, D. (2002). Advances in retort control. In Tucker, G. S. (Ed.), *Second International Symposium, Thermal processing – Thermal Processing: validation chal-lenges*. Session 4:2. Chipping Campden UK: Campden & Chorleywood Food Research Association.

Merson, R. L., Singh, R. P., & Carroad, P. A. (1978). An evaluation of Ball's formula method of thermal process calculations. *Food Technol., 32*(3), 66–72, 75.

Mulley, E. A., Stumbo, C. R., & Hunting, W. M. (1975). Thiamine: a chemical index of the sterilization efficacy of thermal processing. *J. Food Sci., 40*, 985–988.

Myers, T., & Chrai, S. (1980). Basic principles and application of biological indicators. *J. Parenteral Drug Assoc.*, *34*(3), 324–343.

National Bureau of Standards (1940). *Tables of sine, cosine and exponential integrals*, Vol. I. New York.

Naveh, D., Kopelman, I. J., Zechman, L., & Pflug, I. J. (1983a). Transient cooling of conduction heating products during sterilization: temperature histories. *J. Food Process. Preserv.*, *7*, 259–273.

Naveh, D., Pflug, I. J., & Kopelman, I. J. (1983b). Transient cooling of conduction heating products during sterilization: sterilization values. *J. Food Process. Preserv.*, *7*, 275–286.

Naveh, D., Pflug, I. J., & Kopelman, I. J. (1983c). Simplified procedure for estimating the effect of a change in heating rate on sterilization value. *J. Food Protection*, *46*(1), 16–18.

Newman, M., & Holdsworth, S. D. (1989). Methods of thermal process calculation for food particles. *Technical Memorandum No. 321 (revised)*. Chipping Campden, Glos., UK: Campden & Chorleywood Food Research Association.

NFPA (1971). Processes for low-acid canned foods in glass containers. *Bulletin 30-L*. Washington, DC: National Food Processors' Association.

NFPA (1982). Processes for low-acid canned foods in metal containers. *Bulletin 26-L*, (12th edition). Washington, DC: National Food Processors' Association.

Noronha, J., Hendrickx, A., Van Loey, A., & Tobback, P. (1995). New semi-empirical approach to handle time-variable boundary conditions during sterilization of non-conductive heating foods. *J. Food Eng.*, *24*(2), 249–268.

Odlaugh, T. E., Caputo, R. A., & Mascoli, C. C. (1981). Determination of sterilization (F)-values by microbiological methods. *Developments in industrial microbiology*, *22*, 349–354.

Olson, F. C. W., & Stevens, H. P. (1939). Thermal processing of canned foods in tin containers. II. Nomograms for graphic calculation of thermal processes for non-acid canned foods exhibiting straight-line semi-loarithmic heating curves. *Food Res.*, *4*(1), 1–20.

Pagurova, V. I. (1961). *Tables of the exponential integral*. Oxford: Pergamon Press.

Patashnik, M. (1953) A simplified procedure for thermal process calculation. *Food Technol.*, *7*(1), 1–6.

Patino, H., & Heil, J. R. (1985). A statistical approach to error analysis in thermal process calculations. *J. Food Sci.*, *50*, 1110–1114.

Perkins, W. E., Ashton, D. H., & Evancho, G. M. (1975). Influence of z-value of *Clostridium botulinum* on the accuracy of process calculations. *J. Food Sci.*, *40*, 1189–1192.

Pflug, I. J. (1968). Evaluating the lethality of heat processes using a method employing Hick's table. *Food Technol.*, *33*(9), 1153–1156.

Pflug, I. J. (1982). Measuring the integrated time-temperature effect of a heat sterilization process using bacterial spores. In *Food process engineering, AIChE Symp. Series*, *78*(218), 68–75.

Pflug, I. J. (1987). *Textbook for an introductory course in the microbiology and engineering of sterilization processes* (6th Edition). Minneapolis, MN: Environmental Sterilization Laboratory.

Pflug, I. J., & Christensen, R. (1980). Converting an F-value determined on the basis of one z-value to an F-value determined on the basis of another z-value. *J. Food Sci.*, *45*, 35–40.

Pflug, I. J., & Odlaugh, T. E. (1978). Determination of the sterilization value of a process by the 'general method'. In R. B. Sleeth (Ed.), *Introduction to the fundamentals of thermal processing*. Chicago: Institute of Food Technology.

Pflug, I. J., Jones, A. T., & Blanchett, R. (1980a). Performance of bacterial spores in a carrier system in measuring the F_0-value delivered to cans of food heated in a steritort. *J. Food Sci.*, *45*, 940–945.

Pflug, I. J., Smith, G., Holcomb, R., & Blanchett, R. (1980b). Measuring sterilizing values in containers of food using thermocouples and biological indicator units. *J. Food Protection*, *43*(2), 119–123.

Pham, Q. T. (1987). Calculation of thermal process lethality for conduction-heated canned foods. *J. Food Sci.*, *52*(4), 967–974.

Pham, Q. T. (1990). Lethality calculation for thermal process with different heating and cooling rates. *Int. J. Food Sci. Technol.*, *25*, 148–156.

Powers, J. J., Pratt, D. E., Carmon, J. L., Somaatmadja, D., & Fortson, J. C. (1962). Application of extreme-value methods and other statistical procedures to heat-penetration data. *Food Technol.*, *16*(3), 80–89.

Prakash, A., Kim, H. J., & Taub, I. A. (1997). Assessment of microwave sterilization of using intrinsic chemical markers. *J. Microwave Power & Electromagnetic Energy,* *32*(1), 50–57.

Ramaswamy, H. S. (1993). Come-up time effectiveness for process calculations involving thin-packages. *J. Food Eng.*, *19*(2), 109–117.

Ramaswamy, H. S., Awuah, G. B., Kim, H.-I., & Choi, Y.-M. (1996). Evaluation of chemical marker for process lethality measurement at 110°C in a continuous holding tube. *J. Food Proc. Preserv.*, *20*(3), 235–249.

Riedel, L. (1947). The theory of heat sterilization of canned foods. *Mitt. Kältetech. Inst. Karlsruhe, No. 1*, 3–40 [in German].

Robertson, G. L., & Miller, S. L. (1984). Uncertainties associated with the estimation of F_0 values in cans which heat by conduction. *J. Food Technol.*, *19*, 623–630.

Rodriguez, A. C., & Teixeira, A. A. (1988). Heat transfer in hollow cylindrical rods used as bioindicator units for thermal process validation. *Trans. ASAE*, *31*(4), 1233–1236.

Rönner, U. (2002). Validation of heat processes using bio-indicators (polymer beads). In Tucker, G. S. (Ed.), *Second international symposium on thermal processing – Thermal processing: validation challenges* (Session 2:3). Chipping Campden, Glos., UK: Campden & Chorleywood Food Research Association.

Ross, E. W. (1993). Relation of bacterial destruction to chemical marker formation during processing by thermal pulses. *J. Food Process Eng.*, *16*, 247–270.

Sablani, S. S., & Shayya, W. H. (2001). Computerization of Stumbo's method of thermal process calculations using neural networks. *J. Food Eng.*, *47*(3), 233–240.

Schultz, O. T., & Olson, F. C. W. (1940). Thermal processing of canned foods in tin containers. III. Recent improvements in the general method of thermal process calculation. *Food Res.*, *5*(4), 399–407.

Shiga, I. (1970). Temperatures in canned foods during processing. *Food Preserv. Quart.*, *30*(3), 56–58.

Shiga, I. (1976). A new method of estimating thermal process time for a given F value. *J. Food Sci.*, *41*, 461–462.

Simpson, R., Almonacid, S., & Solari, P. (2001). Bigelow's General Method revisited. In J. Welti-Chanes, G. V. Barbosa-Cánovas & J. M. Aguilera (Eds.), *Eighth international conference Engineering and food ICEF8, Vol. 2* (pp. 1815–1820). Lancaster, PA: Technomic Pub. Co.

Simpson, R., Almonacid, S., & Teixiera, A. (2003) Bigelow's General Method revisited: development of a new calculation technique. *J. Food Sci.*, *68*(4), 1324–1233.

Simpson, R. (2005).Generation of isolethal processes and implementation of simultaneous sterilization utilising the revisited general method. *J. Food Eng.*, *67*(1/2), 71–79.

Siriwattanayotin, S., Yoodivhya, T., Meepadung, T., & Ruenglertpanyakul, W. (2006). Simulation of sterilization of canned liquid food using sucrose degradation as an Indicator. *J. Food Eng.*, *73*(4), 307–312, 2006.

Skinner, R. H., Jowitt, R. (1975). A reappraisal of available temperature history models for heat sterilization processes in relation to their suitability for present day computing facilities. In *Proceedings of EFCE Mini-Symposium – Mathematical modelling in food processing* (pp. 203–222). Sweden: Lund University Technology.

Skinner, R. H., & Urwin, S. (1971). Automation of canning process calculations. *Process Biochem.*, *6*(11), 35–38.

Smith, T., & Tung, M. A. (1982). Comparison of formula methods for calculating thermal process lethality. *J. Food Sci.*, *47*, 626–631.

Smout, C., Ávila, I., Van Loey, A. M. L., Hendrickx, M. E. G., & Silva, C. (2000a). Influence of rotational speed on the statistical variability of heat penetration parameters and on the non-uniformity of lethality in retort processing. *J. Food Eng.*, *45*(2), 93–102.

Smout, C., Van Loey, A. M. L., & Hendrickx, M. E. G. (2000b). Non-uniformity of lethality in retort processes based on heat distribution and heat penetration data. *J. Food Eng.*, *45*(2), 103–110.

Spinak, S. H., & Wiley, R. C. (1982). Comparisons of the general and Ball formula methods for retort pouch process conditions. *J. Food Sci.*, *47*, 880–885.

Steele, R. J., & Board, P. W. (1979). Amendments of Ball's formula method for calculating the lethal value of thermal processes. *J. Food Sci.*, *44*, 292–293.

Steele, R. J., Board, P. W., Best, D. J., & Willcox, M. E. (1979). Revision of the formula method tables for thermal process evaluation. *J. Food Sci.*, *44*, 954–957.

Storofos, N. G. (1991). On Ball's formula method for thermal process calculations. *J. Food Process Eng.*, *13*, 255–268.

Storofos, N. G., Noronha, J., Hendrickx, M., & Tobback, P. (1997). A critical analysis in mathematical procedures for the evaluation and design of in-container thermal processes for foods. *Critical Reviews in Food Science and Nutrition, 37*(5), 411–441.

Stumbo, C. R. (1953). New procedures for evaluating thermal processes for foods in cylindrical containers. *Food Technol.*, *7*(8), 309–315.

Stumbo, C. R. (1973). *Thermobacteriology in food processing* (2nd Edition). New York: Academic Press.

Stumbo, C. R., & Longley, R. E. (1966). New parameters for process calculations. *Food Technol.*, *20*, 341–345.

Stumbo, C. R., Purohit, K. S., Ramakrishnan, T. V., Evans, D. A., & Francis, F. J. (1983). *Handbook of lethality guides for low-acid canned foods*. Boca Raton, FL: CRC Press.

Succar, J., & Hayakawa, K. (1982). Prediction of time correction factor for come-up heating of packaged liquid food. *J. Food Sci.*, *47*(3), 614–618.

Sumer, A., & Esin, A. (1992). Effect of can arrangement on temperature distribution in still retort sterilization. *J. Food Eng.*, *15*(4), 245–259.

Tanchev, S. S. (1985). Kinetics of degradation of anthocyanins during sterilization. In *Proc. IUFoST Symposium – Aseptic processing and packaging of foods*. Tylosand, Sweden: SIK.

Teixeira, A. A., Dixon, J. R., Zahradnik, J. W., & Zinsmeister, G. E. (1969). Computer optimization of nutrient retention in thermal processing of conduction heated foods. *Food Technol.*, *23*, 137–142.

Teixeira, A. A., Dixon, J. R., Zahradnik, J. W., & Zinsmeister, G. E. (1969). Computer determination of spore survival distributions in the thermal processing of conduction-heated foods. *Food Technology 23*(3), 78–80.

Teixeira, A. A., Tucker, G. S., Balaban, M. O., & Bichier, J. (1992). Innovations in conduction-heating models for in-line retort control of canned foods with any j-value. In R. P. Singh & A. Wirakarakusumah (Eds.), *Advances in food engineering* (pp. 561–74). Boca Raton, FL: CRC.

Teixeira, A. A., Balaban, M. O., Germer, S. P. M., Sadahira, M. S. Teixeira Neto, R. O., & Vitali, A. A. (1999). Heat transfer model performance in simulation of process deviations. *J. Food Science, 64*(3), 488–493.

Thompson, D. R., Wolf, I. D., Nordsiden, L., & Zottola, E. A. (1979). Home canning of foods: risks resulting from errors in processing. *J. Food Sci.*, *44*, 226–231.

Tobback, P., Hendrickx, M., Weng, Z., Maesmans, G. J., & De Cordt, S. V. (1992). Immobilized peroxidase: a potential bioindicator for evaluation of thermal processes In R. P. Singh & A. Wirakarakusumah (Eds.), *Advances in food engineering*, (pp. 561–74). Boca Raton, FL: CRC Press.

Townsend, C. T., Somers, I. I., Lamb, F. C., & Olson, N. A. (1954). *A laboratory manual for the canning industry*, Washington, DC: National Food Processors' Association.

Tucker, G. S. (1990). Evaluating thermal processes. *Food Manufacture*, June, 39–40.

Tucker, G., & Badley, E. (1990). *CTemp: Centre Temperature Prediction Program for heat sterilization processes (User's Guide)*. Chipping Campden, UK: Campden & Chorleywood Food Research Association.

Tucker, G. S. (1991). Development and use of numerical techniques for improved thermal process calculations and control. *Food Control*, *2*(1), 15–19.

Tucker, G. S., Noronha, J. F., & Heydon, C. J. (1996). Experimental validation of mathematical procedures for the evaluation of thermal process deviations during the sterilization of canned foods. *Food and Bioproducts Processing, Trans. IChemE*, *74*C3, 140–148.

Tucker, G., (1999). A novel validation method: application of time-temperature integrators to food pasteurization treatments. *Food and Bioproducts Processing, Trans. IChemE, 77*C3, 223–231.

Tucker, G. S. (2001a). Validation of heat processes. In P. Richardson (Ed.), *Thermal technologies in food processing* (pp. 75–90). Cambridge, UK.: Woodhead Publishing.

Tucker, G. (2001b). Application of biochemical time-temperature indicators to food pasteurisation treatments. In J. Welti-Chanes, G. V. Barbosa-Cánovas, & J. M. Aguilera (Eds.), *Eighth international conference engineering and food, ICEF8, Vol. 1* (pp. 713–717). Lancaster, PA: Technomic Pub. Co.

Tucker, G. (2002). Capabilities and pitfalls of computer-based prediction software. In Tucker, G. S. (Ed.), *Second international symposium, on thermal processing – Thermal processing: validation challenges.* (Session 4:1). Chipping Campden, Glos., UK: Campden & Chorleywood Food Research Association.

Tucker, G. S. (2004). Validation of heat processes: an overview. In P. Richardson (Ed.), *Improving the thermal processing of foods* (pp. 334–352). Cambridge: Woodhead Publishing.

Tucker, G. S., & Holdsworth, S. D. (1991). Mathematical modelling of sterilization and cooking processes for heat preserved foods–application of a new heat transfer model. *Food and Bioproducts Processing, Trans. IChemE, 69*, C1, 5–12.

Tung, M. A., & Garland, T. D. (1978). Computer calculation of thermal processes. *J. Food Sci., 43*, 365–369.

Uno, J., & Hayakawa, K. (1981). Correction factor of come-up heating based on mass average survivor concentration in a cylindrical can of heat conduction food. *J. Food Sci., 46*, 1484–1487.

Van Loey, M. A. (1996). Enzymic time temperature indicators for the quantification of thermal processes in terms of food safety. PhD Thesis, K. U. Leuven, Belgium.

Van Loey, A., Guiavarc'h, Y., Claeys, W., & Hendrickx, M. (2004). The use of time-temperature indicators (TTIs) to validate thermal processes. In P. Richardson (Ed.), *Improving the thermal processing of foods* (pp. 365–384). Cambridge: Woodhead Publishing.

Van Loey, A. M., Haentjens, T. H., Hendrickx, M. E., & Tobback, P. P. (1997a). The development of an enzymic time temperature integrator to assess the thermal efficacy of sterilization of low-acid canned foods. *Food Biotech., 11*(2), 147–168.

Van Loey, A. M., Haentjens, T. H., Hendrickx, M. E., & Tobback, P. P. (1997b). The development of an enzymic time temperature integrator to assess the lethal efficacy of sterilization of low-acid canned foods. *Food Biotech., 11*(2), 169–188.

Van Loey, A. M., Haentjens, T. H., Hendrickx, M. E., & Tobback, P. P. (1997c). The development of an α-amylase based time temperature integrator to evaluate in-pack pasteurization processes. *Lebensm. -Wiss. u.-Technol., 30*, 94–100.

Van Loey, A. M., Hendrickx, M. E., De Cordt, S., Haentjens, T. H., & Tobback, P. P. (1996). Quantitative evaluation of thermal processes using time-temperature indicators. *Critical reviews in food science and nutrition, 35*(3), 231–262.

Van Loey, A., Ludikhuyze, L., Hendrickx, M., De Cordt, S., & Tobback, P. (1995). Theoretical consideration on the influence of the z-value of a single component time temperature integrator on thermal process impact evaluation. *J. Food Protect., 58* (1), 39–48.

Varga, S., Oliveira, J. C., & Oliveira, F. A. R. (2000a). Influence of the variability of processing factors on the F-value distribution in batch retorts. *J. Food Eng., 44*(3), 155–161.

Varga, S., Oliveira, J. C., Smout, C., & Hendrickx, M. E. (2000b). Modelling temperature variability in batch retorts and its impact on lethality distribution. *J. Food Eng., 4*(3), 163–174.

Vinters, J. E., Patel, R. H., & Halaby, G. A. (1975). Thermal process evaluation by programmable computer calculator. *Food Technol., 29*(3), 42–44.

Wang, Y., Lau, M. H., Tang, G., & Mao, R. (2004). Kinetics of chemical marker M-1 formation in whey protein gels for developing sterilization processes based on dielectric heating. *J. Food Eng., 64*(1), 111–118.

Wang, J., Wolfe, R. R., & Hayakawa, K. (1991). Thermal process lethality variability in conduction heated foods. *J. Food Sci., 56*, 1424–1428.

Warne, D., & Moffit, D. (1985). The effect of container stacking patterns on the delivery of target F_0 values during the thermal processing of canned corned beef. *Inst. Food Sci. Technol. Proceedings (UK), 18*(2), 99–103.

Weng, Z., Hendrickx, M., Maesmans, G., & Tobback, P. (1992). The use of a time–temperature integrator in conjunction with mathematical modelling for determining liquid/particle heat transfer coefficients. *J. Food Eng., 16*, 197–214.

Wiese, K. L., & Wiese, K. T. (1992). A comparison of numerical techniques to calculate broken-line heating factors of a thermal process. *J. Food Proc. Preserv.*, *16*(5), 301–312.

Williams, A. W., & Adams, J. B. (1997). A time-temperature integrator to quantify the effect of thermal processing on food quality. In R. Jowitt (Ed.), *Engineering and food*, *ICEF7* Part 2, §K (pp. 9–12). Sheffield, UK: Sheffield Academic Press.

Yawger, E. S. (1978). Bacteriological evaluation for thermal process design. *Food Technol.*, *32*(6), 59–62.

7
Quality Optimization

7.1. Introduction

The process that is delivered to a canned or packaged food not only inactivates the spoilage micro-organisms, but also cooks the product to an acceptable texture. Canned foods, being convenience foods, are thermally processed to such a degree that they only require reheating prior to consumption. The amount of cooking that a product receives depends on its consistency, the thermal processing conditions and the container or package size. For convection-heating products the internal mixing permits a fairly uniform cook for the whole product; however, for conduction-heating products the heating and, consequently, the degree of cooking vary from the outside to the inside of the food. The product nearest the container wall receives the maximum heat treatment and is consequently the most cooked portion of the food. In general this results in overcooking the outer layers, with a consequent loss of overall quality, especially for products in large-diameter cans. These products are generally perceived as being of lower nutritional value than their fresh or chilled counterparts, which has led to a vast amount of both theoretical and practical work in an attempt to reduce processing conditions. (Holdsworth 1985, 2004).

The methods which have been employed to reduce thermal processes have included altering the geometry of the container, making it thinner, for example, by using plastic trays and pouches; storing the product under chilled conditions; acidification and pasteurization; and using different heating methods, including microwave and ohmic heating.

Some of the earliest work on this subject was empirically based, in particular that of Brown (1950) on canned vegetables and Tischer *et al.* (1953) on canned meat. In the former work, four different processing times and five different temperatures (110–132°C) were used, and it was shown that the long processes at the lower temperatures had an adverse effect on the color and flavor of canned carrots compared with the higher-temperature, shorter-time processes. The texture of carrots, but not of peas, was influenced markedly by the duration of the process. The process conditions had little effect on the levels of carotene or ascorbic acid; however, vitamin B_1 was better preserved by the "high short" processes. The latter work was devoted to a study of the heat processing of beef in 300 × 308 cans in which changes in tenderness and drained juice levels were measured for temperatures in the range 107–124°C. The results were illustrated in three-dimensional graphs, and empirical correlations were developed relating the objective quality measurements to temperature. There is a wealth of work of this type in the literature, but here we are more concerned with optimizing the effects of quality changes and microbial inactivation using more basic definitions. The process engineer should also be aware, in general terms, of the effects of heating on foodstuffs; however, this subject is beyond the scope of this text, and the reader should consult specialized texts for further information, e.g. Hoyem and Kvale (1977), Holdsworth (1979), and food commodity texts.

7.2. Cooking Versus Microbial Inactivation

In the past it was common to use a two-stage process for some ready-meals such as beans in tomato sauce: The product was cooked in the container for a long time at a relatively low temperature and then sterilized at a higher temperature. Such two-stage processes have been superseded by single-stage processes, but they showed that there was an appreciation of the different needs for cooking and sterilization by the production company. Commercial processes are a balance between the requirements for microbiological inactivation and thermal cooking, and the art of canning has been to determine optimum processes for this purpose. In fact, a range of processes can be obtained which have the same F-value but different times and temperatures; such processes are described as *equivalent*. Similarly, there are equivalent cooking processes for different types of processes and conditions (Gillespy 1956; Atherton & Thorpe 1980). For example, Table 7.1 shows some processes for processing beans in tomato sauce, equivalent to a standard process

TABLE 7.1. Minimum equivalent processing times in minutes for sterilization and cooking of beans in tomato sauce in a rotary cooker.

Can size	Process temperature (°C)		
	115.5	121.1	126.6
A1			
sterilization	22	$13\frac{1}{2}$	10
cooking	38	26	19
A2			
sterilization	25	$16\frac{1}{2}$	$12\frac{1}{2}$
cooking	40	28	21
$A2\frac{1}{2}$			
sterilization	27	18	$14\frac{1}{2}$
cooking	41	29	22
A10			
sterilization	35	26	21
cooking	46	34	27

TABLE 7.2. Variation of F- and C-values with different processing temperatures and a time of 1 min (Holdsworth 1992).

Temperature (°C)	$F_{121.1}^{10}$ (min)	$C_{100}^{33.1}$ (min)	$C_{121.1}^{33.1}$ (min)
100	0.00776	1.0	0.23
110	0.0776	2.15	0.46
121.1	1.00	5.05	1.00
130	7.76	10.0	1.85
140	77.6	21.5	3.71
150	776.0	46.4	7.42

of 40 min at 115.5°C in an A2 can. The advantage of using the higher-temperature, shorter-time processes has been to increase the throughput of the canning plant. The main difference between cooking and other thermally degenerate processes, e.g. nutrient destruction, is that the kinetic parameters, z_c- and D_c-values are very different. The z-values are generally much higher for cooking and nutrient degradation, 25–45°C, than those for microbial inactivation, 7–12°C. For detailed values, see Appendix A for microbial inactivation and Appendix B for thermally vulnerable components. In very general terms, for every 10°C rise in temperature the cooking rate doubles, but the sterilization rate increases tenfold.

In Chapter 4 the criterion for cooking was discussed (Mansfield 1962, 1974; Ohlsson 1988). The basic equation for the cook value $C_{T_{\text{ref}}}^{z}$ is given by

$$C = \int_0^t 10^{(T-T_{\text{ref}})/z_c} dt. \tag{7.1}$$

The cook value parameters z_c and T_{ref} differ according to the particular thermolabile component considered. For cooking, the z_c-value chosen is usually 33.1°C and the reference temperature 100°C, and this is designated C_0, i.e. $C_{100}^{33.1}$, although $C_{121.1}^{33.1}$ is often used for comparison with F_0 values. It is important to define the constants z_c and T_{ref} clearly so that there is no misunderstanding.

Table 7.2 shows how F- and C-values vary for ideal circumstances, i.e. instant achievement of temperatures and a time of 1 min. For every 10° rise in temperature, the F-values vary by a factor of 10, the C-values by a factor of approximately 2. Using the lower reference temperature of 100°C, the C-values are higher than using the higher temperatures. In Table 7.3 the sterilizing value is kept constant and the variation of the C-values with processing temperature

TABLE 7.3. Variation of C-values with temperature for a fixed F_0-value and a time of 1 min (Holdsworth 1992).

Temperature (°C)	$F_{121.1}^{10}$ (min)	$C_{100}^{33.1}$ (min)	$C_{121.1}^{33.1}$ (min)
100	4.0	515.3	118.0
110	4.0	103.0	23.8
121.1	4.0	17.35	4.0
130	4.0	4.15	0.96
140	4.0	0.80	0.19
150	4.0	0.17	0.04

is observed. As the temperature rises the amount of cooking, as shown by the C-value, decreases. This is the basis for high-temperature, short-time cooking strategies; ideally at the higher temperatures and with the shorter times less thermal damage will be done to the product. This will only apply if there is no heat transfer lag, and consequently does not apply to larger sizes of container filled with conduction-heated foods. It is, however, applicable to the continuous sterilization of thin liquids, e.g. milk (Holdsworth 1992), and to the processing of convection-heating products. There are, however, serious engineering problems in applying processes of 150°C to canned products. These are the pressure limitations of many conventional retort systems and the problems of the come-up time.

7.3. Process Evaluation

7.3.1. Some Models for Predicting Nutrient and Cooking Effects

For evaluating processes for cooking and other effects, the C-value, known in optimization theory as an *objective function*, determined at the center of the pack, is of less value than the mass-average cook value, C_s. This is because of the uniform distribution of nutrients in the product and the fact that there is a different threshold for contribution to the equivalent of the lethality for heat-vulnerable components. The C_s-value is equivalent to the F_s-value for sterilization, which is the total integrated F-value for the whole volume of the container. It is obtained from the equivalent sterilization definitions,

$$C_s = D_{\text{ref}} \log(c_0/c), \tag{7.2}$$

where D_{ref} is the decimal reduction time for the species being considered, and c and c_0 are the concentrations at times t and 0, respectively.

Silva *et al.* (1992b) have pointed out a way to use the C_s-value that, like the F_s-value, shows a dependence on the D_{ref}-value. The volume-average quality retention value is given by

$$c/c_0 = \frac{1}{V} \int_0^V 10^{-C_c/D_{\text{ref}}} \, dV, \tag{7.3}$$

where

$$C_c = \int_0^t 10^{(T - T_{\text{ref}})/z_c} \, dt, \tag{7.4}$$

McKenna and Holdsworth (1990) have reviewed the published models for determining F_s and C_s. Other work in this area has been presented by Ohlson (1980a,b&c) and Tucker & Holdsworth (1990, 1991).

All the methods outlined in Chapter 6 may be used to evaluate these integrals. Preussker (1970) has given a simplified equation based on Ball's exponential integral equation (Ball & Olson 1957), in which a simple approximate algorithm for the $-Ei(-x)$ function is used, viz., $-Ei(-x) = \ln X^{-1} - 0.6$, which is as follows:

$$C = (L_c/n)[nt - (\ln(T_R - 10)) + 2.06] \tag{7.5}$$

where $L_c = 10^{(T_R - 100)/33}$, T_R is the retort temperature, t is the processing time, n is $h/(15c\rho d)$, h is the heat-transfer coefficient, c is the specific heat, ρ is the density and d the can diameter.

Processes with differing values for T_{ref} and z_c can be inter-converted using the following formulae:

$$C_{T_{ref,1}} = 10^{(T - T_{ref,1})/z_c} t, \tag{7.6}$$

$$C_{T_{ref,2}} = 10^{(T - T_{ref,2})/z_c} t. \tag{7.7}$$

Eliminating t,

$$\log C_{ref,1} - \log C_{ref,2} = (T_{ref,2} - T_{ref,1})/z_c \tag{7.8}$$

or

$$C_{100} = C_{121.1} \times 10^{(121.1 - 100)/z_c}.$$

Similarly, for a sterilization process with a given F-value and a cooking process with a given C-value, it can be shown that

$$\log C = (z/z_c) \log F + (121.1 - 100)/z_c$$

for a process where $t = 1$ min, or

$$\log(C/t) = (z/z_c) \log(F/t) + (121.1 - 100)/z_c \tag{7.9}$$

for a process of duration t min. These equations apply essentially to instantaneous heating and cooling of the product; they do not take into account the realities of heat transfer under normal canning operations.

7.3.2. Some Typical C-Values

The way in which the $C_{121.1}$-value builds up with time is shown in Figure 7.1, and this can be compared with the build-up of the F-value. Relatively few workers have produced target C-values in the same way as for F-values. Table 7.4 contains some typical practical values for processed ready meals (Tucker & Holdsworth 1991). Other workers have produced C-values, in particular Preussker (1970) for static processes and Eisner (1988) for rotary processes.

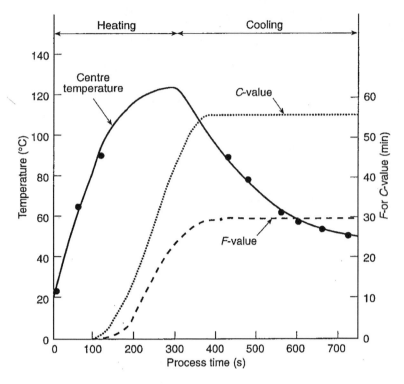

FIGURE 7.1. Graph showing the cumulative increase in F- and C-values.

7.4. Optimization of Thermal Processing Conditions

7.4.1. Graphical Approach

The choice of processing conditions may be determined from a plot of log time versus temperature, on which are drawn two straight lines representing constant F- and C-values, as shown in Figure 7.2. These lines divide the plot into four regions: The line $F_1 O F_2$ marks the boundary between processes that give adequate sterilization and those that do not, while $C_1 O C_2$ marks the boundary between adequate and inadequate cooking. Idealized graphs like this are useful for determining the suitability of various combinations of temperature and time. Table 7.5 lists workers who have used this procedure for process determination for different products. It should be noted that the graphs are based on instantaneous heating followed by instantaneous cooling of the product, in particular to thin films of product. Under more realistic conditions it is necessary to include the effects of heat transfer and dimensions of the object being processed. When this is done the straight lines in Figure 7.2 become curved, as shown in Figure 7.3 (Holdsworth 1985), and the regions have different boundaries.

TABLE 7.4. Some C_0-values for commercial sterilization processes (from Tucker & Holdsworth 1991).

Product	Container size	Process temperature (°C)	Process time (min)	F_0 (heat) (min)	F_0 (total) (min)	f_h (min)	Thermal diffusivity ($\times 10^3$ cm^2 s^{-1})	C_0 (centre) (min)	C_0 (Volume-average) (min)
Beans in tomato sauce	A2	121.1	121	7.0	11.6	64.6	1.46	196.5	331.6
Beans in tomato sauce	UT	121.1	94	5.8	8.3	51.5	1.46	155.0	267.2
Carrot purée	A1	121.1	74	5.5	8.0	37.0	1.58	128.5	210.3
Celeriac purée	A1	121.1	72	4.2	6.0	39.0	1.50	117.2	199.8
Chicken supreme sauce	UT	121.1	86	4.5	6.6	49.0	1.54	138.3	242.1
Chili con carne	UT	121.1	91	4.5	6.6	52.2	1.44	144.0	255.3
Corned beef	300 × 200	121.1	39	4.5	6.6	28.5	1.73	80.9	124.5
Mackerel in tomato sauce	UT	121.1	97	7.0	10.9	50.8	1.49	168.0	280.5
Minced beef	UT	121.1	101	6.0	8.6	55.0	1.37	166.9	287.0
Mushroom soup, cream	A1	115.7	93	3.5	5.8	37.0	1.58	132.8	195.1
Pet food	UT	125.8	84	12.0	20.0	48.0	1.57	180.4	323.6
Stewed steak	UT	121.1	105	9.0	12.0	52.0	1.45	188.3	308.2
Spaghetti hoops in tomato sauce	U8	121.1	41	7.5	11.6	22.2	2.50	96.2	136.0
Spaghetti in tomato sauce	A2$\frac{1}{2}$	121.1	83	6.0	9.0	49.0	2.50	148.5	245.1
White wine sauce	UT	121.1	81	4.5	6.5	46.0	1.64	131.3	229.3

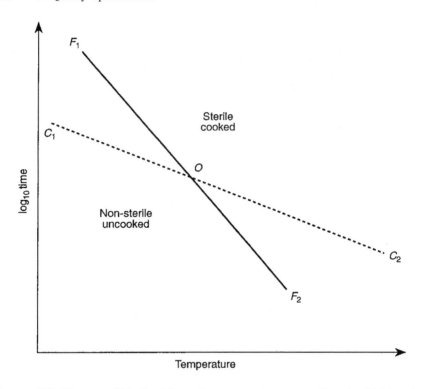

FIGURE 7.2. Diagram of idealized $\log_{10}$ time versus temperature for microbial inactivation ($F_1 O F_2$) and cooking ($C_1 O C_2$) of food product, instantaneously heated.

One of the advantages of this type of representation is that various effects may be plotted on the same graph. In Figure 7.4, for example, the lines for differing percentages of vitamin retention or discoloration have been plotted and appropriate processes selected. While the technique is suited to thin-liquid products, e.g. milk and fruit juices, it is less useful for the average canned products, which show a degree of conduction and contain substantial quantities of particulate material. The advantages of using high-temperature, short-time processes for thin liquid products are clearly demonstrated. For thicker materials, it is necessary to consider alternative methods of heating, e.g. microwave and ohmic heating, to overcome the problems of consistency of canned foods in general. At the present time this can only be achieved in continuous ohmic heating processes with aseptic filling, but in the future the use of microwave in-package sterilization will be realized on a fully commercialized scale.

7.4.2. Optimization Models

7.4.2.1. Simple Methods

One of the earliest mathematical treatments of the optimization of the process conditions for thiamin destruction versus sterilization, in cylindrical cans of

TABLE 7.5. Graphical optimization of microbial versus heat-vulnerable components (after Holdsworth 1985).

	System				
Heat vulnerable component			Microbial		
Description	z_c (°C)	C_{100} (min)	z (°C)	$F_{121.1}$ (min)	Reference
Thiamin destruction in cured meat, 10%, 20% and 50%	–	–	10	0.25	Greenwood *et al.* (1944); Ball and Olson (1957); Jackson *et al.* (1945)
Cooking	33	5–30	10	2–30	Mansfield (1962)
Betanin destruction 5–99%	–	–	10	1.0	Herrmann (1969)
Cooking effect of linear heating	33	0.1–50	10	0.1–50	Preussker (1970)
Enzymes in potatoes	10.3	–	10	2.5	Reichert (1977)
Enzymes	17.5	–	8.9	0.9	Reichert (1977)
Enzymes in green beans	48.9	–	8.9	0.9	Reichert (1977)
Vitamin C	23.2	–	8.9	0.9	Reichert (1977)
Vitamin B$_1$	26.1	–	8.9	0.9	Reichert (1977)
Cooking	25–40	–	8.9	0.9	Reichert (1977)
Sensory	26.5	–	8.9	0.9	Reichert (1977)
Chlorophyll/green beans	87.8	–	8.9	0.9	Reichert (1977)
Cooking	33	10, 36, 52	10.0	1.0	Reichert (1974, 1977)
Cooking peas	29	42, 45, 62	10.0	76.0	Reichert (1974, 1977)
Vitamin B in liver, 5–90% destruction	26.1	–	10.0	5–10	Bauder and Heiss (1975)
Microbial lipase	3.1	–	10.0	2.7	Svensson (1977)
Peroxidase	35	–	10.0	10.0	Svensson (1977)
Thiamin retention 90–99.5%	–	–	10.0	5.0	Ohlsson (1980b)
Thiamin destruction, 1%, 5%, 20% and 50%	–	–	10.0	6.0	Lund (1977)
Thiamin loss in milk, 3%	–	–	10.5	2.0	Kessler (1981)
Lysine loss in milk, 1%	–	–	10.5	2.0	Kessler (1981)
Protease inactivation, 90%	–	–	10.5	2.0	Kessler (1981)
Lipase inactivation, 90%	–	–	10.5	2.0	Kessler (1981)
No discoloration	–	–	10.5	2.0	Kessler (1981)
Enzymes/food particulates	27	–	10.0	3.0	Brown and Ayres (1982)
Anthocyanin in grapes, 90% destruction	23	18	10.0	24.0	Newman and Steele (1978)
Browning and protease destruction	25	–	10.0	4.0	Jelen (1983)

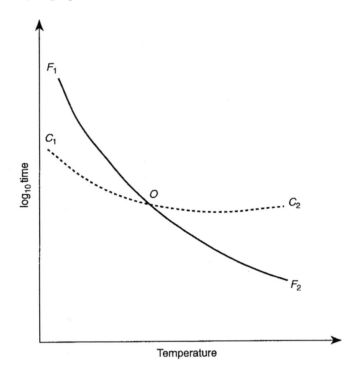

FIGURE 7.3. Diagram of $\log_{10}$ time versus temperature for microbial inactivation ($F_1 O F_2$) and cooking ($C_1 O C_2$) for the central point in a food product.

conduction-heating product, was due to Teixeira *et al.* (1969). A finite-difference method was used for determining the temperature distribution and the corresponding thiamin distribution, employing first-order degradation kinetics. Figure 7.5 shows the percentage thiamin retention plotted against equivalent process times and temperatures, based on 84 min at 121.1°C (Stumbo 1973). The curve shows that the maximum retention of thiamin is obtained for an optimum process of 120°C for 90 min, which is very much a conventional canning process. The curve also shows that for conduction-heating products HTST conditions have an adverse effect on nutrient retention. The model was used for studying the effects of changing variables on the optimum conditions. For heat-vulnerable components with z_c-values lower than for thiamin, the optimum moves to the lower-temperature, shorter-time region and, correspondingly, in the opposite direction for higher z_c-values. Changing the D_c-values had no effect on the position of the maximum but increased the nutrient retention for increasing D_c-values.

In the same year, Hayakawa (1969) published a new procedure, based on dimensionless parameters, for calculating mass-average sterilizing values, F_s, which could be used as C_s-values with the appropriate D_c- and z_c-values (see equation (7.2)). He showed that the method was simpler to use than that of Ball and Olson (1957) and that the results compared more favorably with experimental

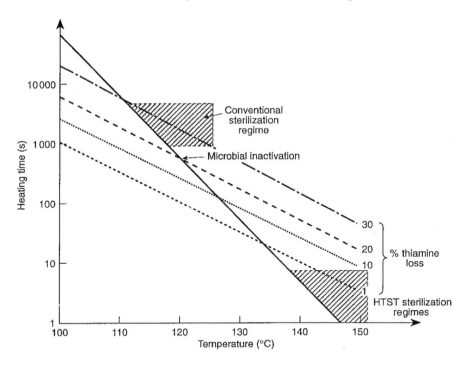

FIGURE 7.4. Degradation of thiamine for different sterilization regimes.

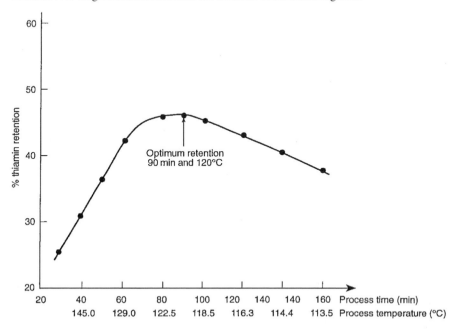

FIGURE 7.5. Optimization curve for percentage thiamine retention ($z = 25°C$) against equivalent process times/temperatures (from Teixeira *et al.*, 1969).

determinations of nutrient retention. A more rigorous method of determining the mass-average value for a physical, chemical or biological factor in food for various simple geometrical shapes was presented by Hayakawa (1971).

Jen et al. (1971) developed Stumbo's method (see Stumbo 1973) further for application to nutrients. In Stumbo's method for obtaining F_s-values a relationship, $F_\lambda - F_c = m\upsilon$ (discussed in Chapter 6), was used for determining F_λ, the F-value at $j = j_c/2$. The probability of microbial survival for values of the volume parameter υ greater than 0.4 was considered negligible. While this is satisfactory for microbial inactivation, it is not so for nutrient degradation. Jen et al. (1971), starting with the equation

$$10^{-F_s/D_{\text{ref}}} = \int_0^V 10^{-F_\lambda/D} dV, \tag{7.10}$$

obtained the equation

$$F_s = F_c + \log\left[1 + AD_{\text{ref}}^{-1}(F_\lambda + F_c)\right], \tag{7.11}$$

which took into account the previously mentioned conditions for nutrient reactions; in equation (7.11)

$$A = -\ln 10/\ln(1 - \upsilon),$$

which, for a value of $\upsilon = 0.19$, gives $A = 10.927$. Newman and Holdsworth (1989) obtained some other values: For a cylindrical container, $A = 10.734$; for a sphere, $A = 9.284$; and for a brick, $A = 11.737$. The method was further extended by producing tables of f_h/U versus g (see Table 6.3) for values of $z = 10$, 24.5 and 25.5°C.

Tables 7.6 and 7.7 show some comparative values of thiamin retention obtained using different methods. The Jen et al. (1971) method produced thiamin retention values close to the analytical values.

Downes and Hayakawa (1977) extended previous approaches to deal with the curvilinear portion of the heating curve and for conditions when f_h was not equal to f_c. New tables were produced for determining temperature distributions and the method compared better than others with analytical results (see Table 7.6).

Teixeira et al. (1975b) studied the effect of container size on thiamin retention in conduction-heating products and showed, as expected, that this had a major influence on nutrient retention. For equal volumes the thiamin retention decreased from 68% to 41% for values of L/D increasing from 0.096 to 1.270 (No. 2 can) and then increased from 43% to 63% for values of L/D increasing from 1.710 to 13.750. Using time-varying surface temperature conditions was found to have little effect on vitamin retention.

Lenz and Lund (1977) showed that for short processing times (less than 20 min), the statistical distribution of C_s-values was normal; however, for longer times the distribution exhibited an increasing standard deviation and a pronounced skewness.

Sjöström and Dagerskog (1977) studied the browning of canned chopped fish at temperatures between 110 and 145°C using $z_c = 33$°C. Using a series of process

TABLE 7.6. Comparisons of thiamin retention predictions using various methods.

Product	Can size	Process temperature (°C)	Process time (min)	Stumbo (1953)	Ball & Olson (1957)	Hayakawa (1969)	Teixeira et al. (1969)	Jen et al. (1971)	Downes & Hayakawa (1971)	Thijssen et al. (1978)	Experimental analysis	Reference
Pea purée	211 × 300	121.1	30	–	75.5	84.1	86.0	89.0	86.1		86.7	Jen et al. (1971)
Pea purée	211 × 300	121.1	60	–	55.2	59.2	61.0	61.4	60.6		65.6	Jen et al. (1971)
Pea purée	211 × 300	115.5	30	–	82.1	91.0	91.5	93.5	92.9		93.6	Jen et al. (1971)
Pea purée	303 × 406	121.1	85	48.7	–	–	49.2	–	–		50.7	Teixeira et al. (1975a)
Pea purée	303 × 406	115.5	136	41.0	–	–	41.6	–	–		42.3	Teixeira et al. (1975a)
Pea purée	303 × 406	126.7	65	47.3	–	–	48.8	–	–		51.3	Teixeira et al. (1975a)
Pea purée	303 × 406	110[a]	–	–	–	–	53.5	–	–		53.5	Teixeira et al. (1975a)
Carrot purée	211 × 300	121.1	60	–	53.9	59.0	–	–	–	63.5	66.8	Thijssen et al. (1978)
Pea purée	211 × 300	121.1	60	–	55.2	59.5	61.4	61.0	–	64.6	65.6	Thijssen et al. (1978)
Pork purée	211 × 300	126.7	40	–	49.2	52.2	–	–	–	57.3	53.0	Thijssen et al. (1978)

[a] + 5.6° step increase every 20 min.

TABLE 7.7. Some theoretical predictions for nutrient retention in canned products.

Product	Can size	Process temperature (°C)	Process time (min)	Initial temperature (°C)	Thiamin	% reduction in experimental results	Chlorophyll	Reference
Conduction	211 × 400	110	80	20	69		85	Savage (1984)
Conduction	211 × 400	110	67	80	68		84	Savage (1984)
Conduction	211 × 400	120	45	20	60		71	Savage (1984)
Conduction	211 × 400	120	35	80	60		70	Savage (1984)
Conduction	211 × 400	130	35	20	64		68	Savage (1984)
Conduction	211 × 400	130	25	80	85		67	Savage (1984)
Simulated product	211 × 300	108.6	44.5	28.9	77.5	79.3	–	Barreiro Mendez et al. (1977)
Simulated product	307 × 409	95.1	59.0	27.9	79	79.8	–	Barreiro Mendez et al. (1977)
Simulated product	211 × 300	117.7	50	25.9	59.6	59.7	–	Barreiro Mendez et al. (1977)
Pork purée	307 × 409	121.11	67	71.11	48 approx.			Banga et al. (1991); Garcia et al. (2006)
Meat paste	144 × 35 mm	112	63			44		Barroso et al. (2001)
Meat paste	144 × 35 mm	115	45			39		Barroso et al. (2001)
Meat paste	144 × 35 mm	118	32			35		Barroso et al. (2001)
Meat paste	144 × 35 mm	120	25			44		Barroso et al. (2001)

times and temperatures equivalent to $F = 7.5$ min, it was shown that the C-value was a minimum at about $127°C$ and 60 min for a position intermediate between the surface and the center. The surface was least discolored at temperatures below $120°C$, and variable temperature heating profiles had little effect on color retention.

Thijssen and Kochen (1980) extended the previously derived short-cut method of process evaluation (Thijssen *et al.* 1978) to the case of variable heating temperature profiles but did not present any experimental data for food products.

Ohlsson (1980a, b, c) conducted an important series of experiments on the determination of C-values for a range of products (fish paste, liver paste, strained beef, strained vegetables, tomato sauce and vanilla sauce) and for a range of different sensory qualities (odor, appearance, taste, consistency, hardness, coarseness and lightness) in different shapes of container. It was shown that for a given F-value, the volume-average cook values for $z = 25°C$ showed minimum values that decreased with increasing temperature and decreasing can size. The optimal processing temperature, between 117 and $119°C$ for a 73×99 mm can, was in agreement with the previous work of Teixiera *et al.* (1969).

Reichert (1980) developed a method of processing known as ΔT-cooking, in which a fixed temperature differential, ΔT, was maintained between the retort temperature and the temperature at the center of the can. This was further extended by Hendrickx (1987), using a numerical algorithm – transmission line matrix (TLM) modeling – to optimize quality retention in conduction-heated foods.

Richardson *et al.* (1988) developed a finite-difference model, for use on an IBM PC-XT to determine nutrient retention in conduction-heating packs. A practical observation from this work was that the results from the experiments were better correlated with the theoretical results when the temperature was measured at the surface of the container than when using the retort thermometer measurements. By this means the effect of the heat-transfer coefficient at the surface of the container, especially during the cooling period, was eliminated. This work was important for modeling the temperature distribution in real time for control purposes.

Alwis *et al.* (1992) used the SPEEDUP[TM] system (Simulating Program for Evaluating and Evolutionary Design of Unsteady-State Processes.) developed by Sargent *et al.* (1982) for optimizing quality and processing effects of a conduction-heating meat product. This showed the need for higher temperatures and shorter times for obtaining better meat quality.

Hendrickx *et al.* (1992c) have used an ANSYS finite element package (De Salvo & Gorman 1989) to determine the maximum surface quality retention as a function of (a) product properties; (b) processing conditions, including geometry and dimensions of the food, surface heat-transfer coefficient, initial product temperature and retort come-up-time; and (c) processing criteria, i.e., target F_o-value.

Kebede *et al.* (1996) studied heat transfer into high-barrier plastic trays containing an 8% bentonite suspension containing ascorbic acid. The containers had a net volume of 460ml and had dimensions $35 \times 97 \times 142$ mm and were made from (i) polypropylene/ethylene vinyl alcohol/polypropylene and (ii) crystallized polyethylene terephthalate. The average retention of ascorbic acid after heating to achieve an F_0 of 11–12 min. in the trays was 83% compared with 75% for the cans. Similarly the trays performed better than the cans during storage at 35°C for 105 days.

Pornchaloempong et al. (2001, 2003) have conducted a theoretical evaluation of the effect of various product and processing factors on the quality retention in products in three conical-shaped containers. The factors studied included retort temperature processing times for equivalent lethalities, z-value, D-values, thermal diffusivity, surface heat transfer and container geometry.

Simpson et al. (2004) have developed a finite–difference model for predicting the quality optimization of Jack mackerel frustrum vacuum-packed in flexible plastic pouches utilizing variable retort temperature profile.

García et al. (2006) have developed retort control profiles for the optimization of thiamine retention in a 307 × 409 can filled with pork purée, using a dynamic optimization technique CVP (control vector parameterization) and the F & C relations of the form of equation 7.3.

Balsa-Canto et al. (2002a, 2002b) have presented a novel and important method for optimization based on proper orthogonal decomposition to give reduced-order models that are easier to solve than the formal non-linear partial differential equations. The method was successfully tested using data for canned pork purée.

Table 7.8 summarizes quality optimization studies.

7.4.2.2. Formal Optimization Techniques

Saguy and Karel (1979) used an elegant multi-iterative mathematical technique, based on the continuous maximum principle theory of Pontryagin et al. (1962), to optimize thiamin retention in pea purée in a 303 × 406 can and pork purée in a 401 × 411 can. The method produced a variable-temperature heating profile that optimized the nutrient retention. A constant heating temperature regime was shown to be almost as good as the theoretically derived profile.

Norback (1980) reviewed the methods of optimization and suggested that the best method was dynamic programming. This could also be applied to process control.

Hildenbrand (1980) developed a two-part approach to solving the problem of optimal temperature control. In the first part, the unsteady-state equation for heat transfer into a finite cylinder was solved using Green's functions. In the second part, a method to ensure that the container received the calculated temperature profile was determined. While the approach seems interesting, no further development appears to have taken place.

Nadkarni and Hatton (1985) examined the previous work and considered that the methods were not sufficiently rigorous to obtain the best optimization results. These workers used the minimum principle of optimal control theory to obtain optimal solutions. Again, simple heating and cooling profiles were better than complex heating profiles.

TABLE 7.8. Some quality optimization studies.

(1) Optimization studies with constant retort temperature		
Static process	Thiamin retention	Lund (1988)
Static process	Effect of packaging	Silva *et al.* (1994a)
Static process	Average & surface quality.	Silva *et al.* (1994b, c, d)
Static & rotary water cascade retorts	Effect of process	Smout *et al.* (2001)
Static process	Thiamin retention	Teixeira *et al.* (1969)
Static process	Thiamin retention	Chen and Ramaswamy (2002b)
Static process	Effect of variability in thermal properties	Baucour *et al.* (2003)
Static & rotary Water cascade retort	Effect of variability in white & green beans	Ávila *et al.* (2006)
Rotary retort	Texture & color of vegetables	Abbatemarco and Ramaswamy (1994)
(2) Optimization studies with variable retort temperature		
(a) Average (volumetric) retention		
Static process	Surface quality: canned salmon	Durance (1997); Durance *et al.* (1997)
Static process	Effect of heating and cooling profiles	Nadkarni and Hatton (1985)
Static process	Pea purée	Saguy and Karel (1979)
Static process	Pea purée	Teixeira *et al.* (1975b)
Static process	Thiamin in cylinders & spheres	Erdögdu and Balaban (2003a).
(b) Optimization of surface quality, process time & energy conservation.		
Static process	Energy optimization	Almonacid-Merino *et al.* (1993)
Static process	Surface quality, minimum process time	Banga *et al.* (1991)
Static process	Energy & surface quality—canned salmon	Durance *et al.* (1997); Noronha *et al.* (1993)
Static process	Theoretical effect on surface quality, process time	Chen and Ramaswamy (2002c)

Banga *et al.* (1991) developed an optimization algorithm, Integrated Control Random Search (ICRS), for three objective functions: maximum overall nutrient retention, maximum retention of a quality factor at the surface of the food, and the minimum process time. It was concluded that the use of a variable-temperature profile was advantageous for preserving optimum surface quality. Similar profiles to those shown in Figure 7.5 were obtained for both overall nutrient retention and surface quality retention. Some other optimization methods are listed in Tables 7.8 and 7.9.

7.4.2.3. Objective Functions

Silva *et al.* (1992a) reviewed and critically examined the objective functions, volume-average retention and volume-average cook value used for optimization purposes. They concluded that for high D-values, e.g. vitamin destruction, the two functions were equivalent; however, for low D-values, e.g. texture and color,

TABLE 7.9. Some optimization techniques applicable to thermal processing.

Technique	Application	Comment	Reference
Continuous minimum principle (Pontryagin et al. 1962)	Cylindrical can; thiamin retention; conduction-heating pack; pea purée	Optimum retort temperature; profile-determined	Saguy and Karel (1979)
General review of techniques	–	Definition of terms, continuous and discrete problems	Norback (1980)
Comparison of optimization techniques	–	Excellent review of method	Nakai (1982)
Continuous minimum principle (Pontryagin et al. 1962)	Cylindrical can; thiamin retention; conduction-heating pack; pork purée	Optimum temperature profiles, optimal control policies	Nadkarni and Hatton (1985)
Super-simplex optimization	–	Modified optimization technique	Nakai et al. (1984); Aishima and Nakai (1986)
Computer program	Batch retorting of canned foods	Optimal temperature profiles for nutrient retention	Kok (1986)
CVP (control vector parameterization)	Thiamin retention; conduction heating pack	Optimal temperature profiles for nutrient retention	Garcia et al. (2006)
Complex method	Batch retorting canned foods	Non-linear constraint technique (Kozmierczak (1996)	Erdögdu & Bablan (2001, 2002, 2003b)
Neural network	Batch retorting canned foods	Prediction of safety & quality	Kseibat et al. (2004)
Proper orthogonal decomposition (POD)	Thiamin retention; canned pork	Retort temperature profiles	Balsa-Canto et al. (2002a, 2002b)
Dynamic optimization g-OPT*	Thiamin retention; canned pork purée	Optimum temperature profiles & good review of methods	Garcia et al.(2006)
Neural networks & genetic algorithms ANN-GA	Idealised system	Effect of variable retort temperature	Chen & Ramaswamy (2002a)

*g-OPT is the dynamic optimization tool of g-PROMS (available from Process Systems Enterprise Ltd., London.)

this is not the case. The volume-average cook value does not take into account the effect of the D-value and consequently underestimates the optimum processing temperature. The effect of finite surface heat-transfer coefficients on the optimal sterilization temperature was found to be significant; however, the initial temperature and the come-up time had little effect (Silva et al. 1992b).

The use of optimization techniques has been discussed in terms of commercial objectives as well as quality and sterilization (Oliveira 2004; Simpson 2004). Business objectives have not been fully studied to date, although Simpson et al. (2003) have produced a comprehensive analysis of the canning operation.

7.4.2.4. Semi-Empirical Equations

Hendrickx *et al.* (1989, 1993) developed a semi-empirical model approach to determining optimal temperatures to minimize surface quality and nutrient losses. This approach was extended to include the effect of the cooling phase and external finite heat transfer, using simple geometrical models, viz. infinite slab, infinite cylinder and sphere (Hendrickx *et al.* 1992). The optimized temperatures were obtained for a minimum surface cook value with the constraint of the sterilization requirement.

The basic empirical formula had the general form

$$T_{op} = a + b \log(F_t/f_h) + c \ln z_q + dT_0, \tag{7.12}$$

where T_{op} is the optimal temperature, F_t is the sterilization value constraint, f_h is the heat penetration factor, z_q is the temperature sensitivity factor for surface quality degradation, and T_0 is the initial temperature of the pack. The regressed values for the various coefficients were $a = 86.68$, $b = 9.73$, $c = 10.46$ and $d = 0.025$. It was shown that variation of the initial temperature and the come-up time had a negligible effect on T_{op}; however, they changed linearly with $\ln z_q$ and $1/Bi$, where Bi is the Biot number (hd/k, in which h is the heat-transfer coefficient, d the characteristic linear dimension, and k thermal conductivity). The effect of water processing with a reduced external heat-transfer coefficient was slightly to increase T_{op}, compared with steam processing. Norohna *et al.* (1993) studied the effects of variable temperature profiles on maximizing the surface quality retention and found that variable-temperature profiles improved the surface quality by up to 20% compared with constant-temperature retort profiles.

This technique was used by Van Loey *et al.* (1994) to optimize the thermal process for canned white beans using both static and rotary processing. The rotary end-over-end process produced the better-quality pack.

Silva *et al.* (1994a, b, c) studied the maximization of the surface quality retention of one-dimensional conduction-heated foods, e.g. pouches. Optimal temperatures were determined for a range of conditions. The semi-empirical approach to determining optimal processing temperatures for quality retention in packaged foods has been shown to be an effective and simple method compared with the complexities of the more formal methods.

Norohna *et al.* (1996a) have optimized the surface quality for three food systems, viz., (i) a meal-set consisting of chili con carne/white rice/peaches slices in syrup in a plastic pouch, (ii) a meal-set consisting of meat/potatoes/spinach in a plastic pouch and (iii) a meal-set comprising a mixture of four vegetables: green beans/peas/corn/carrots in glass. The surface quality retention was studied under conditions of constant and variable retort temperatures, and it was shown that the latter conditions had an advantage. A similar advantage has also been shown in the case of cooked white beans in glass (Norohna *et al.* 1996a, b).

7.4.2.5. Computer Software

Nicolai *et al.* (2001) have developed a number of packages, e.g., CookSim and ChefCad, that deal with a range of food processes, including sterilization operations. The CookSim package is a knowledge-based system that uses a finite-element approach for solving the heat transfer, sterilization and cooking equations. Chef-Cad is used for computer design of complicated recipes consisting of consecutive heating/cooling stages.

7.5. Quality Assessment Through Mass Balance

The hypothesis is that in food-processing applications, quality – in most cases – can be treated by a mass balance. The conditions to carry out a mass balance for a given attribute are the ability to express it in concentration terms (w/w or w/v) and also to account for a kinetic expression (e.g. degradation of quality attribute as a function of time and temperature). Most quality factors, like vitamins, enzymes and color pigments, etc., can be expressed in terms of concentration. In addition, a large amount of work in the area of kinetics for different target attributes has been carried out in the past three decades.

Firstly, the general mass balance equation for the quality of attribute j through the system is:

$$[A_j]_{IN} - [A_j]_{OUT} + [GenerationA_j] + [ConsumptionA_j]$$
$$= [AccumulationA_j]_{SYSTEM}$$

Where

A_j: Mass of Attribute j per unit time (w/t)

As follows, we are presenting the general expression for differential and integral balances, from which we can obtain the adequate balance for the specific thermal processing application.

In general, for an open system in an unsteady-state (transient) condition, the following equation is proposed for a quality balance of factor j in system S.

Differential form (unsteady-state):

$$[FQ]_{j,L} - [F(Q + \delta Q)]_{j,L+\delta L} + \delta V \left[\frac{dQ}{dt}\right]_{j,A} + \delta V \left[\frac{dQ}{dt}\right]_{j,I}$$
$$= \left[\frac{d((\delta V)Q)}{dt}\right]_{j,S} \tag{7.13}$$

Where
F: Flux (L/h)
Q: Quality attribute (kg/L)
V: Volumen (L)

t: Time (h)

δL and δV : small quantity, in this case a very short tube length and a very small volume.

Subindices

L: Position L

A: Activation

I: Inactivation

S: System

Under steady-state:

$$[FQ]_{j,L} - [F(Q + \delta Q)]_{j,L+\delta L} + \delta V \left[\frac{dQ}{dt}\right]_{j,A} + \delta V \left[\frac{dQ}{dt}\right]_{j,I} = 0 \qquad (7.14)$$

Integral form (unsteady-state):

$$[FQ]_{j,i} - [FQ]_{j,o} + V \left[\frac{dQ}{dt}\right]_{j,A} + V \left[\frac{dQ}{dt}\right]_{j,I} = \left[\frac{d(VQ)}{dt}\right]_{j,S} \qquad (7.15)$$

And, under steady-state:

$$[FQ]_{j,i} - [FQ]_{j,o} + V \left[\frac{dQ}{dt}\right]_{j,A} + V \left[\frac{dQ}{dt}\right]_{j,I} = 0 \qquad (7.16)$$

And, for a closed system under unsteady-state:

$$\left[\frac{dQ}{dt}\right]_{j,I} = \left[\frac{d(Q)}{dt}\right]_{j,S}. \qquad (7.17)$$

7.5.1. Demonstration Examples

Examples have been chosen specifically for situations where the classical approach of cooking value concept fails.

7.5.1.1. Quality Evaluation in a Plug Flow Holding Tube (Fluid Without Particles and Constant Temperature)

In a Plug Flow (Figure 7.6) the fluid passes through in a coherent manner, so that the residence time is the same for all fluid elements. The coherent fluid passing through the ideal reactor is known as a plug. As a plug flows through the tube, the fluid is perfectly mixed in the radial direction but not in the axial direction (forwards or backwards). Each plug of differential volume is considered as a separate entity (practically a batch reactor) as it flows down the tubular Plug Flow.

  This is a classical example where it is imperative to do a differential balance. The concentration of the target attribute is changing through the tube length, meaning that we can not select the whole tube as a system. Therefore, to do the balance, first, a differential element at position L and $L + \delta L$ is defined. Because the differential system is very small, we can assume that within this small system the concentration of the target attribute j is homogeneous and equal to $Q_{j,L}$.

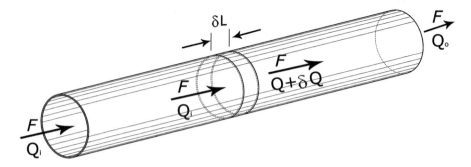

FIGURE 7.6. Quality balance in a Plug Flow system.

According to equation (7.13) and considering steady state condition, equation (7.14), we can write:

$$[FQ]_{j,L} - [F(Q + \delta Q)]_{j,L+\delta L} + \delta V \left[\frac{dQ}{dt}\right]_{j,A} + \delta V \left[\frac{dQ}{dt}\right]_{j,I} = 0$$

Then, considering a first order inactivation kinetics (without an activation term), we obtain:

$$[FQ]_{j,L} - [F(Q + \delta Q)]_{j,L+\delta L} - k_{j,L} Q_{j,L} \delta V = 0. \tag{7.18}$$

Therefore:

$$-F\delta Q_{j,L} - k_{j,L} Q_{j,L} \delta V = 0 \tag{7.19}$$

Where

$$\delta V = A\delta L \tag{7.20}$$

Replacing equation (7.20) into equation (7.19):

$$-F\delta Q_{j,L} - k_{j,L} Q_{j,L} A\delta L = 0 \tag{7.21}$$

Separating variables and integrating from 0 to L and from $Q_{j,i}$ to $Q_{j,o}$:

$$Q_{j,o} = Q_{j,i} e^{-k_j \frac{V}{F}} = Q_{j,i} e^{-k_j \tau} \tag{7.22}$$

Output quality for target attribute j ($Q_{j,o}$) can be evaluted knowing the initial quality ($Q_{j,i}$), process temperature, residence time and reaction constant (k_j). Equation (7.22) can be reformulated to compare with cooking value concept discussed in Chapter 4. In order to relate equation (7.22) to cooking value it is necessary to replace k by D and then express D as a function of temperature.

$$k = \frac{\ln 10}{D} \quad \text{and} \quad D = D_r 10^{\frac{T_r - T}{z}}$$

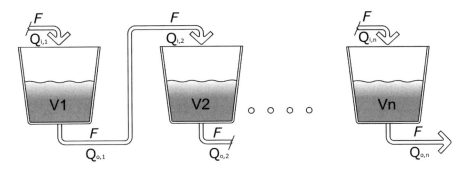

FIGURE 7.7. Quality balance in a series of CSTR reactors.

Expressing $Q_{j,o}$ as a function of $Q_{j,i}$ by $Q_{j,o} = Q_{j,I}/10^x$ and replacing into equation (7.22):

$$C_r = x D_r = 10^{\frac{T - T_r}{zc}} \tau \qquad (7.23)$$

Where, for a constant temperature process, equation (7.23) is similar and comparable with equation (4.26), although equation (4.26) was derived for a closed and unsteady state sytem and equation (7.23) was obtained for an open steady state system.

7.5.1.2. Quality Evaluation in a Series of CSTR Reactors

The application of CSTR, as shown in Figure 7.7, is useful for modeling process systems, since it has good-mixing characteristics.

In this case we are considering each reactor as a system and assuming steady state, first order kinetic for inactivation of target attribute and neglecting the activation term, then, from equation (7.16), we obtain:

First reactor:

$$F Q_{i,1} - F Q_{o,1} - k_1 Q_{o,1} V_1 = 0 \qquad (7.24)$$

Second reactor

$$F Q_{i,2} - F Q_{o,2} - k_2 Q_{o,2} V_2 = 0 \qquad (7.25)$$

By recurrence we are able to derive an equation for the Nth reactor:

$$F Q_{i,N} - F Q_{o,N} - k_N Q_{o,N} V_N = 0 \qquad (7.26)$$

Considering that $Q_{o,j} = Q_{i,j+1}$ and rearranging, we obtain:

$$Q_{o,N} = \frac{F^N Q_{i,1}}{\prod_{j=1}^{N} (F + k_j V_j)} \qquad (7.27)$$

Knowing the quality of the food product at the entrance $(Q_{i,1})$ and system characteristics (F, V_j), it is possible to evaluate quality at the output flux $(Q_{o,N})$.

Although we are not attempting to relate this expression to cooking value concept (equation 4.26), by inspection of equation (7.27), it is not clear that it will be possible to derive such an expression.

7.5.2. Corollary

A general approach to assess quality has been proposed and exemplified through two demonstration examples. The referred approach can be directly extended to assess process safety.

According to the proposed methodology, it is necessary to define a system and carry out a mass balance of the target attribute(s), meaning that it is necessary to express quality in terms of concentration (w/w or w/v) and to have an adequate kinetics expression to account for target attribute(s) degradation. Experience has shown us that most quality attributes can be expressed in concentration terms; in addition, a large amount of work in the area of kinetics has been carried out in the last three decades.

7.6. Conclusions

The subject of determining suitable processing conditions to maximize quality factors in foods undergoing thermal processing is currently being actively investigated. A variety of theoretical models have been developed for predicting optimum temperatures and temperature profiles. While the theoretical aspects have been extensively considered, there is relatively little work on actual food products. It is this aspect that requires further investigation in order to produce equations that may be used by the industry to control or improve the quality of processed products. This is a particularly difficult area in some ways, since it may involve producing a somewhat different product than the consistent brand image required and coveted by marketing people. It is easier to alter the quality attributes, while at the same time altering the geometry and characteristics of the packaging material, e.g. using plastic pouches and cartons. Optimization is an important subject and deserves to be studied in detail. It is not only concerned with defining the conditions for producing certain quality attributes, but can also be used in process and quality control (see Chapter 9).

References

Abbatemarco, C., & Ramaswamy, H. S. (1994). End-over-end thermal processing of canned vegetables: effect of texture and colour. *Food Res. Int.*, 27(4), 327–334.

Abbatemarco, C., & Ramaswamy, H. S. (1993). Heating behavior and quality retention in a canned model food as influenced by thermal processing in a rotary retort. *J. Food Quality*, 16, 273–285.

Aishima, T. & Nakai, S. (1986). Centroid mapping optimization: A new efficient optimization for food research and processing. *J. Food Sci.*, 51(5), 1297–1300, 1310.

Almonacid-Merino, S. F, Simpson, R., & Torres, J. A. (1993). Time-variable retort profiles for cylindrical cans: batch process time, energy consumption, and quality retention model. *J. Food Process Eng., 16*, 271–287.

Alwis, A. A. P. de, Varley, J., & Niranjan, K. (1992). Simulation of thermal food processing operations. In W. Hamm & S. D. Holdsworth (Eds.), *Food engineering in a computer climate* (pp. 253–262). Rugby, U.K: Institution of Chemical Engineers.

Atherton, D. & Thorpe, R. H. (1980). The processing of canned fruit and vegetables. *Technical Bulletin No 4 (revised)*. Chipping Campden, Glos., UK: Campden & Chorleywood Food Research Association.

Ávila, I. M. L. B., Martins, R. C., Ho. P., Hendrickx, M., & Silva, C. L. M. (2006). Variability in quality of white and green beans during in-pack sterilization. *J. Food Eng., 73*(2), 149–156.

Ball, C. O., & Olson, F. C. W. (1957). *Sterilization in food technology–Theory, practice and calculations*. New York: McGraw-Hill.

Balsa-Canto, E., Alonso, A. A., & Banga, A (2002a). A novel, efficient and reliable method for thermal process design and optimization. Part I: theory. *J. Food Eng., 52*(3), 227–234.

Balsa-Canto, E., Alonso, A. A., & Banga, A. (2002b). A novel, efficient and reliable method for thermal process design and optimization. Part II: application. *J. Food Eng. 52*(3), 235–248.

Banga, J. R., Perez-Martin, R. I., Gallardo, J. M., & Casares, J. J. (1991). Optimization of thermal processing of conduction-heated canned foods: study of several objective functions. *J. Food Eng., 14*, 25–51.

Barros, P., Proença, F., Ribeiro, J., Lemos, M. N. D. A., Lemos, F., & Empis, J. (2001). Modelling the change of thiamine concentration in canned meat. In J. Welti-Chanes, G. V, Barbosa-Cánovas & J. M. Aguilera (Eds.), *Eighth international conference on Engineering and food ICEF8, Vol.1* (pp. 683–686). Lancaster, PA: Technomic Pub. Co.

Barreiro Mendez, J. A., Salas, G. R., & Moran, I. H. (1977). Formulation and evaluation of a mathematical model for the prediction of nutrient losses during heat treatment of canned foods. *Archivos Latinamericanos de Nutrición, 27*(3), 325–342.

Barreiro, J., Pérez, C. and Guariguata, C. (1984). Optimization of Energy Consumption During the heat Processing Of Canned Foods. *J. of Food Eng. 3*(1), 27–37.

Baucour, P, Cronin, K., & Stynes, M. (2003). Processing optimization strategies to diminish variability in the quality of discrete packaged foods during thermal processing. *J. Food Eng., 60*(2), 147–156.

Bauder, U., & Heiss, R. (1975). Heat sterilization and quality of food. *Verfahrenstechnik, 9*, 566–573 [in German].

Brown, F. (1950). Effect of time and temperature of processing on the quality and nutritive value of canned vegetables. *Progress report*. Chipping Campden, Glos., UK: Campden & Chorleywood Food Research Association.

Brown, K. L., & Ayres, C. (1982). Thermobacteriology of UHT processed foods. In R. Davies (Ed.), *Developments in food microbiology – 1* (pp. 1219–152). Barking, UK: Elsevier Applied Science.

Chen, C. R., & Ramaswamy, H. S. (2002a). Dynamic modelling of thermal processing using artificial neural networks. *J. Food Process. Preserv., 26*, 91–111.

Chen, C. R., & Ramaswamy, H. S. (2002b). Modeling and optimization of constant retort temperature (CRT) thermal processing using neural networks and genetic algorithms. *J. Food Process Eng., 25*(4). 351–379.

Chen, C. R., & Ramaswamy, H. S. (2002c). Modeling and optimization of variable retort temperature (VRT) thermal processing using coupled neural networks and genetic algorithms. *J. Food Eng., 53*(3), 209–220.

De Salvo, G. J., & Gorman, R. W. (1989). *ANSYS – Engineering Analysis System (rev.4.4)- User's manual.* Houston, PA: Swanson Analysis Systems. Inc.

Downes, T. W., & Hayakawa, K. (1977). A procedure for estimating the retention of components of thermally conducted processed foods. *Lebensm.-Wiss. u.-Technol., 10,* 256–259.

Durance, T. D. (1997). Improving canned food quality with variable retort temperature processes. *Trends in Food Science and Technology, 8,* 113–118.

Durance, T. D., Dou, J., & Mazza, J. (1997). Selection of variable retort temperature processes for canned salmon. *J. Food Process Eng., 20*(1), 65–76.

Eisner, M. (1988). *Introduction to the technique and technology of rotary sterilization.* Milwaukee, WI: private author's edition.

Erdögdu, F., & Balaban, M. O. (2001). Optimization of thermal processing in different container shapes using a complex method. In *Eighth international conference on engineering and food ICEF8 Vol.2* (pp. 1810–1815). Lancaster, PA: Technomic Pub. Co.

Erdögdu, F., & Balaban, M. O. (2002). Non-linear constraint optimization of thermal processing: I Development of a modified algorithm of complex method. *J. Food Process Eng., 25*(1), 1–22.

Erdögdu, F., & Balaban, M. O. (2003a). Non-linear constraint optimization of thermal processing: II. Variable process temperature profiles to reduce process time and improve nutrient retention in spherical and finite cylindrical geometries. *J. Food Process Eng., 26*(4), 303–314.

Erdögdu, F., & Balaban, M. O. (2003b). Complex method of non-linear constrained multi-criterion (multi-objective function). *J. Food Process Eng., 26*(3), 357–375.

García, M. –S., G.Balsa-Canto, E., & Alonso, A. A., & Banga, J. R. (2006). Computing optimal operating policies for the food industry. *J. Food Eng. 74*(1), 13–23.

Gillespy, T. G. (1956). The processing of canned fruit and vegetables. *Technical Bulletin No. 4 (1st Edition).* Chipping Campden, Glos., UK: Campden & Chorleywood Food Research Association.

Greenwood, D. A., Kraybill, H. R., Feaster, J. F., & Jackson, J. M. (1944). Vitamin retention in processed meats. *Ind. Eng. Chem., 36,* 922–928.

Hayakawa, K. (1969). New parameters for calculating mass-average sterilizing values to estimate nutrients in thermally conductive food. *Can. Inst. Food Technol. J., 2*(4), 165–172.

Hayakawa, K. (1971). Mass average value for a physical, chemical, or biological factor in food. *Can. Inst. Food Technol. J., 4*(3), 133–134.

Hendrickx, M. (1987). Time modelling of ΔT-cooking of conduction heated foods. *Berichte der Bundesforschunganstalt, 87*(02), 68.

Hendrickx, M., Van Genechten, K., & Tobback, P. (1989). Optimizing quality attributes of conduction heated foods, a simulation approach. In W. E. L. Speiss & H. Schubert (Eds.), *Engineering and food, ICEF 5, Vol. 2* (pp. 167–176). London: Elsevier Applied Science.

Hendrickx, M. E., Silva, C. L., Oliveira, F. A., & Tobback, P. (1992). Optimization of heat transfer in thermal processing of conduction heated foods. In R. P. Singh & M. A. Wirakartakusumah (Eds.), *Advances in food engineering* (pp. 221–235). Boca Raton, FL: CRC Press.

Hendrickx, M., Silva, C., Oliveira, F., & Tobback, P. (1992c). Optimizing thermal processes of conduction heated foods: generalized equations for optimal processing temperatures. In W. Hamm & S. D. Holdsworth (Eds.), *Food engineering in a computer climate* (pp. 271–276). Rugby, U.K.: Institution of Chemical Engineers.

Hendrickx, M., Silva, C., Oliveira, F., & Tobback, P. (1993). Generalized (semi)-empirical formulae for optimum sterilization temperature of conduction heated foods with infinite surface heat-transfer coefficient. *J. Food Eng.*, *19*(2), 141–158.

Herrmann, J. (1969). Optimization of sterilization processes with regard to destruction of bacteria and chemical changes. *Die Nahrung*, *13*, 639–661 [in German].

Hildenbrand, P. (1980). An approach to solving the optimal temperature control problem for sterilization of conduction-heating foods. *J. Food Process Eng.*, *3*, 123–142.

Holdsworth, S. D. (1979). The effects of heating on fruits and vegetables. In R. J. Priestley (Ed.), *Effects of heating on foodstuffs*. London: Applied Science Publishers.

Holdsworth, S. D. (1985). Optimization in thermal processing. *J. Food Eng.*, *4*, 89–116.

Holdsworth, S. D. (1992). *Aseptic processing and packaging of food products*. London Elsevier Applied Science.

Holdsworth, S. D. (2004). Optimising the safety and quality of thermally processed packaged foods. In P. Richardson (Ed.), *Improving the thermal processing of foods* (pp. 3–31). Cambridge: Woodhead Publishing.

Hoyem, T. & Kvale, O. (Eds.) (1977). *Physical, chemical and biological changes in food caused by thermal processing*. London: Applied Science Publishers.

Jackson, J. M., Feaster, J. F., & Pilcher, R. W. (1945). The effect of canning procedures on vitamins in foods. *Proc. Inst. Food Technologists*, pp. 81–90.

Jelen, P. (1983). Review of basic technical principles and current research in UHT processing of foods. *Can. Inst. Food Sci. Technol. J. 16*, 159–166.

Jen, Y., Manson, J. E., Stumbo, C. R., & Zahradnik, J. W. (1971). A procedure for estimating sterilization of and quality factor degradation in thermally processed foods. *J. Food Sci.*, *36*, 692–698.

Kebede, E., Mannheim, C. H., & Miltz, J. (1996). Heat penetration and quality preservation during thermal treatment in plastic trays and metal cans. *J. Food Eng. 30*(1/2), 109–115.

Kessler, H. G. (1981). *Food engineering and dairy technology*, Chapter 6. Freissing, Germany: Verlag A. Kessler.

Kok, R. (1986). Optimization of the thermal processing of food. In M. Le Mageur & P. Jelen (Eds.), *Food engineering and process applications Vol. 1* (pp. 559–568). London, Elsevier Applied Science.

Kozmierczak, L. (1996). Optimizing complex bioeconomic simulations using an efficient search heiristic. *Research Report No. 704*. Baton Rouge: Louisiana State University.

Kseibat, D. S. Mittal, G. S., & Basir, O. A. (2004). Predicting safety and quality of thermally processed canned foods using a neural network *Trans. Inst. Measurement and Control*, *26*(1), 56–68.

Lenz, M. K., & Lund, D. B. (1977). The lethality-Fourier number method: experimental verification of a model for calculating average quality factor retention in conductive-heating canned foods. *J. Food Sci.*, *42*(4), 997–1001.

Lund, D. B. (1977). Design of thermal processes for maximizing nutrient retention. *Food Technol.*, *31*(2), 71–78.

Mansfield, T. (1962). High-temperature short-time sterilization. In J. Hawthorn & M. Leich (Eds.), *Proc. 1st int. congress food sci. technol Vol. 4* (pp. 311–316). London: Gordon & Breach.

Mansfield, T. (1974). *A brief study of cooking*, San Jose, CA: FMC Corp.

McKenna, A. B., & Holdsworth, S. D. (1990). Sterilization and cooking of food particulates – a review. *Technical Bulletin No. 75*. Chipping Campden, Glos., UK: Campden & Chorleywood Food Research Association.

Nakai, S. (1982). Comparison of optimization techniques for application of food product and process development. *J. Food Sci., 47*, 144–152, 176.

Nakai, S., Koide, K., & Eugster, K. (1984). A new mapping super-simplex optimization for food product and process development. *J. Food Sci., 49*, 1143–1148, 1170.

Nardkarni, M. M,. & Hatton, T. A. (1985). Optimal nutrient retention during the thermal processing of conduction-heated canned foods: application of the distributed minimum principle. *J. Food Sci., 50*, 1312–1321.

Newman, M., & Holdsworth, S. D. (1989). Methods of thermal process calculation for food particles. *Technical Memorandum No. 321 (revised)*. Chipping Campden, Glos., UK: Campden & Chorleywood Food Research Association.

Newman, M., & Steele, D. A. (1984). Methods of thermal process calculations for food particles. *Technical Memorandum, No. 210*. Chipping Campden, Glos., UK: Campden & Chorleywood Food Research Association.

Nicolai, B. M., Verboven, P. & Sheerlinck, N. (2001). Modelling and simulation of thermal processes. In P. Richardson (Ed.), *Thermal technologies in food processing* (pp. 92–112). Cambridge: Woodhead Publishing.

Norback, J. P. (1980). Techniques for optimization of food processes. *Food Technol., 34*(2), 86–88.

Noronha, J., Hendrickx, M., Suys, J., & Tobback, P. (1993). Optimization of surface quality retention during thermal processing of conduction heated foods using variable temperature retort profiles. *J. Food Process. Preserv., 17*, 75–91.

Noronha, J., Van Loey, A., Hendrickx, M., & Tobback, P. (1996a). Simultaneous optimization of surface quality during the sterilization of packed foods using constant and variable retort temperature profiles. *J. Food Eng. 30*(3/4), 283–297.

Noronha, J., Van Loey, A., Hendrickx, M., & Tobback, P. (1996b). An empirical equation for the description of optimum variable retort temperature profiles that maximize surface quality retention in thermally processed foods. *J. Food Process. Preserv., 20*(3), 251–264.

Ohlsson, T. (1980). Temperature dependence of sensory quality changes during thermal processing. *J. of Food Sci.* 45(4), 836–839.

Ohlsson, T. (1980a). Optimal sterilization temperatures for flat containers. *J. Food Sci., 45*, 848–852.

Ohlsson, T. (1980b). Optimal sterilization temperatures for sensory quality in cylindrical containers. *J. Food Sci., 45*, 1517–1521.

Ohlsson, T. (1980c). Optimization of heat sterilization using *C*-values. In P. Linko (Ed.), *Food process engineering Vol. 1* (pp. 137–145), London: Elsevier Applied Science.

Ohlsson, T. (1988). The cook value as a tool for optimizing thermal processes. In *Proc. int. symp. Progress in food preservation processes Vol. 1* (pp. 51–58). Brussels: CERIA.

Oliveira, J. C. (2004). Optimising the efficiency and productivity of thermal processing. In P. Richardson (Ed.), *Improving the thermal processing of foods* (pp. 32–49). Cambridge: Woodhead Publishing.

Pontryagin, L. S., Boltyanskii, V. G., Gamkrelidze, R. V., & Mishchenko, E. F. (1962). *The mathematical theory of optimal processes* (translator K. N. Trirogoff). New York: Wiley-Interscience.

Pornchaloempong, P., Balaban, M. O., & Chau, K. V. (2001). Thermal processing optimization of quality retention in a conical shape. In J. Welti-Chanes, G. V, Barbosa-Cánovas & J. M. Aguilera (Eds.), *Eighth international conference on Engineering and food ICEF8, Vol.1* (pp. 676–682). Lancaster, PA: Technomic Pub. Co.

Pornchaloempong, P., Balaban, M. O., Teixeira, A. A., & Chau, K. V. (2003). Optimization of quality retention in conduction-heating foods of conical shape. *J. Food Proc. Eng., 25*, 557–570.

Preussker, H. (1970). Preservation techniques: selection of temperature and processing times for sterilization and cooking of food. *Die Ernährungswirtschaft/Lebensmittel, 17*(10), 770–788 [in German].

Reichert, J. E. (1974). Optimum sterilization temperatures for prepared foods. *Fleisch-Wirtschaft, 54*, 1305–1313 [in German].

Reichert, J. E. (1977). The *C*-value as an aid to optimization. *Z. Lebensm.-Technol. u.-Verfahr., 28*(1), 1–7 [in German].

Reichert, J. E. (1980). Delta-*T* cooking: a new parameter, *Die Fleischerei, 5*, 478–486 [in German].

Richardson, P. S., Kelly, P. T., & Holdsworth, S. D. (1988). Optimization of in-container sterilization processes. In *Proc. int. symp. Progress in food preservation processes Vol. 2* (pp. 1–11). Brussels: CERIA.

Saguy, I., & Karel, M. (1979). Optimal retort temperature profile for optimizing thiamin retention in conduction-type heating canned foods. *J. Food Sci., 44*, 1485–1490.

Sargent, R. W. H., Perkins, J. D., & Thomas S. (1982). SPEEDUP: Simulation program for the economic evaluation and design of unified processes. In M. E. Lesley (Ed.), *Computer-aided process plant design.* Houston: Gulph.

Savage, D. (1984). Processing sterilized products. In *Profitability of food processing. Inst. Chem. Engrs, Symposium Series No. 84* (pp. 307–316). Rugby, UK: IChemE.

Silva, C., Hendrickx, M., Oliveira, F., & Tobback, P. (1992a). Critical evaluation of commonly used objective functions to optimize overall quality and nutrient retention of heat-preserved foods. *J. Food Eng., 17*(4), 241–258.

Silva, C., Hendrickx, M., Oliveira, F., & Tobback, P. (1992b). Optimal sterilization temperatures for conduction heating foods considering finite surface heat-transfer coefficients. *J. Food Sci., 57*(3), 743–748.

Silva, C. L. M., Oliveira, F. A. R., & Hendrickx, M. (1993). Modelling of optimum processing conditions for sterilization of prepackaged foods. *Food Control, 4*(2), 67–86.

Silva, C., Oliveira, F. A. R., & Pereira, P. A. M. (1994a). Optimum sterilization: A comparative study between average and surface quality. *J. Food Process Eng., 17*, 155–176.

Silva, C. L. M., Oliveira, F. A. R., Lamb, J., Torres, A. P., & Hendrickx, M. (1994b). Experimental validation of models predicting optimal surface quality sterilization temperatures. *Int. J. Food Sci. Technol., 29*, 227–241.

Silva, C. L. M., Oliveira, F. A. R., & Hendrickx, M. (1994c). Quality optimization of conduction heating foods sterilized in different packages. *Int. J. Food Sci. Technol., 29*(5), 515–530.

Silva, C. L. M., Oliveira, F. A. R., & Hendrickx, M. (1994d). Obtaining a well-balanced product quality in thermally processed conduction heating foods by analyzing surface and volume average quality optimum processing conditions. In R. P. Singh & F. A. R. Oliveira (Eds.), *Minimal processing of foods and process optimization.* Boca Raton, FL: CRC Press.

Simpson, R., Almonacid, S., & Teixiera, A. (2003). Optimization criteria for batch retort battery design and operation in food canning plants. *J. Food Process Eng.* 25(6), 515–538.

Simpson, R., Almonacid, S., & Mitchell, M. (2004). Mathematical model development, experimental validations and process optimization: retortable pouches packed with seafood in cone fustrum shape. *J. Food Eng.* 63(2), 153–162.

Simpson, R. (2004). Optimising the efficiency of batch processing with retort systems in thermal processing. In P. Richardson (Ed.), *Improving the thermal processing of foods* (pp. 50–81). Cambridge: Woodhead Publishing.

Simpson, R. (2005). Generation of isolethal processes and implementation of simultaneous sterilization utilising the revisited general method. *J. of Food Eng.* 67(1), 71–79.

Sjöström, C., & Dagerskog, M. (1977). Optimisation of time/temperature relationship for heat sterilization of canned food by computer simulation. In *Proc. mini-symp. Mathematical modelling in food processing* (pp. 227–234). Sweden: University of Lund.

Smout, C., Van Loey, A. M. L., & Hendrickx, M. E. G. (2001). Role of temperature distribution studies in the evaluation and identification of processing conditions for static and rotary water cascading retorts. *J. Food Eng.*, 48(1), 61–68.

Stumbo, C. R. (1953). New procedures for evaluating the thermal process of foods in cylindrical containers. *Food Technol.*, 3, 309–315.

Stumbo, C. R. (1973). *Thermobacteriology in food processing* (2nd Edition). New York: Academic Press.

Svensson, S. (1977). Inactivation of enzymes during thermal processing. In T. Hoyem & O. Kvale (Eds.), *Physical, chemical and biological changes in food caused by thermal processing* (pp. 202–208). London: Applied Science Publishers.

Teixeira, A., Dixon, J., Zahradnik, J. and Zinsmeiter, G. (1969). Computer Optimization of Nutrient Retention in the Thermal Processing of Conduction-Heated Foods. *Food Technol.* 23(6), 845–850.

Teixeira, A. A., Stumbo, C. R., & Zahradnik, J. W. (1975a). Experimental evaluation of mathematical and computer models for thermal process evaluation. *J. Food Sci.*, 40, 653–655.

Teixeira, A. A., Zinsmeister, G. E., & Zahradnik, J. W. (1975b). Computer stimulation of variable retort control and container geometry as a possible means of improving thiamine retention in thermally processed foods. *J. Food Sci.*, 40, 656–659.

Thijssen, H. A. C., Kerkhof, P. J. A. M., & Liefens, A. A. A. (1978). Short-cut method for the calculation of sterilization conditions yielding optimum quality retention for conduction-type heating of packaged foods. *J. Food Sci.*, 43, 1096–1101.

Thijssen, H. A. C., & Kochen, L. H. P. J. M. (1980). Calculation of optimum sterilization conditions for packed conduction-type foods. *J. Food Sci.*, 45, 1267–1272, 1292.

Tischer, R. G., Hurwicz, H., & Zoellner, J. A. (1953). Heat processing of beef. III. Objective measurement of changes in tenderness, drained juice and sterilization value during heat processing. *Food Res.*, 18, 539–554.

Tucker, G. S. & Holdsworth, S. D. (1990). Optimization of quality factors for foods thermally processed in rectangular containers. In R. W. Field & J. A. Howell (Eds.), *Process engineering in the food industry–2*. London, Elsevier Applied Science.

Tucker, G. S., & Holdsworth, S. D. (1991). Mathematical modelling of sterilization and cooking processes for heat preserved foods–application of a new heat transfer model. *Food and Bioproducts Process. Trans. IChemE, 69*, C1, 5–12.

Van Loey, A., Fransis, A., Hendrickx, M., Maesamans, G., Noronha, J. de, and Tobback, P. (1994). Optimizing thermal process for canned white beans in water cascading retorts. *J. Food Sci., 59*(4), 828–832.

8
Engineering Aspects of Thermal Processing

8.1. Thermal Processing Equipment

8.1.1. Introduction

The design, operation and choice of sterilizing and pasteurization systems, known as retorts, cookers or autoclaves, for packaged foods depend on a number of factors.

Product characteristics are the first factors to be considered. Acid products with pH less than 4.5 usually require only a mild treatment, and this can be achieved with steam at atmospheric pressure or boiling water. Consequently, the equipment required is much simpler than for low-acid products with greater than pH 4.5. For pasteurization of acid food products, temperatures of the order of 120–130°C are necessary, and pressure vessels will be required to contain the pressure of the steam or maintain the condition of the hot water or other heating medium. In factories that process a wide range of products, pressurized retorts are often used for acidic food products. It is more usual to find atmospheric equipment operating in factories where fruit products are processed in large volumes and where it is possible to have dedicated systems for retorting.

Many products have high-viscosity components, e.g. meat and vegetable stews, cream-style corn, beans in tomato sauce, macaroni cheese, puddings, sauces with thickening agents, and a variety of cream soups, and the processing of cans of these products is enhanced by using agitating processes. These can involve axial or end-over-end motion, and special retorts or modified retorts are used. Apart from the product aspects, many large sizes of cans of food for the catering food processing industry benefit enormously from agitating processes. The rate of heat penetration is increased, and the problems of the overcooking of the outside layers of the canned product are reduced. In order to ensure adequate sterilization of products undergoing an agitating process, it is necessary to control the headspace in the can carefully, as well as the solid–liquid ratio, consistency and speed of rotation, and other factors such as process temperature, time, pressure and operating cycle.

The nature and form of the packaging material make up the second factor to take into account. It is essential that the conditions applied in the processing operation maintain the integrity of the packaging material. Hermetically sealed containers should not be distorted, and the seams, end(s) and body should not be affected by the process. For all types of sealed container the vapor pressure of the liquid portion of the components increases with increasing temperature, so that as the can is heated externally the internal pressure rises. In the case of pressurized steam there is some counter-balancing between the internal and external pressures, usually, in the case of rigid containers, sufficient to prevent permanent distortion. With larger sizes of can, however, when the process is completed, the can has a residual internal pressure, and this must be counter-balanced during cooling, otherwise there will be a tendency to distort the seams by peaking the ends. There is also a possibility of vacuum formation if the cooling is too rapid, and paneling of the bodies will result. This means that normal steam-operated retorts require a compressed air supply to counter-balance the internal pressure during cooling. While this applies to rigid containers, many of the newer types of plastic containers have seams that are unable to withstand the differential pressures imposed by steam heating. Consequently, it is necessary to use an additional pressure during heating. This can be achieved by the use of steam–air mixtures or pressurized hot water processing. Many new retorts have been devised that can be used with these heating media. The third factor is production throughout. While batch retorts are flexible in relation to container size and process conditions, they have a relatively slow production rate and are labor- and energy-intensive. Consequently, most factories operate high-speed lines with continuous rotary or hydrostatic cookers. Large volume production is necessary to achieve marketing objectives.

The choice of a retorting system is, therefore, quite complex and requires special care. There are, of course, a large number of other factors of a general engineering nature that also have to be considered. These include site location, available floor area and factory layout, as well as many other financial, personnel and marketing aspects, all of which are outside the scope of this text but are dealt with elsewhere (see, for example, Lopez 1987).

During the last two centuries, a large variety of devices have been manufactured and used. In the following discussion only the major types of equipment in current use will be mentioned. The reader seeking historical information should consult Bitting (1916) and Thorne (1986), together with patents and articles in the trade journals, although much of the detail, unfortunately, is not directly available. It would be a useful exercise in mechanical engineering design to document the achievements of individual designers and how they overcame many of the problems involved in continuous cooker design and can handling. Some of the major types of retorting systems currently in use are listed in Table 8.1; they are classified according to the mode of operation and the heat transfer media commonly used. The discussion in this text is confined to the factors which concern the delivery of a safe process, and many aspects of the technology of

TABLE 8.1. Some commercial retorting systems.

System	Heat transfer medium	Condensing saturated steam	Steam–air mixtures	Hot water	Flame
1(a)	Batch – static	Conventional vertical and horizontal retorts	Horizontal retorts: – Lubeca – Lagarde Steristeam – Barriquand	Steriflow – Barriquand [raining water]	—
1(b)	Batch – rotary	Horizontal – Millwall (internal rotation of cans fixed in a cage) – FMC Orbitort	Horizontal retort: – Lagarde	Konservomat – Atmos (Alfa Laval Gruppe) Rotomat – Stock Lubeca	—
2(a)	Continuous – static	Hydrostatic cookers: – Hydroflex – FMC Hydromatic – Stork – Carvallo Hydrostat – M&P Hydron – Mitchell Webster – Hunister	—	—	—
		Horizontal hydrostatic cooker: Hydrolock – LST Storklave – Stork Universal cooker – FMC	—	—	—
	Semi-continuous	Crateless retorts – Malo/Odenberg – FMC			
2(b)	Continuous – rotary	Hydrostatic cookers with rotating carrier bars – FMC – Stork	—	Rotary cooker/ coolers – FMC – Stork Hydroflow	Steriflam – ACB Hydroflam
		Rotary cooker-coolers: reel and spiral Sterilmatic – FMC Steristork – Stork	—	—	—

Manufacturers: ACB, Ateliers et Chantiers de Bretagne, Nantes, France; Atmos-Lebensmittel-technik, Hamburg, Germany; Barriquand SA, Paris, France; ATM Carvallo, Drancy, France; FMC Machinery Europe NV, Sint-Niklaas, Belgium; Hunister, Komplex Hungarian Trading Company, Budapest, Hungary; LST, La Sterilization Thermique, Paris, France; Lubeca Maschinen und Anlagen GmbH, Lübeck, Germany; M&P, Mather and Platt, Bolton, UK (no longer manufactured); Mitchell-Webster, Chisholm-Ryder, USA; Millwall, John Fraser & Sons Ltd., Newcastle-upon-Tyne, UK; Odenberg Engineering Ltd, Dublin, Ireland; Stock, Hermann Stock GmbH, Neumünster, Germany; Stork Amsterdam, Amsterdam, Netherlands.

canning and heat processing will not be dealt with here. The excellent recent texts of Rees and Bettison (1991) and Footitt and Lewis (1995), together with the encyclopedic three-volume work of Lopez (1987), adequately cover many of the technological aspects. More details of the mechanical and technical aspects of the design of retorts are given by Brown (1972), Richardson and Selman (1991), and May (2001).

8.1.2. Batch Retorts

8.1.2.1. Conventional Static Retorts

The conventional batch canning retort is either circular or rectangular in cross-section and arranged either vertically or horizontally. Several retorts are usually arranged in banks, and a sequence of filling, closing, retorting and emptying is operated to give a supply of cans sufficient to maintain continuous labeling and packing. The cans are arranged in baskets, appropriate to the geometry of the retort, either randomly packed or arranged in layers on plastic divider plates. Vertical retorts are usually sunk into the flooring so as to give easy access to the hinged lid at the top. The services to the retort, and indeed any similar type of retort, consists of a steam distribution system with a spreader, cooling water supply, a compressed air supply for pressurized cooling, valves for venting and draining the retort, petcocks for bleeding air from the retort, and an instrument pocket arranged on the side, so as not to interfere with the loading and unloading of the crates of cans. The services and valves are shown in Figure 8.1. The steam and air supplies should not exceed the safe-working pressure of the retort, and should be sufficient to carry out the operations adequately. Batch retorts are operated according to the following general principles estimated for UK practice (Department of Health 1994; Thorpe *et al.* 1975).

1. Baskets of cans are placed in the retort and the lid closed.
2. With the vent, drain and the bypass to the controller valves open, the venting sequence to remove the air from in-between the cans and baskets is started by introducing steam until the temperature in the retort is about 100°C. The drain is then closed.
3. Venting continues for the prescribed time, and the vent valve is then closed.
4. The temperature is brought up to about 5°C below the required processing temperature, and the steam bypass valve closed.
5. The retort is allowed to attain the scheduled temperature using the controller.
6. When the time for processing is complete, the steam valve is closed and the cooling cycle started.
7. Compressed air is introduced to balance the pressure inside the can.
8. While continuing to counter-balance the pressure, water is allowed to run through the overflow/vent valve until cooling is completed, i.e. until the cans have reached a temperature of about 40°C.
9. When the pressure is atmospheric the lid, which should be fitted with a safety bolt, is opened and the baskets removed.

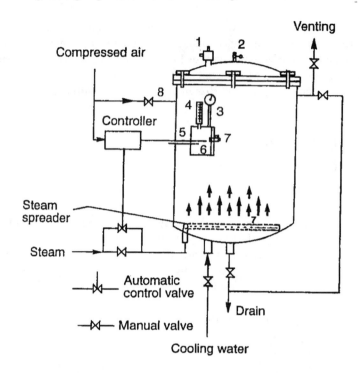

FIGURE 8.1. A vertical batch retort: (1) safety valve; (2) petcocks to maintain a steam bleed from retort during processing; (3) pressure gauge; (4) thermometer; (5) sensing element for controller; (6) thermo-box; (7) steam spreader; (8) air inlet for pressure cooling.

The importance of removing air by venting and using petcocks during processing is paramount to maintaining the correct temperature. Small quantities of air in the steam reduce the temperature significantly, especially at the surface of the cans where condensation is taking place. The venting cycle is established, for all types of retort, by placing thermocouples at strategic points in the crates of cans, and steam is introduced for a period sufficient to make all the thermocouples read the same. It is important to note that the arrangement of cans in the baskets, whether layered or scrambled, affects the heat penetration into the cans, and it is necessary to carry out tests under the conditions that are to be used in production. Sumer and Esin (1992) have attempted to establish a mathematical model based on a model for predicting the come-up time (see Chapter 5, Section 5.2.4 for more details).

A protocol has been published by the Institute for Thermal Processing Specialists determining the temperature distribution for steam heating in batch retorts, which excludes crateless retorts (IFTPS 1992).

The above description is specific to UK operations; other countries use slightly differing procedures. A minimum time is often quoted for the temperature to rise above 100°C for a specific type of retort pipework layout and services, e.g. the vent value shall be open for x minutes until a temperature of at least T is achieved,

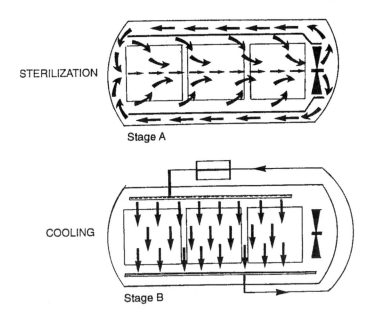

STERILIZATION

Stage A

COOLING

Stage B

FIGURE 8.2. The Lagarde system. Stage A: The sterilizing process uses forced steam circulation. Stage B: The rapid cooling is achieved by spraying of recycled cold water.

after which it is closed and the retort brought up to temperature (Lopez 1987; NFPA 1982).

8.1.2.2. Other Batch Retorts

While steam retorting is convenient for many types of container, several types of plastic container and pouch require overpressure to maintain the integrity of the packaging material. This may be achieved using steam–air mixtures or recirculating pressurized hot water. Hot water is often used for glass jar sterilization. Some typical examples of batch retort are given in Table 8.1. The Lagarde system, developed in 1972, is very versatile and uses steam–air mixtures for heating and water for cooling. The system is illustrated in Figure 8.2. It consists of cans. Rotating mixer blades are used to circulate the steam and to ensure that the mixture is kept uniform; this is particularly important in order to maintain a uniform temperature during processing. The Barriquand system (Steriflow) uses a raining water system to heat the food containers. This uses about 400 l of water, recycled every 9 s at a flow rate of 160 m³ h⁻¹, in a typical four-crate system. The water is heated indirectly with steam using a heat exchanger. The cooling water is separate from the heating system, and may be cooled and recirculated if necessary. The pressure of the system is regulated by using compressed air and is controlled to suit the mechanical requirements of the container. Ramaswamy *et al.* (1993) have evaluated a Lagarde steam/air retort operating in end-over-end

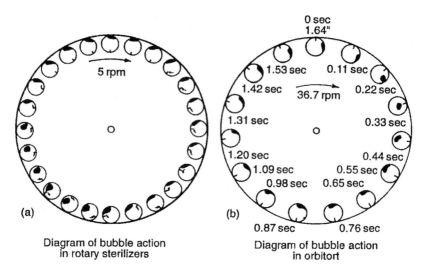

FIGURE 8.3. Typical motion of the headspace bubble in cans being rotated in two different types of rotary cooker: (a) bubble action in rotary sterilizer; (b) bubble action in orbitort (courtesy FMC Corp, St Jose, USA).

mode. A starch suspension (2–4%) was used with 307 × 409 cans at temperatures from 110–130°C and 10–20 rpm.

Mention must also be made of the crateless retort, of which one or two are commercially available. These retorts have top and bottom opening facilities, and the cans are placed directly in the retort, which is filled with water. Steam is introduced when the filling is completed and the water displaced. This is followed by a venting cycle. When the process is finished the cans are cooled from the bottom for the requisite time before being discharged onto a moving belt underwater. The system is of a semi-continuous nature and has the advantages of the flexibility of a batch system and also the high production rate of a continuous cooker. Several retorts, or pots as they are called, are used in a battery system.

A high-speed reel-type sterilizer, known as the Orbitort sterilizer, is used for viscous packs, e.g. cream-style corn and meat and vegetable stews, packed in large-sized catering cans such as 603 × 600 and 603 × 700. The cans rotate around the periphery of a long cylindrical pressure vessel at a speed of 36.7 rpm, such that the headspace bubble is kept continuously moving inside the cans and imparting mixing and agitation to the can contents (Figure 8.3). The inside motion of the cans from the inlet to the outlet is developed by a rotating reel-and-spiral device, which has flights that hold the cans in place and prevent contact during processing. The cans are loaded automatically and are counted until the correct number is in place. The unit is then sealed and steam introduced, followed by venting, processing and cooling.

8.1.2.3. Batch Rotary Systems

The advantages of using agitation of the head-space to increase heat transfer in cans were recognized during the nineteenth century, and many attempts to use the principle were investigated. Agitation may be imparted by either an end-over-end action, e.g. a retort rotating about a central horizontal axis, or by axial rotation of cans in a horizontal plane. While retorts of the former type are occasionally found today, most retorts use the second type of principle. The Millwall retort, for example, has baskets of A10 cans that slide into the system via a pressure lock and fit into an internally rotating cage. The system is ideally suited to catering pack sizes, since these benefit most from having the contents mixed during processing.

The Rotomat, developed by Hermann Stock of Neumünster, Germany, in 1952, is widely used in industry. The Rotomat uses recirculating pressurized hot water, and the temperature pressure differential between the water in the reservoir and the processing vessel provides the necessary overpressure required for plastic and other containers. Retorts of this type are usually operated at about 6–8 rpm; during heating and cooling, however, a faster speed of about 15 rpm is used for the come-up period. Heat transfer depends on the water contacting all the cans in the baskets as well as on the rotational motion. It is necessary to take the nature of the divider plates in the baskets between the cans into account in ensuring that there is adequate distribution of the hot water during the process. A full discussion of these overpressure rotary cookers with regard to determination of temperature distribution and also heat penetration has been presented by Hendrickx (1997).

8.1.3. Continuous Cookers

8.1.3.1. Reel and Spiral Cookers

This type of cooker was originally developed at the end of the nineteenth century; the pressure valve to allow the passage of cans into and out of the various pressurized chambers was developed in 1919 (Ball & Olson 1957). The cooker derives its name from an internal reel-and-spiral arrangement that permits cans to progress along the internal periphery of a cylindrical vessel, allowing them to rotate while keeping them apart. The cylindrical pressure vessels are known as shells, and a typical arrangement consists of preheating, sterilizing and cooling shells, although other arrangements for preheating and cooling are also possible. The cookers are predominantly steam-heated, although it is necessary to avoid condensate flooding the lower part of the sterilizer. The cookers are very flexible inasmuch as the resistance time of the cans, the speed of rotation and the temperature can be adjusted to desired levels (see Figures 8.4 and 8.5).

The most important initial application of the cooker was for sterilizing canned evaporated milk. The first machines, known as AB cookers after the company that first developed them (Anderson and Barngrover, which eventually became a division of the FMC Corporation), were installed at Nestlé Milk Products in the USA for this purpose in 1924. Subsequently various manufacturers made

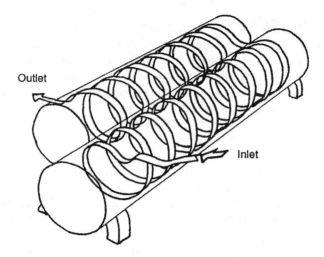

FIGURE 8.4. Path of cans through a reel-and-spiral cooker/cooler.

cookers for use with a variety of canned products. If the valves are well maintained and regularly inspected the cookers behave relatively well, although production difficulties have been reported with valves jamming. Much of the success of this type of cooker depends on the ability to control the pressure inside the valve carefully when the cans are transferred from the heating to the cooling shell. Distorted cans are also a cause of valve jamming.

The CTemp commercial computer program has been used to evaluate the effect of process deviations on the performance of reel-and-spiral cookers (Dobie *et al.* 1994).

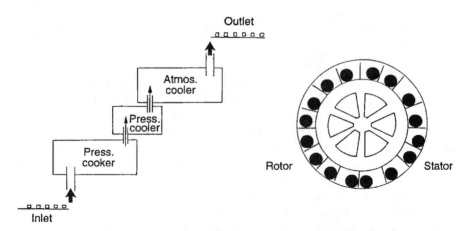

FIGURE 8.5. Arrangement of heating and cooling shells in a typical reel-and-spiral cooker/cooler.

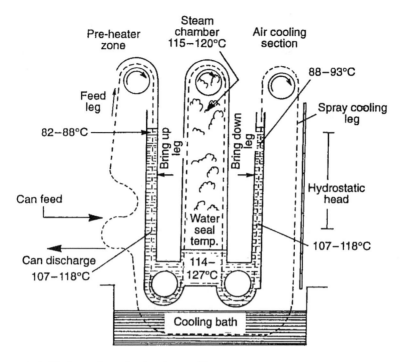

FIGURE 8.6. Hydrostatic cooker/cooler sterilizer.

8.1.3.2. Hydrostatic Cookers

These were first developed for commercial use by Pierre Carvallo in 1948, and subsequently by other manufacturers, e.g. Mitchell-Webster, FMC, Mather and Platt, and Stork. The essential principle was the use of hydrostatic legs of water of sufficient height, about 30 m, to counter-balance the pressure of the steam used in rotary sterilizers. They also required relatively little floor space, being vertically orientated. A typical hydrostatic cooker (Figure 8.6) consists of a preheating leg, a steam sterilizing chamber and one or more cooling legs. The cans or bottles are transported on long horizontal carrier bars (which may or may not impart rotary motion) chained together. The carrier bars enter the preheat leg at the top and progress down until they reach the steam chamber, where they make at least one double pass before emerging upwards through the cooling leg. There are many different arrangements of the legs, depending on the nature of the product being sterilized. There are also canned fruit and fruit juice sterilizers which operate at or near atmospheric pressure.

The Hunister hydrostatic sterilizer has a large number of legs, each operated at a different temperature and pressure, forming a multi-stage process. This arrangement significantly lowers the height to approximately 6 m and can easily be accommodated under normal factory conditions. It can process glass jars and cans at temperatures up to 135°C.

The Hydrolock retort system is a horizontal cooker using the carrier-bar system with a complex rotary pressure-sealing valve of undisclosed design. The carrier bars make multiple passes through the pressure chamber before passing through water at the base of the pressure vessel and discharging to a trough of water below the cooker.

Best *et al.* (1994a, b) have presented a steady-state and a dynamic sterilization of the processing of cans in a Hydrolock sterilizer using zone modeling and a computer program. "Flowpack" (ICI, UK).

8.1.4. Heat Transfer Media

8.1.4.1. Introduction

Condensing saturated steam has the important property of a very high surface heat-transfer coefficient and thus a negligible resistance to heating the surface of the can. The surface of the can is therefore immediately at the temperature of the condensing steam. Given the correct venting conditions, already outlined for batch retorts in Section 8.1.2, then there is no problem with saturated steam. Vigorously boiling water has a similar heat-transfer coefficient to condensing steam. With the requirements for other types of packaging material, steam–air mixtures and pressurized water systems are now commonly in use. These and other heat transfer media do not have the properties of condensing steam and require extra controls to safeguard the process. The heat-transfer coefficients are much lower (see Table 8.2); consequently, there is an appreciable resistance to heat transfer at the surface of the container. This resistance must be taken into account in determining the process requirements (see Chapter 2). In particular, the heat-transfer coefficient is dependent on the velocity of the fluid flow over the surface; consequently, this factor must be controlled by adequate pumping and consistent can orientation and layout in the retorting system. The same applies to cooling water and air, which have low heat-transfer coefficients.

The literature on external heat-transfer coefficients has been reviewed by Holdsworth (1976) in relation to the processing of packages in steam–air mixtures, hot water and fluidized beds.

The relative energy usage for various different types of retorts and cookers is given in Table 8.3 (Holdsworth 1992).

8.1.4.2. Saturated Steam

The properties of dry saturated steam are given in Table 8.4 for a range of temperatures including typical thermal sterilization values. The figures are useful for determining the amount of heat required, including steam supply, for processing operations.

It is important to note that any throttling or incorrectly working valves should be avoided, as they might induce superheating. Superheated steam does not have the latent heat of saturated steam; therefore, superheated steam heats less effectively until the superheat is lost.

TABLE 8.2. Some heat transfer coefficients for heating and cooling media.

Heat transfer media	Process conditions (°C)	Container/product	External heat transfer coefficient ($W\,m^{-2}\,K^{-1}$)	Overall heat transfer coefficient ($W\,m^{-2}\,K^{-1}$)	Reference
(a) Steam – saturated	105.0	Lead cylinder dia. 95 mm	3527	—	Merrill (1948)
	115.5	Lead cylinder dia. 95 mm	4508	—	Merrill (1948)
	121.1	Lead cylinder dia. 95 mm	4995	—	Merrill (1948)
	121.1	No. 2 can/water	—	1417	Merrill (1948)
		No. 3 can/water	—	1474	Merrill (1948)
		No. 2 can/1% bentonite	—	1321	Merrill (1948)
		No. 3 can/1% bentonite	—	1321	Merrill (1948)
	74	Quart jars/water	—	147	Pflug and Nicholas (1961)
	82	Quart jars/water	—	153	Pflug and Nicholas (1961)
	90.5	Quart jars/water	—	158	Pflug and Nicholas (1961)
	99	Quart jars/water	—	158	Pflug and Nicholas (1961)
	107	Quart jars/water	—	170	Pflug and Nicholas (1961)
	74	Pint jars/5% bentonite	—	71	Pflug and Nicholas (1961)
	82	Pint jars/5% bentonite	—	68	Pflug and Nicholas (1961)
	90.5	Pint jars/5% bentonite	—	65	Pflug and Nicholas (1961)
	74	Quart jars/5% bentonite	—	23	Pflug and Nicholas (1961)
	82	Quart jars/5% bentonite	—	23	Pflug and Nicholas (1961)
	90.5	Quart jars/5% bentonite	—	25	Pflug and Nicholas (1961)
	up to	No. $2\frac{1}{2}$, No. 1 tall and lb squat cans/15% bentonite including effect of headspace	—	40–60	Evans and Board (1954)
	115.5	Metallic cylinder	5053	—	Pflug (1964)
	115.5	303 × 406 can/water	—	1093	Pflug (1964)

(cont.)

TABLE 8.2. (Continued)

Heat transfer media	Process conditions (°C)	Container/product	External heat transfer coefficient (W m^{-2} K^{-1})	Overall heat transfer coefficient (W m^{-2} K^{-1})	Reference
(b) Steam/air mixtures 75%	74–107	quart jars/water	—	123–154	Pflug and Nicholas (1961)
	74–90.5	pint jar/5% bentonite	—	57–65	Pflug and Nicholas (1961)
	74–90.5	quart jars/5% bentonite	—	25.5	Pflug and Nicholas (1961)
	74–90.5	solid cylinder	318–812	—	Schmidt and Pflug (1965)
36% 0.45 m/s	74	solid cylinder	193	—	Pflug (1964)
1.14 m/s	74	solid cylinder	250	—	Pflug (1964)
1.78 m/s	74	solid cylinder	255	—	Pflug (1964)
51% 0.45 m/s	82	solid cylinder	415	—	Pflug (1964)
1.14 m/s	82	solid cylinder	425	—	Pflug (1964)
1.78 m/s	82	solid cylinder	442	—	Pflug (1964)
70.5% 0.45 m/s	90.5	solid cylinder	732	—	Pflug (1964)
1.14 m/s	90.5	solid cylinder	823	—	Pflug (1964)
1.78 m/s	90.5	solid cylinder	875	—	Pflug (1964)
75%–	115.5	solid cylinder	851	—	Pflug (1964)
90%–	115.5	solid cylinder	1306	—	Pflug (1964)
75%–	115.5	303 × 406 can/water	—	823	Pflug (1964)
90%–	115.5	303 × 406 can/water	—	948	Pflug (1964)
90%–	121.1	303 × 406 can/water	—	965	Pflug (1964)
55% steam	120	13.5 × 13.5 × 2 cm pouch/40% bentonite	—	410–420	Yamano (1983)
65.9% steam	120	13.5 × 13.5 × 2 cm pouch/40% bentonite	—	360–400	Yamano (1983)
10–50% air	105–125	retort pouch laminate	—	850–2500	Kisaalita et al. (1985)
Positive flow retort: 0% air	105–125	silicone rubber/nylon	—	10310	Tung and Ramaswamy (1986)

15% air	12.1 × 17.8 × 1.9 cm	105–125	—	5480	Ramaswamy and Tung (1988)
35% air	12.1 × 17.8 × 1.9 cm	105–125	—	2360	Tung et al. (1989)
50% air	12.1 × 17.8 × 1.9 cm	105–125	—	1260	Tung et al. (1989)
Lagarde retort: 0% air	12.1 × 17.8 × 1.9 cm	105–125	—	6360	Tung et al. (1989)
15% air	12.1 × 17.8 × 1.9 cm	105–125	—	5220	Tung et al. (1989)
35% air	12.1 × 17.8 × 1.9 cm	105–125	—	4010	Tung et al. (1989)
50% air	12.1 × 17.8 × 1.9 cm	105–125	—		Tung et al. (1989)
Lagarde Retort 0–40% air	122	Retort pouches		9,69-2,500	Britt (2004)
(c) Water – heating					
	303 × 406 can/water	115.5	—	908	Pflug (1964)
	303 × 406 can/water	121.1	—	877	Pflug (1964)
	Lead cylinder	65.5	830	—	Merrill (1948)
	Lead cylinder	87.8	977	—	Merrill (1948)
	Lead cylinder	104.5	1079	—	Merrill (1948)
	Lead cylinder	121.1	1175	—	Merrill (1948)
	307 × 409 can/water	121.1	—	931	Merrill (1948)
	404 × 414 can/water	121.1	—	1079	Merrill (1948)
	307 × 409 can/1% bentonite	121.1	—	636–1079	Merrill (1948)
	404 × 414 can/1% bentonite	121.1	—	880	Merrill (1948)
	300 × 408 can/water	82.2	—	1250–1703	Blaisdell (1963)
	307 × 409 can/citrus juice	26.7	—	1022	Heid and Scott (1937)
	307 × 409 can/citrus segments	26.7	—	176	Heid and Scott (1937)
Flow: in l/s: 0.63	30.4 × 45 × 2.5 pouches/10% bentonite	121.1	—	187	Peterson and Adams (1983)
1.07	30.4 × 45 × 2.5 pouches/10% bentonite	121.1	—	210	Peterson and Adams (1983)
1.82	30.4 × 45 × 2.5 pouches/10% bentonite	121.1	—	250	Peterson and Adams (1983)
3.09	30.4 × 45 × 2.5 pouches/10% bentonite	121.1	—	238	Peterson and Adams (1983)

(cont.)

TABLE 8.2. (Continued)

Heat transfer media	Process conditions (°C)	Container/product	External heat transfer coefficient (W m^{-2} K^{-1})	Overall heat transfer coefficient (W m^{-2} K^{-1})	Reference
4.44	121.1	30.4 × 45 × 2.5 pouches/10% bentonite	—	227	Peterson and Adams (1983)
5.67	121.1	30.4 × 45 × 2.5 pouches/10% bentonite	—	278	Peterson and Adams (1983)
6.93	121.1	30.4 × 45 × 2.5 pouches/10% bentonite	—	272	Peterson and Adams (1983)
	120	Institutional size pouches/8% bentonite	—	186 ± 54	Lebowitz and Bhowmik (1990)
	10–123	13.0 × 18.0 × 2.5 cm pouch/10% bentonite	—	1100–2400	McGinnis (1986)
	120	20.3 × 30.5 × 2.5 cm pouch/8% bentonite	—	178–186	Lebowitz and Bhowmik (1990)
(d) Water – cooling					
	10	305 × 404 can/fish	—	136	Okada (1940a, b)
	10	215 × 404 can/fish	—	57	Okada (1940a, b)
	10	305 × 109 can/fish	—	40	Okada (1940a, b)
	10	303 × 209 × 100 can/fish	—	148	Okada (1940a, b)
	10	215 × 201 × 104 can/fish	—	136	Okada (1940a, b)
	10	412 × 112 × 104 can/fish	—	120	Okada (1940a, b)
	10	401 × 112 × 104 can/fish	—	136	Okada (1940a, b)
86 rpm	26.7	307 × 409 can/citrus segments	—	205	Heid and Scott (1937)
0 rpm	26.7	307 × 409 can/citrus juice	—	216–255	Heid and Scott (1937)
70 rpm	26.7	307 × 409 can/citrus juice	—	500	Heid and Scott (1937)

100 rpm	26.7	307 × 409 can/citrus juice	—	1618	Heid and Scott (1937)
261 rpm	26.7	307 × 409 can/citrus juice	—	823	Heid and Scott (1937)
555 rpm	26.7	307 × 409 can/citrus juice	—	1675	Heid and Scott (1937)
(e) Air heating:					
	145	211 × 400 can/vegetables in brine	—	83	Gillespy (1958)
	145	300 × 410 can/vegetables in brine	—	57	Gillespy (1958)
	145	211 × 400 can/beans in tomato sauce	—	77	Gillespy (1958)
	145	300 × 410 can/beans in tomato sauce	—	51	Gillespy (1958)
	145	300 × 410 fruit/syrup	—	57	Gillespy (1958)
Flow: 4 m/s	145	211 × 400 can/vegetables in brine	—	196	Gillespy (1962)
9 m/s	145	211 × 400 can/vegetables in brine	—	116	Gillespy (1962)
15 m/s	145	211 × 400 can/vegetables in brine	—	159	Gillespy (1962)
Flow: 4 m/s	145	307 × 408 can/vegetables in brine	—	37	Gillespy (1962)
9 m/s	145	307 × 408 can/vegetables in brine	—	54	Gillespy (1962)
15 m/s	145	307 × 408 can/vegetables in brine	—	74	Gillespy (1962)
(f) Air cooling:					
	10	305 × 404 can/fish	—	10.2	Okada (1940a, b)
	10	215 × 404 can/fish	—	7.9	Okada (1940a, b)
	10	305 × 109 can/fish	—	7.9	Okada (1940a, b)
	10	303 × 207 × 100 can/fish	—	10.2	Okada (1940a, b)
	10	215 × 201 × 104 can/fish	—	10.2	Okada (1940a, b)
	10	412 × 115 × 104 can/fish	—	10.2	Okada (1940a, b)
	10	401 × 112 × 104 can/fish	—	10.2	Okada (1940a, b)
(g) Flame/radiation heating:					
Gas flame	approx. 1600	211 × 400 can/water	—	1000	Casimir (1975)
Can surface rotation 10–100	approx. 170	303 × 406 can/silicon fluid	44–47	380–1264	Peralta Rodriguez and Merson (1983)

TABLE 8.3. Relative energy usage for various sterilizers (Holdsworth 1992).

Retort system	Casimir (1971)	Savage (1984)	Jowitt & Thorne (1971)	Ferrua & Col (1975)	Holdsworth (1989)
Batch retort	100	100	100	100	100
Crateless	—	50	—	—	—
Hydrostatic cooker	21.3	25	56	—	56—96
Fluidized-bed	—	—	38	—	—
Microwave sterilizer	1230	—	—	—	—
Continuous rotary cooker	—	—	—	46	45
Continuous atmospheric	—	—	—	64	—
Flame sterilizer (batch)	266	—	—	—	—
Flame sterilizer (continuous)	90	—	—	88	—
Hydrolock cooker	—	—	—	—	54

In using saturated steam the principle outlined in Section 8.1.2 for venting and bleeding the retort must be observed.

Altitude affects the temperature and pressure of saturated steam, and this needs to be kept in mind when canneries are located at high altitudes. The volumes of steam that can flow through orifices of differing sizes are given in Lopez (1987); these are of importance to the steam supply designer. The typical steam demands for some retorts are shown in Table 8.5. With batch retorts, the main steam demand is during the venting, and this falls off progressively after about 10 min to a relatively constant value for the rest of the process. For a vegetable canning operation an overall steam consumption of about 0.8 kg per kilogram of product is required; for a longer process, such as meat canning, the figure rises to about 1 kg per kilogram of product (Holdsworth et al. 1988; Holdsworth 1992). Genta et al. (2001) reported that for canning 20,000 tons of vegetables, the steam consumption was 7 tons/h.

Several models have been proposed for the estimation of energy in retort operations (see Bhowmik & Hayakawa, 1983, 1988; Bhowmik et al. 1985). An interesting optimization study by Guariguata et al. (1984) showed that for a series of processes at different temperatures, corresponding to $F_0 = 2.52$ min, the amount of energy consumed for pea purée in 307 × 409 cans was a minimum of 34.5 kg for a processing temperature of 121.1°C and processing time

TABLE 8.4. Properties of dry saturated steam.

Processing temperature (°C)	(°F)	Absolute pressure (bar)	Latent heat of condensation (kJ kg^{-1})	Specific heat (kJ kg^{-1})	Total heat (kJ kg^{-1})
115.5	(240)	1.72	2216	483	2699
121.1	(250)	2.06	2196	510	2706
126.7	(260)	2.43	2188	532	2720
132.2	(270)	2.90	2166	556	2722
137.8	(280)	3.50	2142	582	2724

TABLE 8.5. Steam requirements for various retorts (Lopez 1987).

Retort type	Steam inlet size (mm)	Maximum demand (kg h^{-1})	Operating demand (kg h^{-1})	Average steam used per case (kg)
Batch	25.4	1134	45–68	2.35
	31.75	1587	45–68	2.35
	38.1	2041	45–68	2.35
	50.8	2721	45–68	2.35
Continuous pressure cooker	50.8	2721	453–680	1.36–1.60

66 min. Bhowmik and Hayakawa (1983) showed that for conduction-heating packs the energy required for sterilization increased with the size of the can, but for convection-heating packs the energy was independent of the can size. In general, steam consumption depends on the method of process operation and the energy-saving measures which are used.

8.1.4.3. Steam–Air Mixtures

Of the various methods of heat-transfer media, less is known about the properties of steam–air mixtures than about any of the others. Although experimental work was carried out as early as the 1930s, non-uniform heat penetration results led to a limited commercial acceptance of the system. In particular, the concept of air in steam, especially with regard to the stringent venting required for batch retorts, was not recommended. However with more extensive investigations and careful retort design, successful sterilizing systems are available (see Table 8.1) which use carefully controlled steam–air mixtures. Table 8.6 gives some equivalent temperatures for different proportions of air mixed with steam. For more extensive

TABLE 8.6. Variation of temperature of steam–air mixtures with pressure (Holdsworth 1976).

% air by volume	Temperature (°C) Steam pressure:			
	1 bar	1.07 bar	1.38 bar	2.06 bar
0	100.0	115.0	126.1	134.4
5	98.3	113.9	124.4	132.8
10	97.2	112.2	122.8	131.1
20	93.9	108.3	118.9	127.2
30	90.6	104.4	115.0	122.8
40	86.1	100.0	110.0	117.8
50	81.7	95.0	105.0	112.2
60	76.1	89.4	98.3	105.6
70	69.4	81.7	90.6	97.2
80	60.6	72.2	80.6	86.7
90	46.1	56.7	63.9	70.0

information see Wasmund (1976). The commercial systems use fans to mix the steam and air, and baffles running the length of the retort to direct the mixture through the baskets holding the cans or packages. The main factors that must be considered to obtain uniform heating are the flow rate, the ratio of steam to air, the direction in which the mixture flows across the surface and the nature of the circulation system. It is necessary to determine suitable values for these and to control them in an appropriate manner. Failure to do this could lead to understerilization or uneven product quality.

Parcell (1930) was one of the first to publish details of steam–air heating of glass containers, where the overpressure was required to retain the metal caps on the glass jars. It was reported that steam–air mixtures heated more slowly and less uniformly than saturated steam at the same temperature. This was directly attributed to a stagnant film of air that was displaced during condensation and formed near the surface of the container. It was found that initially during the coming-up period the temperature distribution within the retort baskets was very wide; however, this disappeared during the heating period. Advantages were also gained by mixing the steam and air prior to entry into the retort.

With the development of sterilizable plastic pouches in the 1960s many workers (see Table 8.2b) investigated the behavior of steam–air mixtures. Extensive studies by Pflug and Nicholas (1961), Pflug and Blaisdell (1961) and Schmidt and Pflug (1965) showed that the f_h values decreased with increasing velocity and temperature of the mixture. It was also shown that the effect of steam–air mixtures was far greater with convection- than with conduction-heating packs.

From Table 8.6 it can be seen that for a given pressure the temperature decreases with increasing quantities of air in the steam. Similarly, the heat-transfer coefficient decreases. The standard method of determining the external heat-transfer coefficient for condensing steam–air mixtures has been to use aluminum blocks or cylinders. These have the advantage that, because aluminum is a rapid heat conductor, the main resistance to heat transfer is the external coefficient, and this can be found without recourse to the internal heat-transfer coefficient. Ramaswamy et al. (1983) derived an empirical equation, based on the formal heat transfer equations (see Chapter 2), for determining heat-transfer coefficients, h, in the temperature range 105–125°C for various percentages of steam, s, in steam–air mixtures:

$$h = 135e^{0.0483s}. \tag{8.1}$$

Tung et al. (1984) studied the effect of flow rate of steam–air mixtures on the heat-transfer coefficient for rectangular metal blocks, using two types of retort – a vertical positive flow retort and a Largarde horizontal forced-circulation retort. In general, the values of h increased with increasing flow rate, due to increasing turbulence. The effect of brick orientation also had an effect on the surface heat-transfer coefficient. The mechanism of heat transfer from a steam–air mixture to a retort pouch laminate has been studied in detail by Kisaalita et al. (1985).

Tung *et al.* (1990) have outlined the historical development of the processing of retort pouches and the use of retorts made by Lagarde and Truxon, which have a forced convection heating system and facilities for rotary processing.

8.1.4.4. Hot Water

The alternative method of processing pouches or plastic containers is to use hot water, which has an overpressure of air. In many retorts (see Table 8.2c) the water is agitated to improve heat transfer, either by recirculation or by bubbling air through the water. The velocity of the water and the way it impacts the containers both affect the heat transfer. As with steam–air processing, it is necessary to place the pouches or semi-rigid plastic containers in a strict geometrical arrangement, normally by using a rigid support system. It is essential that all the containers receive the same treatment, and the insert to support the containers should be designed accordingly. Only by paying strict attention to these design features can the required process be adequately implemented, otherwise there is a danger of non-uniform heating and understerilized packs. Water processing is also used for glass containers, where the excess pressure prevents the metal lids being damaged. Unlike steam, the total heat content is much lower at a given temperature since the latent heat of condensation is not available. However, more recently a very flexible processing system has been developed that uses steam with hot water spray. The FMC Steam Water Spray^{TM} is a horizontal processing vessel with external heat exchangers to control the water temperature. Heating is achieved by a combination of of water sprays using nozzles and direct steam injection. Excellent temperature distribution, efficient heat transfer and good overpressure control are claimed (Vandenberghe 2004).

Some typical values of the heat-transfer coefficients for water processing are given in Table 8.2c.

Peterson and Adams (1983) studied the effect of water velocity of heat penetration into institutional sized pouches, and showed that, using water at 121.1°C, the required process time decreases from 28.5 min for a water velocity of 38 l min^{-1} to 25.7 min for 400 l min^{-1}. This illustrates the need for careful process control to obtain uniform heat penetration. Bhowmik and Tandon (1987) developed a heat transfer correlation,

$$h = ka A^{1/2} \tan A^{1/2}, \tag{8.2}$$

where k is the thermal conductivity, $A = 2.303/\alpha f_h$, α is the thermal diffusivity, f_h is the slope index of the heat penetration curve and a is the half-thickness of the retort pouch. Lebowitz and Bhowmik (1989) used an elegant computer optimization technique to determine heat-transfer coefficients for retortable pouches heated by circulating hot water, The work emphasized the importance of having and maintaining the correct flow patterns during heat processing.

Batch retorts using pressurized hot water cascading retorts, e.g., Barriquand – Steriflow, are an important class of newer retorting systems. Akterian *et al.* (1998) have studied the effect of temperature non-uniformity in batch water cascading retorts with four baskets, using a theoretical model and comparing the results with the sterilization of canned pea purée in 300 × 405 cans. The results showed that the effect of temperature non-uniformity was greater on the F-value than on the thiamin retention.

Similar experimental work has been carried out on overpressure rotary retorts. (Adams & Hart-English 2000; Campbell & Ramaswamy 1992a, b; Lanoiseuille *et al.* 1995; Pak *et al.* 1990; Smout *et al.* 1997, 1998, 2000a, b, 2001; Smout and May (1997) and Varga *et al.* 2000a, b) have produced an important set of guidelines for the correct operation of over-pressure retort systems.

Richardson *et al.* (1997) have studied the distribution of temperature and lethality in a full water immersion system with four retort baskets using PTFE blocks and cans containing water and 5% starch solution.

Varga and Oliveira (2000) have determined the heat-transfer coefficients between bulk medium and packed containers using a Barriquand steriflow retort and cascading water. The containers were filled with 10% bentonite suspension and were arranged in two differing stacking patterns. The calculated heat-transfer coefficients varied from 162–203 W m^{-2} °C^{-1} for heating and 75–142 W m^{-2} °C^{-1} for cooling in the middle of the stack of packs. The values were lower for the top of the packs compared with the bottom.

Ramaswamy *et al.* (1991) made a study of the temperature distribution in a single-crate water cascade retort to establish the uniformity of the process.

8.1.4.5. Water Cooling

The cooling of cans and packages is important in reducing the temperature to a level so that thermophilic organisms do not germinate and cause spoilage. The process is generally rather slow because of the low heat-transfer coefficients, especially in static water. In continuous cookers, e.g. hydrostatic cookers, it is often necessary to have several cooling legs to achieve adequate cooling within the cooker. While heat transfer in the steam chamber is highly effective, it is necessary to remember that the hot bars also have to cool, and they themselves shield the cans from the water to some extent. Other types of continuous cooker, e.g. reel-and-spiral pressure cookers, have their own cooling shell, and the cans are rapidly cooled. Relatively little work has been done on the engineering aspects of cooling, and few data are available in the literature on heat-transfer coefficients.

8.1.4.6. Air Heating

For some years there was interest in dry processing methods to conserve water and possibly reduce leaker spoilage problems. Ekelund (1961) described a rotating hot air sterilizer which used air at temperatures of 140–165.5°C and velocities of 7–10 m s^{-1}. The method was shown to have application to the high-temperature short-time in-pack sterilization of convection-heating products. The rate of heating is controlled almost entirely by the rate of external heat transfer to the can. Ekelund (1961) obtained heat-transfer coefficients of the order of

70 W m^{-2} K^{-1} for the above-quoted velocities. Gillespy (1958, 1962) published extensive details of trials using a prototype cooker, with small sizes of can, which would withstand the internal pressure generated, in the absence of a superimposed pressure. The method does not appear to have been developed for commercial use.

8.1.4.7. Fluidized Bed

Piggot (1963a, b) described experimental work on the application of fluidized beds to the heating of cans and obtained high heat transfer rates. For example, for the heating of a 307×201 can of salmon at 117°C, a center temperature of 115°C was observed after 36 min processing. This shows that fluidized bed heating could be as effective as steam heating. Like hot air processing, the method, being unpressurized, could only be used for small containers. The method was extensively investigated by Jowitt and Thorne (1972), who showed that for conduction-heating packs, processing times were almost the same as for steam heating, but for convection-heating packs processing times were longer under comparative conditions to achieve the same F-value. Using a theoretical heat transfer model, it was shown that heat-transfer coefficients were in excess of 600 W m^{-2} K^{-1} (Thorne & Jowitt 1972; Jowitt & Mynott 1974). The method has not so far been developed for commercial use.

8.1.4.8. Flame Sterilization

The commercial success of the flame sterilizer was due to Max Beauvais, in France, who developed the process in the 1950s (Beauvais et al. 1961). This continuous process involved rolling the cans over a series of gas flames and subsequently cooling the cans in water. The main problem was to prevent the can lids from peaking due to the unrestrained pressure inside the cans. This was subsequently overcome by making stronger ends for the containers. The development of the process has been well documented by Richardson (1987), who discussed the problems of temperature measurement in the container during flame sterilization, heat transfer, container requirements, product quality and economics of using flame sterilizers. Most of the work on heat transfer uses the complex models for radiation heat transfer outlined in chapter 2 (Peralta Rodriguez & Merson 1982, 1983). Teixeira Neto (1982) discussed the rate of heat transfer to convection-heating liquid packs and determined the effect of headspace, rotational speed and viscosity. He showed that nearly 90% of the resistance to heat transfer was due the external heat-transfer coefficient. Cummings and Wright (1984) developed a thermocouple system that had a special rotating adaptor and support, and Paulus and Ojo (1974) have reported on the internal and external temperature distribution during flame processing. The main advantages of this method of high-temperature short-time process were ones of economy and improved product quality (Leonard et al. 1975a, b, 1976, 1977, 1984a, b; Noh et al. 1986).

8.2. Total and Transient Energy Consumption in Batch Retort Processing

A detailed description will be given according to the work done by Simpson *et al.* (2006). The transient energy balance for a system defined as the retort including cans without their contents, and the steam and condensate in the retort requires no work term. The heat transfer terms – between the system and its environment – include radiation and convection to the plant cook room environment and heat transfer to the food within the cans. Equations should be solved simultaneously and the heat transfer equation for the food material can solved numerically using, e.g. an explicit finite difference technique. Correlations valid in the range of interest (100°C through 140°C) should be used to estimate the thermodynamic properties of steam, condensate, and food material. The thermal process is divided into three steps: (a) venting period, (b) period after venting to reach process temperature, and (c) holding time. Cooling step is not analyzed because no steam is required. First, the mathematical model for the food material is presented and then a full development of the energy model for the complete thermal process.

Nomenclature

A: Area (m^2)
Cp: Specific heat (J/kg K)
E : Energy (J)
g_c: Universal conversion factor; 1 (Kg m/N s^2)
$\underline{H}$: Enthalpy (J/Kg)
h: Heat-convection coefficient (W/m^{2}K)
$\dot{m}$: Mass flow rate (kg/s)
M: Mass (kg)
P: Pressure (Pa)
Pm: Molecular weight (kg/kmol)
$\dot{Q}$: Thermal energy flow (W)
R: Ideal gas constant 8.315 (Pa m^3/kmol °K); (J/kmol K)
t: Time (s)
$t*$:Time required to eliminate air from retort
T: Temperature (K)
T_0: Initial temperature
$\overline{T}$: Average product temperature (K)
v : Velocity (m/s)
V: Volume (m^3)

Greek symbols

ρ: Density (kg/m^3)
ε: Surface emissivity of retort shell at an average of emitting and receiving temperatures (dimensionless)
γ: Ratio of specific heat at constant pressure to specific heat at constant volume (dimensionless)
σ: Stefan-Boltzmann constant, 5.676×10^{-8} (W/m^2 K^4)

Subscripts

a: Air
b: Bleeder
amb: Ambient
c: Convection
cv: Condensed vapor
cr: Critical value
cw: Cooling water
e: Metal container
in: Insulation
p: Food product
r: Radiation
rt: Retort
s: Steam
sl: Saturated liquid
sv: Saturated vapor
rs: Retort surface
t: Time
v: Vapor
w: Ccondensed water

8.2.1. Mathematical Model for Food Material

The food material is assumed to be homogeneous, isotropic, and contained in a cylindrical format; therefore, the heat conduction equation for the case of a finite cylinder could be expressed as (other geometries can be considered):

$$\frac{1}{r}\frac{\partial T}{\partial r} + \frac{\partial^2 T}{\partial r^2} + \frac{\partial^2 T}{\partial z^2} = \frac{1}{\alpha}\frac{\partial T}{\partial t} \tag{8.3}$$

Where T is a function of the position (r, z) and time (t). The respective boundary and initial conditions are as follows: $T(food\ material, 0) = T_0$; where T_0 is a known and uniform value through the food material at time 0.

To estimate the temperature at food surface at any time t, a finite energy balance is developed at the surface.

$$-kA\frac{\partial T}{\partial r} + hA\partial T = MCp\frac{\partial T}{\partial t} \tag{8.4}$$

In most practical cases, it can be assumed that the Biot number is well over 40, meaning that the temperature of the surface of the food material could be equalized, at any time, with retort temperature (Teixeira *et al.* 1969; Datta *et al.* 1986; Simpson *et al.* 1989; Almonacid-Merino *et al.* 1993; Simpson *et al.* 1993). The aforementioned statement is not necessarily applicable for processing retortable pouch or semi rigid trays and bowls (Simpson *et al.* 2003). The model (equation 8.4) considers the possibility of a Biot number less than 40, but is also suitable for a Biot number equal to or larger than 40.

8.2.2. Mass and Energy Balance During Venting

Before expressing the energy balance, it is necessary to define the system to be analyzed: Steam-air inside the retort – at any time $t\ (0 \leq t \leq t^*)$, during venting – is considered as the system (Figure 8.1).

Global mass balance

$$\dot{m}_s - \dot{m}_{sv} - \dot{m}_a = \frac{dM}{dt} \tag{8.5}$$

Mass balance by component

$$Air: -\dot{m}_a = \frac{dM_a}{dt} \tag{8.6}$$

$$Vapor: \dot{m}_s - \dot{m}_{sv} - \dot{m}_w = \frac{dM_{sv}}{dt} \tag{8.7}$$

$$Condensed\ water: \dot{m}_w = \frac{dM_w}{dt} \tag{8.8}$$

Where:

$$M = M_a + M_{sv} + M_w;\ \dot{m} = \dot{m}_{sv} + \dot{m}_a \tag{8.9}$$

General energy balance

$$[\underline{H}_s \dot{m}_s]_{IN} - [\underline{H}_{sv} \dot{m}_{sv} + \underline{H}_a \dot{m}_a]_{OUT} + \delta \dot{Q} - \delta \dot{W} = \frac{dE_{system}}{dt} \qquad (8.10)$$

Where:

$$\delta \dot{Q} = \delta \dot{Q}_c + \delta \dot{Q}_r + \delta \dot{Q}_p + \delta \dot{Q}_e + \delta \dot{Q}_{rt} + \delta \dot{Q}_{in} \qquad (8.11)$$

and

$$\delta \dot{W} = 0 \qquad (8.12)$$

Replacing the respective terms into equation (8.11), the term δQ in equation (8.10) can be quantified as:

$$\delta \dot{Q} = hA\,(T_{in} - T_{amb}) + \sigma\varepsilon A\left(T_{in}^4 - T_{amb}^4\right) + M_p Cp_p \frac{d\bar{T}_p}{dt} + M_{rt} Cp_{rt} \frac{d\bar{T}_{rt}}{dt}$$

$$+ M_e Cp_e \frac{d\bar{T}_e}{dt} + M_{in} Cp_{in} \frac{d\bar{T}_{in}}{dt} \qquad (8.13)$$

The following expression shows how the cumulative term of equation (8.10) was calculated. Because of the system definition, changes in potential energy as well as kinetic energy were considered negligible:

$$\frac{dE_{system}}{dt} = M_{sv}\frac{d\underline{H}_{sv}}{dt} + \underline{H}_{sv}\frac{dM_{sv}}{dt} - P_{sv}\frac{dV_{sv}}{dt} - V_{sv}\frac{dP_{sv}}{dt} + M_a\frac{d\underline{H}_a}{dt} + \underline{H}_a\frac{dM_a}{dt}$$

$$- P_a\frac{dV_a}{dt} - V_a\frac{dP_a}{dt} + M_w\frac{d\underline{H}_w}{dt} + \underline{H}_w\frac{dM_w}{dt} - P_w\frac{dV_w}{dt} - V_w\frac{dP_w}{dt}$$

$$(8.14)$$

The mass flow of condensate was estimated from the energy balance as:

$$\dot{m}_w\left(\underline{H}_{sv} - \underline{H}_{sl}\right) = \delta \dot{Q} = hA\,(T_{in} - T_{amb}) + \sigma\varepsilon A\left(T_{in}^4 - T_{amb}^4\right) + M_p Cp_p \frac{d\bar{T}_p}{dt}$$

$$+ M_{rt} Cp_{rt}\frac{d\bar{T}_{rt}}{dt} + M_e Cp_e\frac{d\bar{T}_e}{dt} + M_{in} Cp_{in}\frac{d\bar{T}_{in}}{dt} \qquad (8.15)$$

Therefore:

$$\dot{m}_w$$

$$= \frac{hA\,(T_{in} - T_{amb}) + \sigma\varepsilon A\left(T_{in}^4 - T_{amb}^4\right) + M_p Cp_p \frac{d\bar{T}_p}{dt} + M_{rt} Cp_{rt}\frac{d\bar{T}_{rt}}{dt} + M_e Cp_e\frac{d\bar{T}_e}{dt} + M_{in} Cp_{in}\frac{d\bar{T}_{in}}{dt}}{\left(\underline{H}_{sv} - \underline{H}_{sl}\right)}$$

$$(8.16)$$

Therefore the steam mass flow demand during venting should be obtained replacing equations (8.7), (8.8), (8.9), (8.13), (8.14), and (8.16) into equation (8.10).

8.2.3. Mass and Energy Consumption Between Venting and Holding Time (To Reach Process Temperature)

As was mentioned before, first, it is necessary to define the system to be analyzed: Steam and condensed water inside the retort were considered as the system (Figure 8.1).

$$\text{Global mass balance: } \dot{m}_s - \dot{m}_b = \frac{dM}{dt} \tag{8.17}$$

$$\text{Vapor: } \dot{m}_s - \dot{m}_b - \dot{m}_w = \frac{dM_{sv}}{dt} \tag{8.18}$$

$$\text{Condensed water: } \dot{m}_w = \frac{dM_w}{dt} \tag{8.19}$$

Energy balance on the bleeder: System, steam flow through the bleeder, considering an adiabatic steam flow:

$$\left[\underline{H}_{sv}\dot{m}_b\right]_{IN} - \left[\left(\underline{H}_b + \frac{v^2}{2g_c}\right)\dot{m}_b\right]_{OUT} = 0 \tag{8.20}$$

Where the bleeder is assumed to be operating in steady state condition, with no heat, no work, and negligible potential energy effects, the energy balance around the bleeder reduces to (Balzhiser 1972):

$$\left(\underline{H}_b - \underline{H}_{sv}\right) + \frac{v_b^2 - v_{sv}^2}{2g_c} = 0. \tag{8.21}$$

For a gas that obeys the ideal gas law (and has a Cp independent of T):

$$\left(\underline{H}_{sv} - \underline{H}_b\right) = Cp\left(T_{sv} - T_b\right) \tag{8.22}$$

Neglecting v_{sv}^2 in relation to v_b^2, and replacing equation (8.20) into equation (8.19), it reduces to:

$$v_b^2 = -2g_c Cp T_{sv}\left(\frac{T_b}{T_{sv}} - 1\right). \tag{8.23}$$

Considering an isentropic steam flow in the bleeder which obeys the ideal gas law, equation (8.21) could be re-written as:

$$v_b^2 = -\frac{2g_c P_{sv}}{\rho_{sv}}\left(\frac{\gamma}{\gamma - 1}\right)\left(\left(\frac{P_b}{P_{vs}}\right)^{\left(\frac{\gamma-1}{\gamma}\right)} - 1\right) \tag{8.24}$$

Where the continuity equation is:

$$\dot{m}_b = \rho v_b A. \tag{8.25}$$

Therefore, combining equation (8.24) and equation (8.25):

$$\dot{m}_b = \frac{P_s A_b}{\sqrt{\frac{RT_s}{\gamma}}} \left(\frac{P_{\text{amb}}}{P_s}\right) \sqrt{\left(\frac{2}{\gamma - 1}\right)\left(1 - \left(\frac{P_{\text{amb}}}{P_s}\right)^{\left(\frac{\gamma-1}{\gamma}\right)}\right)}. \qquad (8.26)$$

The maximum velocity of an ideal gas in the throat of a simple converging nozzle is identical to the speed of sound at the throat conditions. The critical pressure is Pc (Balzhizer 1973):

$$Pc = P_{\text{amb}}\left(\frac{2}{\gamma + 1}\right)^{\frac{\gamma}{\gamma-1}} \qquad (8.27)$$

Then equation (8.26) will be valid for P_s in the following range:

$$\left(\frac{2}{\gamma + 1}\right)^{\frac{\gamma}{\gamma-1}} \leq \frac{P_{\text{amb}}}{P_s} < 1. \qquad (8.28)$$

If P_s is bigger than P_c, substituting equation (8.27) into equation (8.28), the expression for the mass flow is as follows:

$$\dot{m}_b = \frac{P_s A_b}{\sqrt{\frac{RT_s}{\gamma}}}\left(\frac{2}{\gamma + 1}\right)^{\frac{\gamma+1}{2(\gamma-1)}}; \quad \text{for} \quad \frac{P_{\text{amb}}}{P_s} < \left(\frac{2}{\gamma + 1}\right)^{\frac{\gamma}{\gamma-1}}. \qquad (8.29)$$

8.2.4. Mass and Energy Balance During Holding Time

System: Steam inside the retort (Figure 8.1).

$$\text{Global Mass balance: } \dot{m}_s - \dot{m}_b = \frac{dM}{dt} \qquad (8.30)$$

$$\text{Vapor: } \dot{m}_s - \dot{m}_b - \dot{m}_w = \frac{dM_{sv}}{dt} \qquad (8.31)$$

$$\text{Condensed water: } \dot{m}_w = \frac{dM_w}{dt} \qquad (8.32)$$

Energy balance on the bleeder: The steam flow through the bleeder was estimated as previously mentioned, therefore:

$$\dot{m}_b = \frac{P_s A_b}{\sqrt{\frac{RT_s}{\gamma}}}\left(\frac{P_{\text{amb}}}{P_s}\right)\sqrt{\left(\frac{2}{\gamma - 1}\right)\left(1 - \left(\frac{P_{\text{amb}}}{P_s}\right)^{\left(\frac{\gamma-1}{\gamma}\right)}\right)}; \quad \text{if} \left(\frac{2}{\gamma + 1}\right)^{\frac{\gamma}{\gamma-1}} \leq \frac{P_{\text{amb}}}{P_s} < 1$$

$$\dot{m}_b = \frac{P_s A_b}{\sqrt{\frac{RT_s}{\gamma}}}\left(\frac{2}{\gamma + 1}\right)^{\frac{\gamma+1}{2(\gamma-1)}}; \quad \text{if} \frac{P_{\text{amb}}}{P_s} < \left(\frac{2}{\gamma + 1}\right)^{\frac{\gamma}{\gamma-1}}$$

$$\dot{m}_w = \frac{hA(T_{in} - T_{\text{amb}}) + \sigma\varepsilon A(T_{in}^4 - T_{\text{amb}}^4) + M_p Cp_p \frac{dT_p}{dt} + M_{rt} Cp_{rt} \frac{dT_{rt}}{dt} + M_e Cp_e \frac{dT_e}{dt} + M_{in} Cp_{in} \frac{dT_{in}}{dt}}{(H_{sv} - H_{sl})}$$

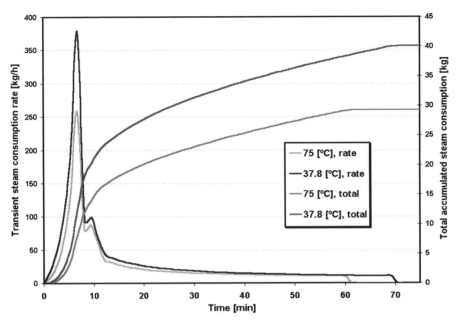

FIGURE 8.7. Total and transient steam consumption for two different initial food temperatures (37.8 and 75 [°C]).

Therefore the steam mass flow was estimated as:

$$\dot{m}_s = \dot{m}_w + \dot{m}_b \tag{8.33}$$

8.2.5. Numerical Results

As an example, Figure 8.7 compares the total and transient steam consumption for two different initial food temperatures (37.8 and 75°C). Figure 8.7 shows that for the high temperature, the maximum peak and the total steam consumption could be reduced as much as 30% and 29% respectively. This example shows that initial temperature of food material is an important variable to reduce the peak of energy consumption.

8.3. Pressures in Containers

8.3.1. Development of Internal Pressures

When a hermetically sealed container of food is heated the internal pressure rises due to increasing vapor pressure of the liquid portion; air and other gases in the headspace expanding; and expansion of the food mass. The rise is partially counter-balanced, in the case of a can, by expansion of the can and the outward movement of the can ends, which are designed with expansion rings. The situation is different with semi-rigid packaging materials that have only a limited elastic expansion, which can damage the material or the seams.

The headspace is also very important, not only in the case of agitated packs but also in all types of rigid container. The headspace has the function of allowing for the liquid and gaseous expansion.

The filling temperature also has an effect on the pressure: the higher the filling temperature the lower the pressure developed internally. This also produces a lower vacuum in the can when it is cooled. It is important to eliminate the air after filling. This was formerly done by clinching the lid on the can and "exhausting" the can by heating and then seaming. It is now done by blowing steam across the can before seaming, in a process known as steam flow closing. The removal of air from the headspace also reduces the amount of oxygen in the can and reduces the likelihood of product–container interactions.

8.3.2. Internal Pressure Calculation

A general equation for calculating the pressure developed in a container is

$$P_T = P_a + P_w + C_1 + C_2,$$ (8.34)

where P_T is the total is the total pressure, P_a is the partial pressure of the air in the headspace, P_w is the saturated vapor pressure of water (or oil) at the temperature of the food surface, C_1 is a correction factor for the expansion or shrinkage of the package and C_2 is a correction factor for the thermal expansion of the food.

The partial pressure of the air depends on both the temperature and the volume of air, and is calculated using Charles's Law

$$(P_i V_i)/T_i = (P_f V_f)/T_f,$$ (8.35)

where P is the pressure, V is the volume and T the temperature; the subscripts i and f refer to the initial and final conditions.

The strain pressure P developed is given by

$$P = P_a + P_w - P_{\text{ext}},$$ (8.36)

where the pressure of air and water vapor are the final pressures at the processing temperature and P_{ext} is the external pressure of steam (and air). Strain pressures in cans vary in the range 1–2 bar, with ambient external pressure and $\frac{1}{2} - 1$ bar under processing conditions.

The pressure of water vapor may be estimated from the equation

$$P_w = e^{(16.0 - 4967/T)}.$$ (8.37)

A more complex relation has been derived for plastic containers of liquid, where the effects of expelled non-condensable gases are taken into account (Beck 1985). The pressure–volume relationship for flexible pouches has been derived for flexible packaging materials (Davis *et al* 1960). The pressure developed during the processing of pouches of food has been examined by Yamano (1983) for steam–air processing. The internal pressure profile has been examined by Patel *et al.* (1991) for the processing of semi-rigid food packages in steam–air mixtures.

8.3.3. Processing Requirements

As previously mentioned, the internal pressure in rigid containers is counter-balanced to a certain extent by the pressure of the steam or hot water in the heating stage. However, at the end of the heating stage, if the pressure on the can is suddenly released while still hot, there is a danger of can distortion, known as peaking, which may weaken the seam or cause the can to jam in a valve or a subsequent process, such as labeling. The effect is more serious with the larger-diameter cans, and it is usual to prevent such distortion by superimposing a pressure during the cooling cycle, for sufficient time for the can to have cooled and the internal pressure to have reduced to a comfortable level. The situation is more complicated in continuous reel-and-spiral pressure cookers, where pressure must be maintained in the transfer valve during the transfer of cans from the cooker to the cooler. These valves require careful design in order to prevent can distortion and jamming. From the design point of view it is necessary to know the mechanical characteristics of the cans being processed, as well as the pressure–temperature relationship in the can/container for the processing conditions. Can manufacturers supply tables of strain pressures that the cans and seams can withstand to enable the determination of safe working pressures.

An important part of the technology of can handling concerns the bacteriological status of cooling water. For the purposes of this book, it should be mentioned that the seams of a can are vulnerable to ingress of spoilage microorganisms at the cooling stage, when a vacuum is developing in the headspace of the can. It is, therefore, necessary to chlorinate cooling water so that it has a residual of 1 ppm free chlorine before it is used. This topic is dealt with in standard canning technology books. See also Thorpe and Gillespy (1968) and CCFRA (1985, 1997).

8.3.4. Semi-Rigid Containers

The development of semi-rigid containers and sterilizable flexible packaging materials resulted in the need for superpressure processing. This has been achieved using steam–air mixtures and air-pressurized water. For a half-litre semi-rigid pack, containing 100 cm^3 of air, the pressure in the container, originally about 1 bar, is increased to about 3.2 bar at 121.1°C. This includes the effect of the air and the water vapor present in the pack, and gives an estimate of the overpressure required to prevent seam and container damage.

The retort profiles for this type of package vary depending on the type of package and its construction. A typical come-up profile is a linear ramp from 70 to 121°C in 15 min and a cooling period from 121 to 40°C in the same time.

Campbell and Ramaswamy (1992a) have studied the heating rate, lethality and cold-spot location in air entrapped in retort pouches. Massaguer *et al.* (1994) have discussed the selection of heat-transfer media, including air/steam, mixtures and pressurized hot water

8.4. Mechanical Agitation and Rotation of Cans

8.4.1. End-Over-End Agitation

This type of motion creates a stirring of the contents in the can and improved heat transfer. The result of this type of motion is to enable processes to be shortened, with the accompanying attributes of improved quality characteristics. More recently Tucker (2004) has considered all the factors that are involved in the effectiveness of rotation in improving heat transfer to and in containers of food.

As shown in Table 8.7a, the types of product which benefit from this kind of motion are convection packs containing solids or particulates. During the rotation the headspace bubble moves through the food mixture, causing a continuous stirring action; the degree of mixing depends on the speed of rotation and the size of the headspace. It is necessary, in using this type of technique, to ensure that both these factors are carefully and correctly controlled.

The time for the contents of the can to reach processing temperature was found to decrease with increasing speed of rotation. There is, however, a limiting speed, the optimum speed of rotation, beyond which no further reduction is observed. This is due to centrifugal force, which induces a static condition of the food in the can. The agitation of the can's contents will cease when the gravitational force equals the centrifugal force, i.e. when the maximum speed of rotation, $N_{\max} = 0.498/R_r^{1/2}$, where R_r is the radius of rotation of the can. For a retort of 66 cm, this means that $N_{\max}$ should be less than 36.8 rpm. The factors affecting the heat penetration have been studied in detail by Anantheswaran and Rao (1985a, b), who showed that the rate of heat transfer was independent of the size of can for headspace volume 3–9% and a radius of rotation up to 15 cm. Heat transfer correlations for both Newtonian and non-Newtonian foods were discussed in Chapter 2.

8.4.2. Axial Rotation and Spin Cooking

Axial rotation in cans has been studied up to 500 rpm (for the higher range above 180 rpm the process is usually described as spin processing). The degree of mixing depends on the size of the headspace, and this and the speed of

TABLE 8.7. Some heat transfer studies with canned foods processed in agitating cookers.

Type of motion/retort system	Rotation speed (rpm)	Can size	Products	Heat transfer factors	References
(a) End-over-end	0–50	307 × 409	Cut green beans, peas, diced beets and carrots, corn kernels, lima beans, and mixed vegetables	f_h	Roberts and Sognefast (1947)
	0–200	Wide range	Water, peas, carrots, beets, asparagus, cabbage, mushroom soup, evaporated milk	cut	Clifcorn et al. (1950)
	10, 20 and 30	300 × 208; 300 × 401	Water, suet, pea purée	f_h	Niinivaara et al. (1968)
	15–90	307 × 409	Water, tomato purée, corn kernels, peas, and orange concentrate	hpc	Conley et al. (1951)
	27–144	303 × 700	5% bentonite	hpc	Parchomchuk (1977)
	20–40	Special cans	Water, organic liquids, sugar solution, salt solution	htc	Duquenoy (1980)
	20–120	Special cans	84° brix glucose syrup	htc	Naveh and Kopelman (1980)
	40–140	74 × 80 mm	5–6% carboxymethylcellulose	hpc	Hotani and Mihori (1983)
	0–37	Special can	Water and sucrose, glycerine and guar gum solutions	htc	Anantheswaran and Rao (1985a, b)
	0–100	300 × 306; 307 × 408	Six oils of different viscosity	f_h	Javier et al. (1985)
	0–20	307 × 409	Plastic particles	htc	Sablani and Ramaswamy (1993;1996;1997); Sablani et al. (1997); Ramaswamy et al. (1993;1997)

(cont.)

TABLE 8.7. (Continued)

Type of motion/retort system	Rotation speed (rpm)	Can size	Products	Heat transfer factors	References
	0–20	[307 × 407]	Nylon spheres/guar gum	f_h	Krishnamurthy et al. (2001)
	10–15	73 × 109.5 mm cans	3.5% cereal starch	f_h	Denys et al. (1996)
	10–15	81 × 172 mm glass jars	3.5% cereal starch	f_h	Denys et al. (1996)
	10–50	cans	Colflo 67 starch	f_h: rpm	Emond and Tucker (2001)
	0–5	303 × 406	Guar gum 0.4,0.5,0.75%	Htc	Price and Bhomik (1994)
	10	307 × 409	Ny; pn spheres/CMC	htc;F_0	Mang and Ramaswamy (2005)
(b) Axial	0–50	307 × 409	Cut green beans, peas, diced beets and carrots, corn kernels, lima beans and mixed vegetables	f_h	Roberts and Sognefast (1947)
	0–250	401 × 411	Peaches	hpc	Van Blaricom (1955)
	20–120	Special cans	84° brix glucose syrup	htc	Naveh and Kopelman (1980)
	40–140	74 × 80 mm	5–6% carboxymethylcellulose	hpc	Hotani and Mihori (1983)
	0–100	300 × 306; 307 × 408	Six oils of different viscosity	f_h	Javier et al. (1985)
(c) Spin	0–150	Range	Water and silicone oils	htc	Soule and Merson (1985)
	0–400	Various	Orange and citrus juice products	f_h	Casimir (1961)
	0–400	202 × 214	Fruits and fruit juice products; tomato juice, paste and soup	cut	Pruthi et al. (1962)
	0–500	603 × 700	Papaya purée and passion fruit juice	hpc	Wang and Ross (1965)
	0–420	Various	Sucrose solution, CMC, tropical fruit juices and nectars	htc	Quast and Siozawa (1974)

(d) FMC Steritort	0–8	303×406; 608×700	Water/sucrose solutions; spheres/water, spheres/sucrose solution	htc	Lenz and Lund (1978)
	5–11	303×406	Cream style corn formulations	f_h	Berry et al. (1979)
	3–10	Various	cream of celery soup	F_0	Berry and Bradshaw (1980)
	1–15	Various	Whole kernel corn in brine, and vacuum-packed corn	f_h	Berry and Dickerson (1981)
	0–9	Various	Mushrooms, sliced	f_h	Berry and Bradshaw (1982)
	3–11	Various	Milk-based products	f_h	Berry and Kohnhorst (1985)
	2–8	303×406; 603×700	Water and glycerine, sugar and guar gum solutions	htc	Rao et al. (1985)
	2–6	303×406	Snap beans	htc	Fernandez et al. (1988)
	1–10	Various	White beans in brine	F_0	Deniston et al. (1991)
	0–10	221×300	Tomato concentrate	htc	Deniston et al. (1992)
	0–10	603×700	Tomato concentrate	htc	Deniston et al. (1992)
(e) FMC Orbitort	25–42	Various	Cream of celery soup	F_0	Berry and Bradshaw (1980)
	10–642	Various	Whole kernel corn in brine, and vacuum-packed corn	f_h	Berry and Dickerson (1981)
2(f) Water Cascade Retort	0 & 10	Glass jars	Green & white beans	f_h	Ávila et al. (1999, 2006) and Smout et al. (2000a)
	0–25	Glass jars 600ml	White beans (Phaseolus vulgaris)	f_h	Denys et al. (1996, 2000)
	0	Trilaminate pouch	Silicon rubber bricks	f_h	Campbell and Ramaswamy (1992b)
	0–20	Glass Jars 370ml	White beans	htc	Van Loey et al. (1994)

cut = come-up time; hpc = heat penetration curve; htc = heat transfer coefficient.

rotation have to be controlled. Most interest in this type of cooking has been shown in the processing of tropical fruit juices with delicate flavors. The high-speed processes have been shown to give better flavor retention and overall quality. Table 8.7 b, c gives brief details of some of the products that have been studied.

8.4.3. Steritort and Orbitort Processes

The Steritort is a pilot model of the FMC Sterilmatic continuous reel-and-spiral pressure cooker–cooler, and is used to determine suitable rotational speeds and heat penetration for establishing processes for the full-scale model. Speeds of operation are usually in the range 10–42 rpm, and the cooker can be used for a wide range of products that benefit from axial rotation. Some experimental work with this type of system is indicated in Table 8.7d.

The Orbitort, unlike the Sterilmatic cooker, does not allow cans to roll during processing. The usual speed of rotation is 35 rpm. Experimental work has shown that, for a product such as whole kernel corn, the size of the headspace bubble volume and the filled weight are important in obtaining a consistent and satisfactory process. Rao and Anantheswaran (1988) have given a full review of the heat transfer aspects of agitated and rotary cooking processes.

8.4.4. Shaka^{TM} Retort Process

Packages in the retort are vigorously agitated (shaken) at a frequency of 100 to 200 cycles per minute. The shaken of the food product is achieved through a horizontal movement of the baskets. Packages agitation allows an increase of heat transfer rate. Then, processing time is greatly reduced compared to standard retort systems.

Reduction of process time has a high positive impact on quality of end-product (color, taste, vitamin retention, etc.) compared to quality retention obtained with standards retorts. One of the main advantages of this fairly new system (SHAKA^{TM} technology) is that can be used with all types of packages from rigid cans, glass jars and trays to flexible packaging such as retortable pouches. Products such as sauces, soups, babyfood, vegetables or petfood have been positively tested.

In addition, a significant reduction in process time greatly improves the utilization of the sterilization equipment (production capacity). In general, manufacturers claim that 4 to 5 sterilization cycles per hour can be performed on the SHAKA^{TM} system.

8.5. Commercial Pasteurizers

Pasteurization temperatures are usually of the order of 90–100°C, and hot water or sprayed water is the usual heating medium. While the process may be carried out in normal canning retorts, it is more usual to use tunnel pasteurizers with moving belts operating at atmospheric pressure, and zones of sprayed water at different temperatures. For a detailed treatment of the pasteurization of food products see Buckenhskes *et al.* (1988).

A typical industrial pasteurizer for glass jars of an acidic product would be of the order of 12m in length and would have (i) a preheating section at 70°C, 5 mins, (ii) a pasteurization section at 90°C, 30 min., (iii) a cooling section 70°C, 12 min., (iv) a cooling section at 50°C, 7 min., and (v) a cooling section at 20°C, 7 min. The conveyor belt would move the jars at 0.003 m/s. Each zone would be thermostatically controlled and include heat recovery systems.

For packs that benefit from agitation, rotary atmospheric cooker-coolers may be used. Using the system with various heating and cooling zones, it is possible to employ gentle thermal treatment of glass bottles and jars and other packaging systems that may be damaged with thermal shock.

Plazl *et al.* (2006) have modeled the 3-D temperature distributions in glass jars of a tomato product being sterilized in a 5-stage pasteurizer using a finite difference technique.

Horn *et al.* (1997) have modeled an industrial tunnel pasteurizer for bottled beer, (0.33 l Vichy botttle DIN 6075). Temperature measurements were made inside the bottle in two positions: one in the bottom zone and the other in the body of the fluid. The model showed good agreement with the experimentally determined results.

Guerreri (1993) developed a simulation model involving heat transfer and fluid dynamics for an 8-stage continuous pasteurizer; the first four sections had hot water temperatures of 25, 36, 47 and 67°C, respectively, followed by four cooling sections. A comparison was made with a commercial plant involving 20 cl & 100 cl bottles, filled with water; the improved simulation gave PUs which agreed with the achieved values. The model also examined the effect of altering the shape and dimensions of the bottles

8.6. Computer Simulation of Fluid Dynamics Heat Transfer

A detailed study using a CFD of the experimental visualization and mathematical simulation of the isothermal flow of both a Newtonian liquid (Corena Oil 27) and also a non-Newtonian shear-thinning liquid (Ketrol) has been reported by Hughes *et al.* (2003). This involved a study of the headspace bubble movement inside a closed cylindrical container during off-axis rotation in order to simulate axial and end-over-end agitation. It was shown that under certain conditions of rotational speed and off-axial positions the headspace bubble moves through the fluid in the can and so enhances the mixing inside.

8.7. Batch Processing and Retort Scheduling

Batch processing has been widely practiced but little analyzed in the context of canned food plants. Although high speed processing with continuous rotary or hydrostatic retort systems can be found in very large canning factories, such systems are not economically feasible in the majority of small to medium-sized canneries (Simpson *et al.* 2003).

Batch processing will be analyzed in a retrospective and a prospective view. Firstly, batch processing problem structure will be defined in relation to canned food plants. Then, batch processing in canned food plants will be discussed. To bridge the gap between thermal processing and industrial engineering in optimizing design and operation of food canning plants we will discuss and present specifics procedures e.g. retort scheduling.

Finally, we will try to discuss and analyze this large and diverse field, in which there should be plenty of room for surprises, particularly for those who take time to look closely enough with an open and speculative mind.

8.7.1. Batch Processing Problem Structure in Canned Foods

Batch processing with a battery of individual retorts is a common mode of operation in many food-canning plants (canneries). Although high speed processing with continuous rotary or hydrostatic retort systems can be found in very large canning factories (where they are cost-justified by high volume throughput), such systems are not economically feasible in the majority of small to medium-sized canneries (Norback & Rattunde 1991). In such smaller canneries, retort operations are carried out as batch processes in a cook room in which the battery of retorts is located. Although the unloading and reloading operations for each retort are labor intensive, a well designed and managed cook room can operate with surprising efficiency if it has the optimum number of retorts and the optimum schedule of retort operation.

This type of optimization in the use of scheduling to maximize efficiency of batch processing plants has become well known, and is commonly practiced in many process industries. Several models, methods and implementation issues related to this topic have been published in the process engineering literature (Rippin 1993; Kondili *et al.* 1993; Reklaitis 1996; Barbosa & Macchietto 1993; Lee & Reklaitis 1995a, b). However, specific application to retort batteries in food canning plants has not been addressed in the food process engineering literature. Food canneries with batch retort operations are somewhat unique in that the cannery process line as a whole is usually a continuous process in that unit operations both upstream and downstream from the retort cook room are normally continuous (product preparation, filling, closing, labeling, case packing, etc.). Although retorting is carried out as a batch process within the cook room, unprocessed cans enter and processed cans exit the cook room continuously at the same rate (see Figure 8.8). Since the entire process line operates continuously, food canneries are often overlooked as batch process industries. The focus of this work was to apply these batch process optimization techniques only to the retort operations within the cook room, and not the entire process line of the cannery.

Food processing, and thermal processing in particular, are industries confronted with strong global competition. Continuous innovation and improvement of processing procedures and facilities is needed. Although the literature in food

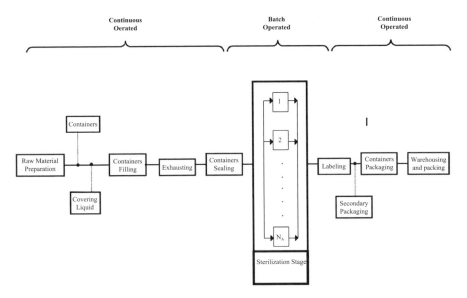

FIGURE 8.8. General simplified flow diagram for a canning plant.

science and thermal processing is very extensive, most of the references deal with the microbiological and biochemical aspects of the process or with engineering analysis of a single unit process operation, and rarely analyze the processing operations in the context of manufacturing efficiency. The early stages of a project usually involve studies of alternative processes, plant configurations and types of equipment. Among problems confronted by canned food, plants with batch retort operations are peak energy/labor demand, underutilization of plant capacity and underutilization of individual retorts.

8.7.2. Batch Processing in Canned Food Plants

In batch retort operations, maximum energy demand occurs only during the first few minutes of the process cycle to accommodate the venting step, while very little is needed thereafter in maintaining process temperature. Likewise, peak labor demand occurs only during loading and unloading operations, and is not required during the holding time at processing temperature. In order to minimize peak energy demand it is customary to operate the retorts in a staggered schedule, so that no more than one retort is venting at any one time. Similar rationale applies to labor demand, so that no more than one retort is being loaded or unloaded at any one time. Too few retorts in a battery can leave labor unutilized, while too many will leave retorts unutilized. The optimum number will maximize utilization of labor and equipment, thus minimizing on-going processing costs. Alternatively, the optimum number of retorts may be based upon maximizing the economic rate of return on the capital investment in the project measured in terms of net present value, which takes many additional factors beyond processing costs into account.

In the case of maximizing output from a fixed number of retorts for different products and container sizes, iso-lethal processes can be identified for each of the various products (alternative combinations of retort temperature and process time that deliver the same lethality), from which a common set of process conditions can be chosen for simultaneous processing of different product lots in the same retort (see 8.8).

8.7.3. The Hierarchical Approach

The hierarchical approach consists of successive refinements, and the design procedure is similar to the hierarchical planning strategy discussed in the artificial intelligence (AI) literature (Douglas 1988).

In contrast to normal true batch processes, canned food plants are operated with just one stage functioning in a batch mode. During normal operation of the sterilization stage (Figure 8.8), the various retort units are filled with cans, perform the retorting process for a specified period and then shut down, and the cycle is repeated. As previously mentioned, in canned food plants, all units, with the exception of retorts, operate continuously. The distinction between batch and continuous processes are sometimes somewhat "fuzzy" (Douglas 1988). According to the literature, when a plant has one or two batch operations with large production rates that otherwise operate continuously; they are normally referred to as a continuous process. Although most of the food science and food engineering literature refers to a canning plant as a batch plant, when the sterilization stage is operated in batch mode, and the hierarchical approach is applied, it is assumed that it is better to classify it as a continuous process.

The design effort will be to decide whether a concept is sufficiently promising from an economic point of view so that a more detailed study could be justified. In our specific case the flow scheme of the process is presented in Figure 8.8. Although some exceptions to this flow scheme could be justified, the following analysis will consider it as a general flow scheme for canned foods plants. The main target in the following sections will be to decide the optimum number of retorts that can be allocated in a canned food plant. The approach will be to identify a general procedure that can be applied to canned food plants.

Finally, to decide and optimize canned food plant design and operation the Net Present Value (or Net Present Worth) profitability evaluation method should be utilized.

8.7.4. Retort Scheduling

Batch processing in food canneries consists of loading and unloading individual batch retorts with baskets or crates of food containers that have been filled and sealed just prior to the retorting operation. Each retort process cycle begins with purging of all the atmospheric air from the retort (venting) with inflow of steam at maximum flow rate, and then bringing the retort up to operating pressure/temperature, at which time the flow rate of steam falls off dramatically to the relatively low level required to maintain process temperature. The retort is

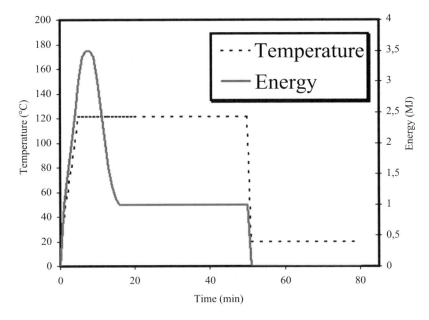

FIGURE 8.9. Energy consumption in a batch cycle.

then held at the process temperature for the length of time calculated to achieve the target lethality (F_0 value) specified for the product. At the end of this process time, steam to the retort is shut off, and cooling water is introduced to accomplish the cool down process, after which the retort can be opened and unloaded.

One of the factors that should be considered when retort scheduling is the energy demand profile during sterilization processing (Almonacid et al. 1993). In batch retort operations, maximum energy demand occurs only during the first few minutes of the process cycle to accomplish the high steam flow venting step. Very little steam is needed thereafter to compensate for the bleeder (and convection and radiation losses) in maintaining process temperature (Bhowmik et al. 1985; Barreiro et al. 1984). A typical representation of the energy demand profile during one cycle of a retort sterilization process is shown in Figure 8.9. As shown, at the initial stage of the process a high peak of energy consumption occurs (venting before reaching the retort temperature), later decreasing dramatically, and finally reaching a low and constant value (convection, radiation and bleeder). Thus, the energy demand for the whole plant will be conditioned upon this acute venting demand in the sterilization process of each retort operating cycle. To minimize the boiler capacity and maximize energy utilization, it is necessary to determine adequate scheduling for each individual retort.

Likewise, peak labor demand occurs only during loading and unloading operations, and is not required during the holding time at processing temperature. Therefore, a labor demand profile would have a similar pattern to the energy demand profile. In order to minimize the peak energy and labor demands, the

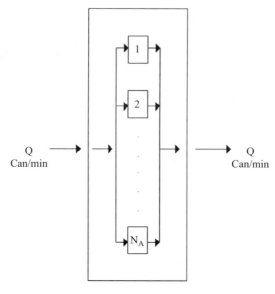

Sterilization Step

FIGURE 8.10. Diagram for operation of a battery with optimum number (N_A) of retorts such that the cook room system operates with continuous inflow and outflow of product.

retort must operate in a staggered schedule so that no more than one retort is venting at any one time, nor being loaded or unloaded at any one time. When a battery consists of the optimum number of retorts for one labor crew, the workers will be constantly loading and unloading a retort throughout the workday, and each retort will be venting in turn, one at a time. Under these optimum circumstances, unprocessed product will flow into and processed product will flow out of the retort battery system as though it were a continuous system as shown in Figure 8.10, while the energy profile will appear as in Figure 8.11.

The optimum number of retorts in the battery will maximize utilization of labor and equipment, thus minimizing unit-processing costs. Too few retorts in a battery can leave labor unutilized, while too many will leave retorts unutilized. A Gantt chart showing the temporal programming schedule of the battery retort system (see Figure 8.12) can be used as a first step in determining the optimum number of retorts. Optimum operation of the retort battery can be achieved if the loading step of the last retort starts at the same time as the first retort finishes its cycle and is ready for unloading. This means that the loading time multiplied by the number of retorts must fit within the total time to load, process, and unload one retort. This relationship can be expressed mathematically:

$$t_c + t_p + t_d = t_c N_A$$

Where N_A is number of retorts and t_c, t_p, and t_d are loading, process, and unloading times, respectively. Considering that loading and unloading times are

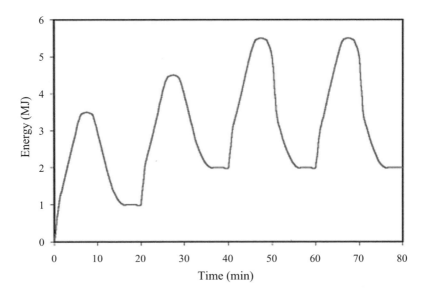

FIGURE 8.11. Energy demand profile from retort battery operating with optimum number of retorts and venting scheduling.

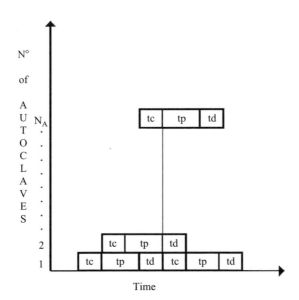

FIGURE 8.12. Gantt chart showing temporal programming schedule of the battery retort system operation.

equal ($t_c = t_d$), we get:

$$N_A = 2 + \frac{t_p}{t_c} \tag{8.38}$$

Therefore:

$$3 \leq N_A \leq \infty,$$

and the minimum number of retorts for optimum operation under this criterion is 3. The number of retorts for any given situation will depend upon the ratio of process time to loading/unloading time.

Moreover, according to the operation scheme presented in Figure 8.12, the following mathematical relationships can relate the plant production capacity (Q) to loading time and retort size:

$$Qt_c = KV \tag{8.39}$$

Rearranging equation (8.39) and replacing t_c from equation (8.38) it is possible to obtain an expression for production capacity (Q) as a function of processing time (t_p) and retort number (N_A) as follows:

$$Q = \frac{KV(N_A - 2)}{t_p}. \tag{8.40}$$

From equation (8.40) it is possible to infer that production capacity is directly influenced by process temperature because the higher the process temperature, the shorter the process time and the higher the production capacity (more batches per day).

8.8. Simultaneous Sterilization of Different Product Lots in the Same Retort

Simultaneous sterilization applies, mainly, to small canneries with few retorts that are frequently required to process small lots of different products in various container sizes that normally require different process times and retort temperatures. In these situations, retorts often operate with only partial loads because of the small lot sizes, and are under-utilized. The proposed approach to this optimization problem is to take advantage of the fact that, for any given product and container size, there exists any number of alternative combinations of retort temperature (above the lethal range) and corresponding process time that will deliver the same lethality (F_0 value). These can be called isolethal processes. They were first described by Teixeira et al. (1969) to find optimum isolethal process conditions that would maximize nutrient retention (thiamine) for a given canned food product, and later confirmed by others (Lund 1977; Ohlsson 1980). Barreiro et al. (1984) used a similar approach to find optimum isolethal process conditions that would minimize energy consumption.

Important to this analysis, is the fact that the differences found in the absolute level of quality retention are relatively small over a practical range of isolethal process conditions. This relative insensitivity of quality over a range of different isolethal process conditions opens the door to maximizing output from a fixed number of retorts for different products and container sizes. Isolethal processes can be identified for each of the various products; from which a common set of processes and conditions can be chosen for simultaneous sterilization of different product lots in the same retort.

8.8.1. Simultaneous Sterilization Characterization

In terms of analysis, a range of isolethal processes for selected products and container sizes should be obtained from experimental work. Heat penetration tests should be conducted on each product in order to establish process time at a reference retort temperature to achieve target lethality (F_0 values). A computer program can be utilized to obtain the equivalent lethality processes according to the following specifications:

- Two F_0 values should be considered for each product ($F_{0\,min}$ and $F_{0\,max}$). The referred values are product-related, but in general, $F_{0\,min}$ is chosen according to a safety criterion and $F_{0\,max}$ according to a quality criterion.
- For each F_0 value ($F_{0\,min\,j}$ and $F_{0\,max\,j}$) isolethal processes at retort temperatures of TRT_1, TRT_2, TRT_3, ..., TRT_N should be obtained for each product.
- The discrete values that define each process per product at different temperatures will be transformed as a continuous function through the cubic spline procedure (for both $F_{0\,min}$ and $F_{0\,max}$, per product). Obtaining a set of two continuous curves per product (Figure 8.13).

In addition, the following criteria should be established for choosing the optimum set of process conditions for simultaneous sterilization of more than one product:

- The total lethality achieved for each product must be equal or greater than the pre-established $F_{0\,min}$ value for that specific product,
- The total lethality for each product must not exceed a pre-established maximum value ($F_{0\,max}$) to avoid excessive over processing.

8.8.2. Mathematical Formulation for Simultaneous Sterilization

Let us assume we have $n \geq 2$ products, say $P_1, \ldots, P_n$, which are processed at the plant location. Let us consider the index set

$$X = \{1, 2, \ldots, n\}, \tag{8.41}$$

Considering the temperature interval $[T_{min}, T_{max}]$, which denotes the temperature capabilities of the process.

Each product P_j, for $j \in X$, has attached two strictly decreasing continuous functions, say

$$m_j, M_j : [T_{min}, T_{max}] \to (0, +\infty) \tag{8.42}$$

Region restricted for the maximum and minimum iso-lethal curves

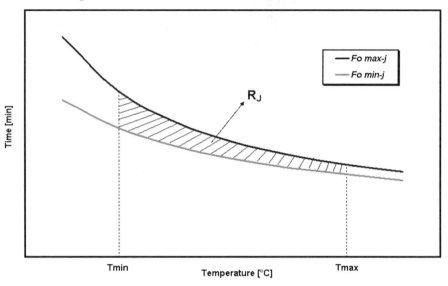

FIGURE 8.13. Region restricted for the maximum and minimum isolethal curves ($F_{0\,min-j}$ and $F_{0\,max-j}$) for the j-nth product.

Where $m_j(T) \leq M_j(T)$, for each $T \in [T_{min}, T_{max}]$. The meaning of $m_j(T)$ (respectively, $M_j(T)$) is the minimum time (respectively, maximum time) needed to process the product P_j at temperature T.

Defining the region R_j:

$$R_j = \{(T, t) : T_{min} \leq T \leq T_{max}, m_j(T) \leq t \leq M_j(T)\} \tag{8.43}$$

The interpretation of R_j is that the product P_j can be processed at temperature T with time t if and only *if* $(T, t) \in R_j$ (Figure 8.13).

It is clear that a sub-collection of products will be:

$$P_{j_1}, \ldots, P_{j_r} \tag{8.44}$$

Where

$$1 \leq j_1 \leq j_2 \leq \cdots \leq j_r \leq n$$

Can be simultaneously processed at temperature T and time t if and only if

$$(T, t) \in R_{j_1} \cap R_{j_2} \cap \cdots \cap R_{j_r}$$

It follows that in order to obtain all possible sub-collection of products that can be simultaneously processed is equivalent to finding all possible subsets

$$Q = \{j_1, \ldots, j_r\} \subset X$$

$$r > 0,$$

$$1 \leq j_1 \leq j_2 \leq \cdots \leq j_r \leq n,$$

For which it holds that

$$I_Q = R_{j_1} \cap R_{j_2} \cap \cdots \cap R_{j_r} \neq \emptyset$$

8.8.3. Computational Procedure

In the practical sense, we have the products $P_1, \ldots, P_n$ and the temperature interval $[T_{min}, T_{max}]$.

(1) We choose a positive integer $k \in \{1, 2, 3, \ldots\}$ and a partition $P = \{T_0 = T_{min}, T_1, \ldots, T_k = T_{max}\}$
Where $T_m < T_{m+1}$ for $m = 0, \ldots, k-1$.
(2) For each product P_j we compute the values $m_j(T_m), M_j(T_m), m = 0, 1, \ldots, k$
(3) For each $m \in \{0, 1, \ldots, k\}$, we define the values $m_p(T_m) = $ Maximum $\{m_p(T_m): j \in P\}; M_p(T_m) = $ Minimum $\{M_p(T_m): j \in P\}$.
(4) If for some $m \in \{0, 1, \ldots, k\}$ we have that $m_p(T_m) \leq M_p(T_m)$, then we observe that products $P_1, \ldots, P_n$ can be simultaneously processed at temperature T_m with time $t \in [m_p(T_m), M_p(T_m)]$.

8.8.4. Expected Advantages of the Implementation of Simultaneous Sterilization

The opportunity to carry out simultaneous sterilization and the possibility of employing alternative processes (same F value) provides flexibility to optimize retort utilization. Within a pre-established range of F values, it was possible to obtain all the combinations for simultaneous sterilization. This procedure is of special relevance for small companies that normally work with many different products at the same time. Practical implementation of this flexibility will require close attention to batch record-keeping requirements of the FDA Low Acid Canned Food regulations, and the need to file with the FDA each of the alternative processes as an acceptable scheduled process for each product.

According to Simpson (2003) an important finding indicates that the lower the process temperature the higher the possibility to attain simultaneous sterilization for more products at the same time. Independent of the selected process temperature, several combinations of simultaneous sterilization can be attained.

Simultaneous sterilization provides flexibility to optimize retort utilization. This procedure is of special relevance for companies that normally work with many different products at the same time.

References

Adams H. W., & Hardt-English, P. (1990). Determining temperature distribution in cascading water retorts. *Food Technol., 44* (12), 110–112.

Akterian, S. G., Smout, M. C., Hendrickx, M. E., & Tobback, P. P. (1998). Application of sensitivity functions for analysing the impact of temperature non-uniformity in batch sterilizers. *J. Food Eng., 37*(1), 1–10.

Almonacid-Merino, S. F., Simpson, R., & Torres, J. A. (1993). Time-variable retort temperature profiles for cylindrical cans: batch process time, energy consumption, and quality retention model. *J. Food Proc. Eng., 16*(4), 171–187.

Anantheswaran, R. C., & Rao, M. A. (1985a). Heat transfer to model Newtonian liquid foods in cans during end-over-end rotation. *J. Food Eng., 4*, 1–19.

Anantheswaran, R. C., & Rao, M. A. (1985b). Heat transfer to model non-Newtonian liquid foods in cans during end-over-end rotation. *J. Food Eng., 4*, 21–35.

Ávila, I. M. L. B., Martins, R. C., Ho. P., Hendrickx, M., & Silva, C. L. M. (2006). Variability in quality of white and green beans during in-pack sterilization. *J. Food Eng., 73*(2), 149–156.

Ávila, I. M. L. B., Smout, C., Silva. C. L. M., & Hendrickx, M. (1999). Development of a novel methodology to validate optimum sterilization conditions for maximizing texture quality of white beans in glass jars. *Biotechnology Progress, 15*, 565–572.

Ball, C. O., & Olson, F. C. W. (1957). *Sterilization in food technology – Theory, practice and calculations.* New York: McGraw-Hill.

Balzhizer, R. E., Samuels, M. R., & Eliassen, J. D. (1972). *Chemical Engineering Thermodynamics.* Prentice Hall, New Jersey.

Barbosa, A. P., & Macchietto, S. (1993). Optimal design of multipurpose batch plants 1. Problem formulation. Computers and Chemical engineering, *17*, S33–S38.

Barreiro, J., Pérez, C., & Guariguata, C. (1984). Optimization of Energy Consumption During the heat Processing Of Canned Foods. *J. of Food Eng., 3*(1): 27–37.

Beauvais, M., Cheftel, M., & Thomas, G. (1961). A new method for heat processing canned foods. *Food Technol., 15*(4), 5–8.

Beck, R. E. (1985). Autoclaving of solutions in sealed containers: theoretical pressure–temperature relationships. *Pharm. Manf., No. 6*, 18–23.

Berry, M. R., & Bradshaw, J. G. (1980). Heating characteristics of condensed cream of celery soup in a Steritort. *J. Food Sci., 45*, 869–74, 879.

Berry, M. R., & Bradshaw, J. G. (1982). Heat penetration for sliced mushrooms in brine processed in still retort and agitating retorts with comparisons to spore count reduction. *J. Food Sci., 47*, 1698–1704.

Berry, M. R., & Dickerson, R. W. (1981). Heating characteristics of whole kernel corn processed in a Steritort. *J. Food Sci., 46*, 889–895.

Berry, M. R., & Kohnhorst, A. L. (1985). Heating characteristics of homogeneous milk-based formulas in cans processed in an agitating retort. *J. Food Sci., 50*, 209–214, 253.

Berry, M. R., Savage, R. A., & Pflug, I. J. (1979). Heating characteristics of cream-style corn processed in a Steritort: effects of head space, reel speed and consistency. *J. Food Sci., 44*, 831–836.

Best, R. J., Bewaji, E. O., & Johns, W. R. (1994a). Dynamic simulation for waste reduction in hydrostatic canning retort operations. In *Proc. 4th. Bath conference food process engineering Vol. 1* (pp. 261–268). Rugby, UK: Institution of Chemical Engineers.

Best, R. J., Bewaji, E. O., Johns, W. R., & Kassim, H. O. (1994b). Steady-state and dynamic simulation of food processes including spatial effects. *Food and Bioproducts Processing, Trans. IChemEng., 72*C2, 79–85.

Bhowmik, S. R., & Hayakawa, K. (1983). Influence of selected thermal processing conditions on steam consumption and on mass average sterilizing values. *J. Food Sci., 48*, 212–216, 225.

Bhowmik, S. R., & Hayakawa, K. (1988). Quality retention and steam consumption of selected thermal process. *Lebensm.-Wiss. u.-Technol., 21*, 13–19.

Bhowmik, S. R., & Tandon, S. (1987). A method for thermal process evaluation of conduction heated foods in retortable pouches. *J. Food Sci., 52*, 201–209.

Bhowmik, S. R., Vichnevetsky, R., & Hayakawa, K. (1985). Mathematical model to estimate steam consumption in vertical still retort for thermal processing of canned foods. *Lebensm.-Wiss. u.-Technol., 18*, 15–23.

Bitting, A. W. (1916). Processing and processing devices. *Bulletin No. 9*. Washington, DC: National Food Processors' Association.

Blaisdell, J. L. (1963). Natural convection heating of liquids in unagitated food containers. PhD thesis, Michigan State University.

Bratt, L. (1995). Heat Treatment. In R. J. Footitt & A. S. Lewis (Eds.), *The canning of meat and fish*. Glasgow: Blackie Academic and Professional.

Britt. I. (2004). Trends with innovative packaging. In G. S. Tucker (Ed.), *Third international symposium Thermal processing – Process and package innovation for convenience foods* (Session 1:1). Chipping Campden, Glos., UK: Campden & Chorleywood Food Research Association.

Brown, A. C. (1972). Engineering aspects of pasteurization and sterilization of foods by heat. In *The mechanical engineer's contribution to process engineering in the food industry* (pp. 1–5). London: Institution of Mechanical Engineers.

Buckenhskes, H., Gierschner, K., & Hammes, W. P. (1988). *Theory & practice of pasteurization* (in German). Braunschweig, Germany: Die Industrielle Obst- & Gemusevertung.

Campbell, S., & Ramaswamy, H. S. (1992a). Heating rate, lethality and cold-spot location in air entrapped retort pouches during over-processing. *J. Food Sci., 57*(2), 485–490.

Campbell, S., & Ramaswamy, H. S. (1992b). Distribution of heat transfer rate and lethality in a single basket water cascade retort. *J. Food Process Eng., 15*(1), 31–48.

Casimir, D. J. (1961). New methods of sterilizing by heat can rotation during thermal processing. In *Proc. 4th. int. congress – Canned foods* (pp. 354–359). Paris: CIPC.

Casimir, D. J. (1971). *Economics of Flame Sterilization*. Food Research Report No. 54. Australia: CSIRO, Division of Food Research.

Casimir, D. J. (1975). Flame sterilization. *CSIRO Food Research Quarterly, 35*(2), 34–38.

CCFRA. (1985). *Hygienic design of post process can handling equipment. Technical Manual No. 8*. Chipping Campden, Glos. UK: Campden & Chorleywood Food Research Association.

CCFRA. (1997). *Microbiological control in food industry process waters: chlorine dioxide and bromine as alternatives to chlorine. Technical Report*. Chipping Campden, Glos. UK: Campden & Chorleywood Food Research Association.

Clifcorn, L. E., Peterson, G. T., Boyd, J. M., & O'Neil, J. M. (1950). A new principle for agitating in processing canned foods. *Food Technol., 4*, 450–460.

Conley, W., Lawrence, K., & Schuhmann, L. (1951). The application of end-over-end agitation to the heating and cooling of canned food products. *Food Technol., 5*, 457–460.

Cummings, D. B., & Wright, H. T. (1984). A thermocouple system for detecting centre temperature in cans processed by direct flame sterilization. *Can. Inst. Food Sci. Technol. J., 17*(3), 152–156.

Datta, A. K., Teixeira, A. A., & Manson, J. E. (1986). Computer based retort control logic for on-line correction of process deviations. *J. of Food Sci., 51*(2), 480–483, 507.

Davis, E. G., Karel, M., & Proctor, B. E. (1960). The pressure volume relationship in film packages during heat processing. *Food Technol., 14*, 165–169.

Deniston, M. F., Kimball, R. N., Pedersen, L. D., Parkinson, K. S., & Jones, M. C. (1991). Effects of steam/air mixtures on a convection-heating product processed in a Steritort. *J. Food Sci., 56*(1), 27–30.

Deniston, M. F., Kimball, R. N., Storofos, N. G., & Parkinson, K. S. (1992). Effect of air/steam mixtures on thermal processing of an induced convection heating product tomato concentrate in a Steritort. *J. Food Process Eng., 15*(1), 49–64,

Denys, S., Noronha, J., Stoforos, N. G., Hendrickx, M., & Tobback, P. (1996). A semi-empirical approach to handle broken-line heating: determination of empirical parameters and evaluation of process deviations. *J. Food Process. Preserv., 20*(4), 331–146.

Denys, S., Noronha, J., Stoforos, N. G., Hendrickx, M., & Tobback, P. (1996). Evaluation of process deviations, consisting of drops in rotational speed, during thermal processing of foods in rotary water cascading retorts. *J. Food Eng., 30*(3/4), 327–338.

Department of Health (1994). *Guidelines for the Safe Production of Heat Preserved Foods*. London: HMSO.

Dobie, P. A. P., Tucker, G. S., & Williams, A. (1994). The use of "CTemp" model to assess the effects of process deviations in reel and spiral cooker coolers. *Technical Memorandum No. 697*. Chipping Campden, Glos., UK: Campden & Chorleywood Food Research Association.

Douglas, J. M. (1988). *Conceptual Design of Chemical Processes*. McGraw-Hill International Editios. Chemical Engineering Series.

Duquenoy, A. (1980). Heat transfer to canned liquids. In P. Linko (Ed.), *Food Process Engineering Vol. 1* (pp. 483–489). London: Applied Science Publishers.

Ekelund, E. A. (1961). New methods of sterilizing by heat – hot air sterilization and heat transfer. In *Proc. 4th int. congress. Canned foods* (pp. 10–14). Paris: CIPC.

Evans, H. L., & Board, P. W. (1954). Studies in canning process. I. Effect of headspace on heat penetration in products heating by conduction. *Food Technol., 8*, 258–262.

Fernandez, C. L., Rao, M. A., Rajavasireddi, S. P., & Sastry, S. K. (1988). Particulate heat transfer to canned snap beans in a Steritort. *J. Food Process Eng., 10*, 183–198.

Ferrua, J. P., & Col, M. H. (1975). Energy consumption rates for sterilization equipment. *Canner/Packer*, Jan., 44–46.

Footitt, R. J., & Lewis, A. S. (1995). *The canning of meat and fish*, Glasgow: Blackie Academic and Professional.

Genta, M. L., Soria, S. J., Jandula, A. M., & Alvarez, N. V. (2001). Canning: continuous processing. In J. Welti-Chanes, G. V. Barbosa-Cánovas, & J. M. Aguilera, (Eds.), *International conference – Engineering and food ICEF8 Vol. 1* (pp. 692–696). Lancaster, PA: Technomic pub. Co., Inc.

Gillespy, T. G. (1958). Eklund hot air cooker experimental trials I, *Technical Memorandum No. 28*, Chipping Campden, Glos., UK: Campden & Chorleywood Food Research Association.

Gillespy, T. G. (1962). Eklund hot air cooker experimental trials II, *Technical Memorandum No. 48*. Chipping Campden, Glos., UK: Campden & Chorleywood Food Research Association.

Guariguata, C., Perez, C. R., & Barreiro, J. A. (1984). Optimization of energy consumption during processing of canned foods. *J. Food Process Eng., 3*(1), 27–39.

Guerrei, G. (1993). Pasteurization of beverages: unsteady state heat transfer model and process analysis. *Food & Bioproducts Processing, Trans. IChemE, 71*,C2, 67–76.

Heid, J. L., & Scott, W. C. (1937). Processing and cooling of canned foods – some heat transfer problems. *Fruit Products J., 17*, 100–104.

Hendrickx, M. (1997). A new approach to heat distribution and heat penetration studies for thermal processing of foods in modern batch-type overpressure rotary retorts. In M. Hendrickx (Coordinator), *Thermal processing of foods – Recent developments in process validation*. Belgium: K. U. Leuven.

Holdsworth, S. D. (1976). Heat transfer to containers of food during sterilization: a literature review. *Technical Bulletin No. 31*. Chipping Campden, Glos., UK: Campden & Chorleywood Food Research Association.

Holdsworth, S. D. (1989). *Data Base of Energy Usage in the UK Canning Industry.* Chipping Campden, Glos., UK: Campden & Chorleywood Food Research Association.

Holdsworth, S. D. (1992). Energy usage in the canning industry. In Y. H. Hui (Ed.), *Encyclopedia of food science and technology, Vol. 2* (pp. 714–721). New York: Wiley.

Holdsworth, S. D., Dennis, C., Mallalieu, B., & Powell, B. (1988). *Energy Monitoring and Targeting Manual for the Canning Industry. Technical Manual No. 23.* Chipping Campden, Glos., UK: Campden & Chorleywood Food Research Association.

Horn, C. S., Franke, M., Blakemore, F. B., & Stannek, W. (1997). Modelling and simulation of pasteurization and staling effects during tunnel pasteurization of bottled beer. *Food and Bioproducts Processing, Trans. IChemE, 75*(C1), 23–33.

Hotani, S., & Mihori, T. (1983). Some thermal engineering aspects of the rotation method of sterilization. In T. Motohiro & K. I. Hayakawa (Eds.), *Heat sterilization of food* (pp. 121–129). Tokyo: Koseicha-Koseikaku.

Hughes, J. P., Jones, T. E. R., & James, P. W. (2003). Numerical simulation and experimental visualization of the isothermal flow of liquid containing a headspace bubble inside a closed cylinder during off-axis rotation. *Food and Bioproducts Processing, Trans. IChemE, 81*(C2), 119–128.

IFPS. (1992). Temperature distribution protocol for processing in steam still retorts, excluding crateless retorts. Fairfax, VA: The Institute for Thermal Processing Specialists.

Javier, R. A., Naveh, D., Perlstein, E., & Kopelman, I. J. (1985). Convective heating rate parameters of model solutions in an agitating retort simulator. *Lebensm.-Wiss. u. -Technol., 18*, 311–315.

Jowitt, R., & Mynott, A. R. (1974), Fluidized bed processing of convection-heating canned foods. *Proc. 4th int. congress food science technology, 4*, 41–44.

Jowitt, R., & Thorne, S. N. (1971). Evaluates variables in fluidised-bed retorting. *Food Eng., 43*(11), 60–62.

Jowitt, R., & Thorne, S. N. (1972). Continuous heat processing of canned foods, *Food Trade Rev., 42*(6), 9–13.

Kisaalita, W. S., Lo, K. V., Staley, L. M., & Tung, M. A. (1985). Condensation heat and mass transfer from steam/air mixtures to a retort pouch laminate. *Can. Agric. Eng., 27*(2), 137–145.

Kondili, C., Pantelides, R. and Sargent, H. (1993). A general algorithm for short-term scheduling of batch operations- I. MILP formulation. *Computers and Chemical engineering, 17,* 211–227.

Krishnamurphy, H., Ramaswamy, H. S., Sanchez, G., Sablani, S., & Pandey, P. K. (2001). Effect of guar gum concentration, rotation speed, particle concentration, retort temperature, and diameter of rotation on heat transfer rates during end-over-end processing of canned particulate non-Newtonian fluids. In J. Welti-Chanes, G. V. Barbosa-Cánovas & J. M. Aguilera (Eds.), *Eighth international conference Engineering and food, ICEF8 Vol. 1* (pp. 665–670). Lancaster, PA: Technomic Pub. Co.

Lanoiseuille, J.-L., Canadau Y., & Debray, E. (1995). Predicting internal temperature of canned foods during thermal processing using a linear recursive model. *J. Food Sci., 60*(4), 833–835, 840.

Lebowitz, S. F., & Bhowmik, S. R. (1989). Determination of reportable pouch heat-transfer coefficients by optimization method. *J. Food Sci.*, *54*, 1407–1412.

Lebowitz, S. F., & Bhowmik, S. R. (1990). Effect of retortable pouch heat-transfer coefficients of different thermal processing stages and pouch material. *J. Food Sci.*, *55*(5), 1421–1424, 1434.

Lee, B., & Reklaitis, G. V. (1995a). Optimal scheduling of cyclic batch processes for heat integration-I. Basic formulation, *Computers and Chemical engineering, 19*(8), 883–905.

Lee, B., & Reklaitis, G. V. (1995b). Optimal scheduling of cyclic batch processes for heat integration-II. Extended problems, *Computers and Chemical engineering, 19*(8), 907–931.

Lenz, M. K., & Lund, D. B. (1978). The lethality-Fourier number method: heating rate variations and lethality confidence intervals for forced-convection heated foods in containers. *J. Food Process Eng.*, *2*, 227–271.

Leonard, S., Marsh, G. L., Merson, G. L., York, G. K., Buhlert, J. E., Heil, J. R., & Wolcott, T. (1975a). Quality evaluation of canned fruit cocktail experimentally processed by Steriflamme. *J. Food Sci.*, *40*, 257–258.

Leonard, S., Marsh, G. L., Merson, G. L., York, G. K., Buhlert, J. E., Heil, J. R., & Wolcott, T. (1975b). Chemical, physical and biological aspects of canned whole peeled tomatoes thermally processed by Steriflamme. *J. Food Sci.*, *40*, 254–256.

Leonard, S., Marsh, G. L., York, G. K., Merson, R. L., Heil, J. R., Wolcott, T., & Ansar, A. (1976). Flame sterilization of tomato products and fruits in 630 × 700 cans. *J. Food Sci.*, *42*, 828–832.

Leonard, S. J., Marsh, G. L., York, G. K., Heil, J. R., & Wolcott, T. (1977). Evaluation of tomato canning practices using flame sterilization. *J. Food Process. Preserv.*, *1*(4), 313–323.

Leonard, S J., Heil, J. R., Carroad, P. A., Merson, R. L., & Wolcott, T. K. (1984a). High vacuum sterilization fruits – influence of can type on storage stability of vacuum packed peach and pear slices. *J. Food Sci.*, *49*, 263–266.

Leonard, S., Osaki, K., & Heil, J. (1984b). Monitoring flame sterilization processes. *Food Technol.*, *38*(1), 47–50, 83.

Lopez, A. (1987). *A complete course in canning, Vol. 1: Basic information on banning*, Baltimore, MD: The Canning Trade, Inc.

Lund, D. B. (1977). Design of Thermal Processes for Maximizing Nutrient Retention. *Food Technol., 31*, 71–78.

Massaguer, P. R., Cardelli, C. F., & Aguilera, H. G. (1994). Selection of heating media for thermal processing retort pouches. In T. Yano, R. Matsuno, & K. Nakamura (Eds.), *Developments in food engineering. ICEF6 Part 2* (pp. 710–712). London: Blackie Academic & Professional.

May, N. S. (2001). Retort technology. In P. Richardson (Ed.), *Thermal technologies in food processing* (pp. 7–28). Cambridge: Woodhead Publishing.

McGinnis, D. S. (1986). Surface heat transfer distribution in a weir type pressurized water retort for processing foods in flexible retort pouches. *Can. Inst. Food Sci. Technol. J.*, *19*(2), 45–52.

Meng Y., & Ramaswamy H. S. (2005). Heat transfer coefficients associated with canned particulate/non-Newtonian fluid (CMC) system during end over end rotation. *Food and Bioproducts Processing, Trans. IChemE, 83*(C3), 229–237.

Merrill, D. G. (1948). Heating rates of food in glass and other containers. *Ind. Eng. Chem.*, *40*(12), 2263–2269.

Naveh, D., & Kopelman, I. J. (1980). Effect of some processing parameters on the heat-transfer coefficients in a rotating autoclave. *J. Food Process. Preserv.*, *4*, 67–77.

NFPA (1982). Thermal processes for low-acid canned foods in metal containers. *Bulletin No. 26L*. Washington, DC: National Food Processors' Association.

Niinivaara, F. P., Laine, J. J., & Heikkunen, E. (1968). The importance of head space and of can rotation in the sterilization of canned products. *Die Fleischwirtschaft*, *48*(4), 431–436 [in German].

Noh, B. S., Heil, J. R., & Patino, H. (1986). Heat transfer study on flame pasteurization of liquids in aluminum cans. *J. Food Sci.*, *51*(3), 715–719.

Norback, J and Rattunde, M. (1991). Production planning when batching is part of the manufacturing sequence. *J. of Food Proc. Eng.*, *14*, 107–123.

Ohlsson, T. (1980). Temperature dependence of sensory quality changes during thermal processing. *J. of Food Sci.*, *45*(4), 836–839.

Okada, M. (1940a). Cooling of canned foods. *Bull. Jap. Soc. Sci. Fish*, *9*, 64–66.

Okada, M. (1940b). Cooling of rectangular canned foods. *Bull. Jap. Soc. Sci. Fish*, *9*, 208–210.

Parcell, J. W. (1930). Investigations of retorting of glass containers. *Canning Age*, *11*, 475–479.

Parchomchuk, P. (1977). A simplified method of processing canned foods. *J. Food Sci.*, *42*, 265–268.

Park, D. J., Cabes, L. J. Jr, & Collins, K. M. (1990). Determining temperature distribution in rotary full-immersion, hot-water sterilizer. *Food Technol.*, *44*(12), 113–118.

Patel, P. N., Chandarana, D. I., & Gavin, A. III (1991). Internal pressure profile in semi-rigid food packages during thermal processing in steam/air. *J. Food Sci.*, *56*(3), 831–834.

Paulus, K., & Ojo, A. (1974). Heat transfer during flame-sterilization. *Proc. 4th int. congress – Food Science & Technology*, *4*, 443–448.

Peralta Rodriguez, R. D., & Merson, R. L. (1982). Heat transfer and chemical kinetics during flame sterilization. In *Food process engineering*, AIChE Symp. Series, No. 218 (pp. 58–67).

Peralta Rodriguez, R. D., & Merson, R. L. (1983). Experimental verification of a heat transfer model for simulated liquid foods undergoing flame sterilization. *J. Food Sci.*, *48*, 726–733.

Peterson, W. R., & Adams, J. P. (1983). Water velocity and effect on heat penetration parameters during institutional size retort pouch processing. *J. Food Sci.*, *48*, 457–459, 464.

Pflug, I. J. (1964). *Evaluation of heating media for producing shelf stable food in flexible packages Part 1*. Natick, MA: US Army Laboratories.

Pflug, I. J., & Blaisdell, J. L. (1961). The effect of velocity of steam–air mixtures on the heating of glass containers. *Quart. Bulletin. Michigan Agr. Expt. Station*, *44*(2), 235–244.

Pflug, I. J., & Nicholas, R. C. (1961). Heating rates at pasteurization temperatures as affected by heating mediun and product in glass containers. *Quart. Bulletin, Michigan Agr. Expt. Station*, *44*(1), 153–165.

Piggott, G. M. (1963a). Fluidised bed retort proposed at IFT. *Canner/Packer*, *132*(7), 27–29.

Piggott, G. M. (1963b). Fluidised bed heat processing of canned foods. *Food Processing*, *24*(11), 70–72.

Plazl, I., Lakner, M., & Koloini, T. (2006). Modelling of temperature distributions in canned tomato based dip during industrial pasteurization. *J. Food Eng.*, *75*(3), 400–406.

Price, R. B., & Bhowmik, S. R. (1994). Heat transfer in canned foods undergoing agitation. *J. Food Eng.*, *23*(4), 621–629.

Pruthi, J. S., Satyanarayana Rao, N. S., Susheela, R., & Satyanarayana Rao, B. A. (1962). Studies on the spin pasteurization of some important tropical canned fruit products. *Proc. 1st int. congress Food Science & Technology, 4*, 253–270.

Quast, D. G., & Siozawa, Y. Y. (1974). Heat transfer rates during axially rotated cans. *Proc. 4th int. congress of food Science & Technology, 4*, 458–468.

Ramaswamy, H. S., & Tung, M. A. (1988). Heating rates of thin-profile packages in two pilot-scale steam/air retorts. *Can. Inst. Food Sci. Technol. J.*, *21*(3), 316–322.

Ramaswamy, H. S., & Tung, M. A. (1983). A method to measure surface heat transfer from steam to air mixtures in batch retorts. *J. Food Sci.*, *48*, 900–904.

Ramaswamy, H. S., Abbatemarco, C., & Sablani, S. S. (1993). Heat transfer rates in canned food model as influenced by processing in an end-over-end rotary steam/air retort. *J. Food Process. Preserv., 17*(4), 367–389.

Ramaswamy, H., Campbell, S., & Passey, C. (1991). Temperature distribution in a 1-basket water-cascade retort. *J. Canadian Inst. Food Sci. Technol.*, *24*(1/2), 19–26.

Ramaswarmy, H. S., & Sablani, S. S. (1997). Particle motion and heat transfer in cans during end-over-end rotation: influence of physical properties and rotational speed. *J. Food Process. Preserv., 21*(2), 105–127.

Rao, M. A., & Anantheswaran, C. (1988). Convective heat transfer in fluid foods in cans. *Advances Food Research, 32*, 40–84.

Rao, M. A., Cooley, H. J., & Anantheswaran, C. (1985). Convective heat transfer to canned liquid foods in a Steritort. *J. Food Sci.*, *50*, 150–54.

Rees, J. A. G., & Bettison, J. (Eds.) (1991). *Processing and packaging of heat preserved foods*. Glasgow: Blackie Academic & Professional.

Reklaitis, G. V. (1996). Overview of scheduling and planning of batch process operations. In *"Batch Processing System Engineering"*. (G. V. Reklaitis, A. K. Sunol, D. W. Rippin, O. Hortacsu, eds) pp. 660–705, Springer, Berlin.

Richardson, P. S. (1987). Review: flame sterilization. *Int. J. Food Sci. Technol.*, *22*, 3–14.

Richardson, P. S., & Selman, J. D. (1991). Heat processing equipment. In J. A. G. Rees & J. Bettison (Eds.), *Processing and packaging of heat preserved foods*. Glasgow: Blackie Academic & Professional.

Richardson, P. S., Heydon, C. J., May, N. S., & Tucker, G. S. (1997). Lethality distribution for in-pack foods thermally processed in batch systems. In R. Jowitt, (Ed.), *Engineering and food, ICEF7 §K* (pp 5–8). Sheffield, UK: Sheffield Academic Press.

Rippin, D. W. (1993). Batch process system engineering: A retrospective and prospective review. *Computer and Chemical Engineering, 17*, S1–S13.

Roberts, H. L., & Sognefast, P. (1947). Agitating processes for quality improvement in vacuum-packed vegetables. *Canner, 104*, 20–24.

Sablani, S. S., & Ramaswamy, H. S. (1993). Fluid/particle heat transfer coefficients in cans during end-over-end processing. *Lebens. –Wiss. u.–Technol.*, *26*(6), 498–501.

Sablani, S. S., & Ramaswamy, H. S. (1996). Particle heat transfer coefficients under various retort operation conditions with end-over-end rotation. *J. Food Process Eng., 19*(4), 403–424.

Sablani, S. S., & Ramaswamy, H. S. (1997). Heat transfer to particles in cans with end-over-end rotation: influence of particle size and concentration(%v/v). *J. Food Process Eng., 20*(4), 265–283.

Sablani, S. S., Ramaswamy, H. S., & Mujumdar, A. S. (1997). Dimensionless correlations for convective heat transfer to liquid and particles in cans subjected to end-over-end rotation. *J. Food Eng., 34*(4), 453–472.

Savage, D. (1984). Processing sterilized products. *Proc. symp. Profitability of food processing, No. 84*. Rugby, UK: Institution of Chemical Engineers.

Schmidt, E. D., & Pflug, I. J. (1965). A study of the variables that affect heat penetration rates in glass containers. *Quart. Bulletin, Michigan Agr. Expt. Station, 48*(3), 397–410.

Simpson, R., Aris, I., & Torres, J. A. (1989). Sterilization of conduction-heated foods in oval-shaped containers. *J. Food Sci., 54*(4), 1327–1331, 1363.

Simpson, R., Almonacid, S., & Torres. J. A. (1993). Mathematical models and logic for the computer control of batch retorts: Conduction-heated foods. *J. Food Eng., 20*(3), 283–295.

Simpson, R. S. (2005). Generation of isolethal processes and implementation of simultaneous sterilisation utilising the revisited general method. *J. Food Eng., 67*(1/2), 71–79.

Simpson, R., Almonacid, S., & Teixiera, A. (2003). Optimization criteria for batch retort battery design and operation in food canning plants. *J. Food Process Eng., 25*(6), 515–538.

Simpson, R., Cortes, C., & Teixeira, A. (2006). Energy consumption in batch processing; model development and validation. *J. Food Eng., 73*(3), 217–224.

Smout, C., & May, N. (Eds.) (1997). *Guidelines for establishing heat distribution in batch over-pressure retort systems. No. 17*. Chipping Campden, Glos., UK: Campden & Chorleywood Food Research Association.

Smout, C., Van Loey, A., Hendrickx, M., & Tobback, P. (1997). Heat distribution studies in overpressure (rotary) retorts. In R. Jowitt (Ed.), *Engineering and food ICEF7* (pp. K25–K28). Sheffield, UK: Sheffield Academic Press.

Smout, C., Van Loey, A., & Hendrickx, M. (1998). Heat distribution in industrial scale water cascading (Rotary) retort. *J. Food Sci., 63*(5), 882–886.

Smout, C., Ávila, I., Van Loey, A. M. L., Hendrickx, M. E. G., & Silva, C. (2000a). Influence of rotational speed on the statistical variability of heat penetration parameters and on the non-uniformity of lethality in retort processing. *J. Food Eng., 45*(2), 93–102.

Smout, C., Van Loey, A. M. L., Hendrickx, M. E. G. (2000b). Non-uniformity of lethality in retort processes based on heat distribution and heat penetration data. *J. Food Eng., 45*(2), 103–110.

Smout, C., Van Loey, A. M. L., & Hendrickx, M. E. G. (2001). Role of temperature distribution studies in the evaluation and identification of processing conditions for static and rotary water cascading retorts. *J. Food Eng., 48*(1), 61–68.

Smout, C., Ávila, I., Van Loey, A. M. L., Hendrickx, M. E. G., & Silva, C. (2000a). Influence of rotational speed on the statistical variability of heat penetration parameters and on the non-uniformity of lethality in retort processing. *J. Food Eng., 45*(2), 93–102.

Soule, C. L., & Merson, R. L. (1985). Heat-transfer coefficients to Newtonian liquids in axially rotated cans. *J. Food Process Eng., 8*, 33–46.

Sumer, A., & Esin, A. (1992). Effect of can arrangement on temperature distribution in still retort sterilization. *J. Food Eng., 15*, 245–259.

Teixeira, A., Dixon, J., Zahradnik, J., & Zinsmeiter, G. (1969). Computer Optimization of Nutrient Retention in the Thermal Processing of Conduction-Heated Foods. *Food Technol., 23*(6), 845–850.

Teixeira Neto, R. O. (1982). Heat transfer rates to liquid foods during flame-sterilization. *J. Food Sci., 47*, 476–481.

Thorne. S. (1986). *History of food preservation*. London: Parthenon.

Thorne, S. N., & Jowitt, R. (1972). Fluidised bed heating of canned foods: the influence of the outside heat transfer coefficient on the sterilizing value of the process. In *Proc. 1st. int. conf. Heat and mass transfer in food engineering*, Paper B3. Netherlands: University of Wageningen.

Thorpe, R. H., Atherton, D. A., & Steele, D. A. (1975). *Canning retorts and their operation. Technical Manual No. 2*. Chipping Campden, Glos., UK: Campden & Chorleywood Food Research Association.

Thorpe, R. H., & Gillespy T. G. (1968). *Post-process sanitation in canneries. Technical Manual No. 1*. Chipping Campden, Glos. UK: Campden & Chorleywood Food Research Association.

Tucker, G. S. (2004). Improving rotary thermal processing. In P. Richardson (Ed.), *Improving the thermal processing of foods* (pp. 124–137). Cambridge: Woodhead Publishing.

Tung, M. A., & Ramaswamy, H. S. (1986). Steam/air media for retort pouch processing. In M. Le Mageur and P. Jelen (Eds.), *Food engineering and process applications Vol. 1* (pp. 521–546). London: Elsevier Applied Science.

Tung, M. A., Ramaswamy, H. S., Smith, T., & Stark, R. (1984). Surface heat transfer coefficients for steam/air mixtures in two pilot scale retorts. *J. Food Sci., 49*, 939–943.

Tung, M. A., Morello, G. F., & Ramaswamy, H. S. (1989). Food properties, heat transfer conditions and sterilization considerations in retort processes. In R. P. Singh & A. G. Medina (Eds.), *Food properties and computer-aided engineering of food processing systems* (pp. 49–71). London: Kluwer Academic Publishers.

Tung, M. A., Britt, I. J., & Ramaswamy, H. S. (1990). Food sterilization in steam/air retorts. *Food Technol., 44*(12), 105–109.

Van Blaricom, L. O. (1955). Design and operating characteristics of a new type, continuous, agitating cooker and cooler as applied to the processing of freestone peaches. *Bulletin 429*. Clemson, SC: South Carolina Agricultural Expt. Station.

Van Loey, A., Fransis, A., Hendrickx, M., Maesmans, G., Noronha, J. De, & Tobback, P. (1994). Optimizing thermal process for canned white beans in water cascading retorts. *J. Food Sci., 59*(4), 828–832.

Vandenberghe, M. (2004). Thermal processing solutions for innovative packages. In G. S. Tucker (Ed.), *Third international symposium on thermal processing – Process and package innovation for convenience foods* (Session 4:1). Chipping Campden, Glos., UK: Campden & Chorleywood Food Research Association.

Varga, S., & Oliveira, J. C. (2000). Determination of the heat transfer coefficient between bulk medium and packed containers in a batch retort. *J. Food Eng., 44*(4), 191–198.

Varga, S., Oliveira, J. C., & Oliveira, F. A. R. (2000a). Influence of the variability of processing factors on the F-value distribution in batch retorts. *J. Food Eng., 44*(3), 155–161.

Varga, S., Oliveira, J. C., Smout, C., & Hendrickx, M. E. (2000b). Modelling temperature variability in batch retorts and its impact on lethality distribution. *J. Food Eng., 44*(3), 163–174.

Wang, J.-K., & Ross, E. (1965). Spin processing for tropical fruit juices. *Agric. Eng., 46*(3), 154–156.

Wasmund, R. (1976). The effect of air in steam heating on heat transfer. *Zuckerind., 26*(1), 13–18 [in German].

Yamano, Y. (1983). Suitability of steam-and-air retort for sterilization of reportable pouches and analysis of heat penetration parameters. In T. Motohiro and K.-I. Hayakawa (Eds.), *Heat sterilization of food*. Tokyo: Koseicha-Koseikaku.

9
Retort Control

9.1. Process Instrumentation

9.1.1. Introduction

The main objective of a control system is to ensure that an adequate thermal process is delivered to the product. For this purpose a sensor is required for each variable, an indicator and recording system, control valves and a system for operating the valves correctly. The systems may be entirely manually operated or automatically controlled; alternatively, as in many cases with batch retorts, just one variable is automatically controlled, e.g. the temperature.

Process instrumentation is concerned with the equipment with which a canning process is controlled, either manually or automatically. It includes the measurement, recording and manipulation of process variables. The main variables with which the process operator is concerned are:

(a) The temperature of the heating and cooling process;
(b) The time to achieve the desired process;
(c) The pressure to maintain the required process temperatures during heating and cooling, as well as maintaining the package integrity;
(d) The water level, especially during the cooling operation;
(e) The rotation speed of each container in any agitating process.

Manual operation of batch retorts can be difficult, and it requires a skilled process operator to maintain an adequate flow of cans from a bank of batch retorts. In particular, the start of the cooling operation needs particular attention, since the pressure will be changing rapidly during the initial stages. After the steam is turned off and the cooling water turned on it is necessary, in some cases, to introduce air and control the pressure of the system to match the differential pressure on the can, to prevent irreversible distortion. This is a particular case where pneumatically or electrically operated control valves are useful.

The accuracy of measurements must be considered and the effects of a change of temperature appreciated. For example, if the thermometer is incorrectly calibrated and gives a reading of 121°C when the actual temperature is only 120°C, then the process will be of 20% lower lethal value. Similarly, failure to measure the process temperature correctly can lead to errors. In sterilization operations any shortfall in the time allowed can lead to a sub-lethal process, since most of the lethality accrues during the latter part of the process.

The advantages of automatic control for the heating and cooling process are obvious. The need to be aware of fluctuations in the temperature of the process and to maintain a constant temperature are paramount to canning operations.

The UK recommendations (Department of Health 1994) require the following minimum level of instrumentation on each retort or processing system:

(a) Master temperature indicator (MTI) – platinum resistance thermometer or mercury-in-glass of suitable specification;
(b) Temperature controller – accuracy -0.5 to $+1°C$;
(c) Temperature/time recording device – usually associated with (b), and should agree with (a);
(d) Pressure gauge – accuracy $±1\%$ of full-scale deflection;
(e) Indicating timing device – calibrated to 1 min intervals and to facilitate the correct operation of the above;
(f) Bleeds in the retort lid and the instrument pocket to ensure that air is eliminated from the system and does not interact with the steam to give a potentially lower temperature.

National codes of practice for other countries are available, e.g. for the United States, the Food and Drug Administration Regulations (FDA 1983). While these give details for many different types of retort, the principles are essentially the same for each country.

While there have been many changes in the last decade, with the development of newer types of electronic devices, it is necessary to remember that these must be calibrated to identifiable and traceable standards. The systems should also have a "fail-safe" policy associated with their operation.

9.1.2. Temperature Measurement

We now consider the main methods of measuring temperature in canning retorts.

9.1.2.1. Mercury-in-glass thermometers

This type of thermometer has been used on canning retorts since the beginning of retorting. The main requirements are that the stem should have a scale not less than 150 mm, which should be graduated at intervals of not more than 1°C (2°F) over a range of about 50°C (100°F), and readable to 0.5°C (1°F). The longer the mercury tailpiece, the more sensitive and responsive the instrument, but the more difficult to install. The minimum length is about 90 mm. In selecting this type of thermometer it is essential to make sure that the instrument has a rapid response time. Some instruments, by virtue of their design and bulb shielding, have very slow response times, and for this reason are unsatisfactory. The mercury column and the calibrations should be easily seen by the operator. Calibration of these instruments is not easy. For some time it was customary to use an oil bath and to calibrate against a calibrated NPL (National Physical Laboratory, UK) thermometer. However, it was ultimately established that the base and brass fittings of these thermometers were subject to different heat-transfer conditions when mounted on a hot radiating retort. The bulk of the thermometer when immersed in an oil bath was sufficiently great for heat to be conducted away from the oil bath at a greater rate than it could be supplied. It was therefore recommended that these thermometers should be calibrated in a steam pipe with fittings similar to those in a retort. The standard NPL thermometer was also mounted in a similar case so that the mountings of both the test thermometer and the standard were similar. This is now the accepted method of calibration in the UK, and has been proved to be most satisfactory in operation. It is important that all thermometers of this type are calibrated frequently and all the records preserved. The instrument pocket on the side of the retort should be continuously bled with a petcock during the heating operation, so that the steam is free from air.

Another system using constant temperature blocks has been developed by Cossey (1991). This can also be used on-site in factories, since it is portable.

9.1.2.2. Thermocouples

Thermocouples are widely used in the food industry for determining the temperatures in cans and for carrying out venting trials in retorts. The most common type in use is the type T, which has a temperature range from 200 to 400°F and consists of copper and copper/nickel wires. Thermocouples are unsuitable for use as MTIs, the main problems resulting from the nonlinear temperature–voltage relationship and the problems of cold junction compensation. The practical problems of conduction errors in thermocouples have been discussed by Withers *et al.* (1993).

9.1.2.3. Platinum Resistance Thermometers (PRTs)

These devices consist of a sensing resistor within a protective sheath, fitted with internal conducting wires and external terminals. These are connected to a digital display unit. The sensor is a precision resistor of known temperature coefficient and consists either of a coil of wire wound on a former or a film device where platinum is deposited on a substrate. The sensing element is mounted in a stainless steel sheath, which is insulated from the sensor. The accuracy' of these sensors must conform to certain standards, e.g. BS 1904: 1964 and 1984, and DIN 43 760: 1980. In general the resistor should have a resistance of $100\,\Omega$ at $0°C$ and $138.5\,\Omega$ at $100°C$. The main advantages of PRTs over mercury-in-glass thermometers are:

(a) Digital displays are less prone to reading errors;
(b) The ability to measure temperatures in remote places and display the readings locally;
(c) The ability to replay to a secondary display, recorder, controller or control computer;
(d) System components are interchangeable within overall measurement tolerances;

(e) The elimination of mercury and glass from the food-processing environment;
(f) Improved measurement resolution and overall measurement tolerances.

Extensive research (Bown & Slight 1980, 1982; Bown 1984) has shown that the PRTs are entirely suitable for canning retort operation. It has taken a considerable period of time, however, to convince the industry. A standard for PRTs for use in the food industry was developed by Richardson and Bown (1987). A system of calibration, using constant temperature blocks, has the advantage of being portable and can be used in factories, where the sensors may be in remote situations (Cossey 1991). Every retort is required to have a device fitted which continuously records the temperature and time during the retort operation, so that a permanent record is available for future reference, should the need arise. Various devices for doing this are available, most of them having a feedback control system to control an automatically operated control valve. Electronic recording systems, including computer control systems, are replacing older types of instruments. Whatever type of system is operated, the MTI is the absolute standard for processing operations. It is recommended, in comparing the MTI with the recorder controller, that the temperature difference should not exceed more than $\pm 0.5°C$ for a total of more than 1 min.

9.1.3. Pressure Measurement

Legislation for pressure vessels demands that every pressure vessel is fitted with a pressure-indicating device. For canning purposes the pressure gauge is used to ensure that the pressure is correct in relation to the steam temperature, so that air in the steam may be detected. It should never be used to control the temperature of the steam heating operation, but only for assisting in applying the correct cooling overpressure protocol to prevent can "peaking."

Pressure gauges are usually of the Bourdon tube type with a range of 0–2 bar, with the scale divided into 0.1 bar divisions. The pressure gauge is usually mounted adjacent to the MTI.

Pressure transducers are used with computer control systems; these are either of the strain gauge or piezo-electric crystal type. It is essential in this type of operation that the transducers are sufficiently robust to withstand the cannery environment.

Closely associated with this topic are pressure-relief valves, which must be fitted to the retort, and a steam distribution system to ensure that the safe working pressure of the retort cannot be exceeded under any operating conditions. This applies to steam, water and compressed air supplies.

9.1.4. Water Level

For manual operation of batch retorts, a sight glass is used to detect the level of water in the retort during the cooling operation. For automatic or electronic operation conductivity or capacitance, probe sensors are used. The signals from these devices are used to determine the initiation of the cooling cycle by controlling the water in venting, drain and overflow valves. Level control is also necessary in many types of cooker/cooler, especially hydrostatic cookers, where the level of water in the hydrostatic legs has to be controlled, and in hot water sterilizing retorts, where separate tanks are used to heat the water and store it during circulation.

9.1.5. Rotation Monitors

For the satisfactory operation of rotary retorts it is necessary to have devices for sensing and controlling their rotation or movement when agitated-pack processing is being used. It is particularly important during such processes that the speed of rotation is rigidly maintained at a constant rate, otherwise there is a danger of understerilization. Withers (1991) devised a can rotation counter for use in reel and spiral sterilizers (see also Saguy & Kiploks 1989; Withers & Richardson 1992).

9.1.6. Lethality Measurement

The use of direct on-line measurement of lethality has only been realized with the development of suitable transducers containing programmed microchips. It had been realized for many years that if it were possible to measure the temperature inside a can during processing, then the process could be continuously controlled on this signal by determining the lethality. This could only be achieved if the temperature could be measured remotely, since the possibility of leading wires into batch retorts during operation was decidedly not a practical proposition. The use of radio-telemetry and other techniques was actively progressed for many years, with regrettably limited success. A method progressed by Holdsworth (1974), based on work on analog computation of sterilization values by Overing-ton (1975), made use of the retort temperature profile and subsequent conversion of this into a lethality value using a mathematical model (see Figure 9.1). This was developed further by Bown (1982, 1983) using a microprocessor system; however, implementation was slow, in terms of lethality control, because of the problems of computing the lethality in real time on the available facilities.

A transducer for the direct measurement of lethal rates has been devised by David and Shoemaker (1985); the transducer, a log/antilog amplifier, has the ability to convert the voltage signal generated by a thermocouple into an exponential relationship of the lethality type. This is a simple alternative to using elaborate PCs. Many commercial retorts and experimental retort simulators have built-in computerized systems for programmed operation of the retort sequences and the simultaneous determination of achieved lethality. Current trade literature shows the ever increasing use of microelectronics for the safe operation of retort systems.

Data acquisition units have already been mentioned for determining heat penetration into cans (see Chapter 5). They may also be used for determining environmental temperatures in retorts, especially in continuous cookers. For a range of current devices, see May and Withers (1993). The design of these devices

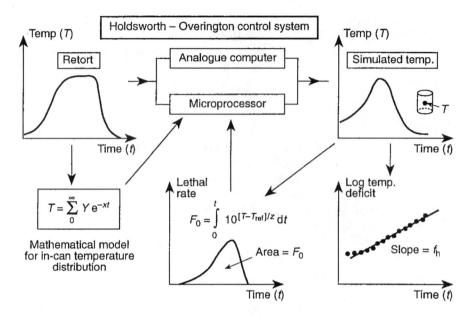

FIGURE 9.1. Retort control system using process mathematical model.

changes so rapidly with the development of new microelectronic devices that it is difficult to keep up with the new developments.

Teixeira and Tucker (1996) have discussed the development of on-line computer simulation control for sterilization equipment and have outlined the critical points to ensure the systems perform correctly. In particular they have stressed the need to look carefully at the factors that determine the performance of the model, e.g., the apparent thermal diffusivity, of the system.

9.2. Process Control

9.2.1. Introduction

A process control system may be entirely manual, with the process operator manipulating the control valves; semi-automatic, with, for example, the stream being controlled and the other operations manual; or fully automatic. While some batch retort operations are semi-automatically controlled, most processes are now continuously controlled, with operator supervision at a computer terminal.

There are three elements of process control: a sensor, giving either an analog or a digital output, often with a visual display of the measured variable; a control valve, often described as a final control element; and a controller to produce an output signal from a sensor input. In this section it is only possible to give a general review of the application of control systems in canning. The subject is developing very rapidly, and readers requiring more detailed knowledge should consult books on process control, e.g. Stephanopoulos (1984) and Smith and Corripio (1985).

9.2.2. Control Valves and Actuators

Control valves for automatic systems are modifications of the standard types of valve with either an electric motor or a pneumatic system to actuate the motion of the valve spindle. The valve motion may be up and down or rotary, depending on the type of response required. A large number of retorts are fitted with the simplest type of pneumatic valve with compressed air applied to a diaphragm. The method of imparting motion to a valve is known as actuation and the device as an actuator. The air signal applied to the actuator's diaphragm comes from a pneumatic controller. Electrical valve actuators often impart a rotary motion and are controlled by an electrical signal from an electric or electronic controller.

Valves may be either "on–off," e.g. gate valves, or continuously modulated types, e.g. globe valves, depending on the type of operation. For the control of the steam to a retort the valves are usually continuously modulated, so that the temperature in the retort may be finely controlled. If compressed air actuated valves are used, then it is necessary to convert the electrical input signal to a compressed air output signal. The electrical output from the controller is usually a 4–20 mA current loop and I/P (current I to air pressure P) converters are used. The output pressures are 0.2–1.00 bar, corresponding to 4–20 mA electrical current. A pressure greater than the minimum applied to the actuator diaphragm overcomes the initial force in the actuator return spring, and causes the valve to open. If the pressure fails then the valve closes and shuts down the operation. Baines (1985) reviewed the available types of valves and actuators available for the control of steam. Cossey (1988, 1990) made a practical examination of the performance of control valves used on canning retorts over a 15 month period. This showed that there was little change in the performance of the valves and system.

9.2.3. Interfaces

The interfaces between the control systems and the control valve fit into 4 basic categories:

(a) Digital inputs – on–off devices, e.g. relays, thermostats;
(b) Digital outputs – on–off devices, e.g. pilot solenoid valves;
(c) Analog inputs – modulating devices, e.g. temperature sensors;
(d) Analog outputs – I/P converters.

For the automatic control of retorts a selection of the above types of interface are required so that the incoming signals to the controller can be converted into the appropriate outgoing signals to actuate the valves. Analog to digital (A/D) converters are widely used to convert signals into an appropriate form.

9.2.4. Control Systems

The standard type of pneumatic controller, widely used on canning retorts, consists of a comparator device which compares the measured signal with the desired signal, and a method of regulating the output to conform with the requirements of the deviant or error signal. Several types of controller action are possible:

1. On–off control is the simplest type, but tends to produce an oscillating response.
2. Proportional (P) action has an output signal proportional to the deviation, and produces a control value which has an inherent offset and oscillating response.
3. Integral (I) action (also known as automatic reset) has an output signal proportional to the time integral of the error, and is generally used with proportional action to reduce the offset.
4. Derivative (D) action (also known as rate control) has an output proportional to the derivative of the error.

Types 3 and 4 are used in conjunction with type 2, as follows:

(a) P + I: This reduces the offset inherent in P-only control;
(b) P + D: This reduces oscillations as well as offset;
(c) P + I + D: This reduces oscillations, offset and returns the control value to a stable condition fastest.

Each of the control values can be adjusted to match the process load and conditions. The required setting values can be calculated from the theory of process control, e.g. Ziegler–Nichols procedure, or obtained by trial and error.

There are electric and electronic versions of the pneumatic controller that have the same capabilities.

9.2.5. Computer Control

9.2.5.1. Direct Digital Control

In this type of control, conventional controllers are replaced by signals generated by a digital computer. The required set points are entered into the computer directly by the operator or are created via a computer program. This type of system is becoming more widespread, but requires knowledge of the dynamics of the process to be operated.

Microprocessors have been widely used in the control of process plants; these do not have the computational power of large computers but are ideally suitable for process control of batch systems (Bown 1985a, b).

9.2.5.2. Supervisory Digital Control

This is used for more complex systems, and would include not only the retort operation but the whole of the factory operations. Again some knowledge of the dynamic behavior of the operation is required in the form of a mathematical or empirical model. The system can be used with existing types of controller or with a direct digital control system.

9.2.6. Process Dynamics

The process dynamics of a system may be studied by altering the level of one of the inputs, e.g. temperature of steam to a retort, and studying the change in temperature of the system with time, i.e. the response. This can then be modeled and an equation developed that will predict the behavior when process disturbance occurs. The nature of the change of input may be a Dirac impulse, step-function, linear ramp or a sinusoidally varying function. In general the step-function behavior is used.

One of the earliest studies was carried out by Biro (1978), who developed differential equations for the heat and material balances for a batch retort, in which jars were heated in water by steam injection. A Dirac function was used as the system disturbance in the steam flow, and the response determined using Laplace transformations (see Chapter 2). Mulvaney (1987) used a Hydrolock simulator (see Chapter 8) with a microprocessor-based control system (Mulvaney & Rizvi 1984), to study the step-response of the retort. The PI control model used was derived as a discrete algorithm, rather than in the classical linear form. The results of the simulation were shown graphically; the simulation model developed a PI control, which was controlled to within $\pm 0.1°C$ (see Mulvaney et al. 1990). More recently Alonso et al. (1993a, b) have extended the Mulvaney PI control model to quality control and retort cooling. Ryniecki and Jayas (1993) also used the step-function method to study the dynamic response of a retort with respect to the lethality delivered to the processed cans. The method was sufficiently accurate for the on-line prediction of lethality for use in process control. One notable aspect of the development of dynamic modeling is the effect of venting procedures and the come-up time. From venting experiments it has been shown that the response of the batch retort corresponds to a series of two first-order systems or an overdamped second-order system (Valle & Soule 1987a, b). A simple relation for the temperature during the come-up period of the venting schedule was used by Sumer and Esin (1992).

A computer program was developed from a mathematical analysis of a Barriquand 1300 Steriflow retort in order to simulate the dynamic properties. This enables a control protocol to be established for the heat exchange system, which consists of an external heat exchanger and a horizontal over-pressure retort vessel (Loucka 1997).

A number of models for the energy distribution in retorts have also been developed, e.g. Singh (1977) and Bhowmik et al. (1985), but without studying the dynamic response.

9.3. Retort Control

9.3.1. Control of Batch Retorts

9.3.1.1. Conventional Control

The main operating stages have already been listed in Chapter 8, and the typical lay-out of process valves is given in Thorpe *et al*. (1975). The simplest system consists of a control valve on the steam supply with a manual by-pass valve for venting. This is usually described as a semi-automatic system. The temperature of the retort is continuously measured in the pocket of the retort and the signal fed to the controller. The system may be extended by introducing further automatic valves on the steam by-pass, vent and air pressure for cooling with a single controller. Hughes (1971a, b) gives a variety of different methods of controlling retorts with increasing complexity. The problem with increasing the degree of control on batch retorts is that the cost of additional valves increases with the number of retorts and this is prohibitive for many canners; consequently, smaller canneries tend to have only the simplest steam control.

One of the important parts of the retort cycle is the control of pressure in the retort during the cooling of the cans in order to prevent the deformation of the cans, especially those with large diameters. Various strategies for controlling the pressure have been presented by Alonso *et al*. (1994). A microcomputer system was developed for batch retort control using digital PI (proportional + integral) control. and six retort temperature profiles were studied including 3 isothermal, sinusoidal, step and ramp retort temperature profiles (Fastag *et al*. 1996).

The use of oxygen analyzers (Pederson 1984; Tracey 1982) and thermostatic air vents (Felmingham & Leigh 1966) for controlling venting have been investigated; however, they are not yet recommended for commercial use.

The developments in the control of retort systems have been reviewed by Holdsworth (1983, 1985, 1988, 1989) and Richardson and Holdsworth (1989).

9.3.1.2. Computer Control

Most of the developments in the application of computer control have taken place in the 1980s with the availability of microprocessors and minicomputers. Table 9.1 shows some of the work that has been done on different aspects of computer control, mainly in the form of laboratory demonstrations. One of the first systems to be used in a UK cannery was due to Bown (1982, 1983, 1985a, b, 1986) and colleagues (see Bown & Duxbury 1986). The temperature and pressure signals from three banks of eight retorts were linked respectively with three microprocessor units (MACSYM 350). The data from the sensors were displayed on graphics terminals and were refreshed every 3 s. Data could also be printed out for reference purposes. This commercial system was shown to be very effective in the factory environment.

The **LOG-TECH**[TM]**-E System** is marketed by FMC of St. Jose, CA, and used with their retorts. LOG-TECH includes a dedicated controller system with

TABLE 9.1. Some investigations into computer control of sterilizing retorts.

Author(s)	System/study	Comment
Earle and Navankasattusas (1974)	Laboratory batch retort	Electrical analog of sterilization value
Navankasattusas and Lund (1978)	Laboratory batch retort	On-line measurement of accomplished lethality
Getchel (1980)	General	Discussion of principles
Steele (1980a, b)	Horizontal retort	General principles
Hildenbrand (1980)	Theoretical study	Optimal temperature control
Teixeira and Manson (1982)	Theoretical study	Discussion of basic principles
Bown (1985a, b)	Three-crate batch retort	Microprocessor application
Datta et al. (1986)	Theoretical study	On-line correction of process deviations
Lappo and Povey (1986)	Laboratory batch retort	Temperature deviations and errors investigated
Richardson et al. (1988)	Theoretical study	Lethality model
Gill et al. (1989)	Thermal process simulator	On-line data acquisition: process deviations, lethality calculations
Bermudez de Castro and Martinez Varela (1989)	Theoretical study	Optimal control theory
Laurenceau et al. (1991)	Hydrolock simulator	Lethality control model
Mihori et al. (1991)	Theoretical study	Control strategy for 1D heat transfer
Simpson et al. (1993)	Theoretical study	Control strategy for conduction-heating products
Ramesh et al. (1994)	Theoretical study	Optimal sterilization

recipe-selection system, automatic temperature deviation correction, using the Ball formula calculation and numerical estimation of the F_0-value of the process. This is linked to a host PC screen with options, including a fully-graphic representation of the retort system and data logging facilities. The NumeriCAL^{TM} thermal process calculation model is also incorporated which can be used on-line for real-time predictive modeling (see Section 9.3.5.) (Martens & Clifton 2002).

9.3.2. Efficient and General On-Line Correction of Process Deviations in Batch Retort

The following on-line control analysis has been summarized from the work done by Simpson et al. (2007). The aim of this research study was the development of a safe, simple, efficient and easy to use procedure to manage on-line corrections of unexpected process deviations in any canning plant facility. Specific objectives were to:

1. Develop strategy to correct the process deviation by an alternative "proportional-corrected" process that delivers no less than final target lethality, but with near minimum extended process time at the recovered retort temperature.
2. Demonstrate strategy performance by comparing "proportional-corrected" with "commercial-corrected" and "exact-corrected" process times.

3. Demonstrate consistent safety of the strategy by exhaustive search over an extensive domain of product and process conditions to find a case in which safety is compromised.
4. Estimate potential economic impact of chosen on-line correction strategy in terms of gains in manufacturing efficiency or productivity of a typical food cannery.

9.3.2.1. Scope of the Strategy

To reach the objectives stated above, the approach to this work was carried out in four tasks, one in support of each objective. Task 1 consisted of developing the strategy for on-line correction of process deviations with minimum extended process time using the method of "proportional correction." Task 2 consisted of choosing appropriate mathematical heat-transfer models for construction of the equivalent lethality curves or "look-up tables" needed for use with each respective strategy and for determining the final lethality and quality retention for each of the thousands of cases simulated in the study. Task 3 included the complex optimization search routine that was carried out to demonstrate validity and consistent safety of the strategy. The fourth and final task was to estimate the impact on plant production capacity of alternative corrected process times used in response to the same frequency and type of deviations occurring annually in a typical cannery. Methodology employed in carrying out each of these tasks is described in greater detail below.

Task 1 – Proportional Correction Strategy Development

The objective for the strategy required in this task was to accomplish an on-line correction of an unexpected retort temperature deviation by an alternative process that delivers final target lethality, but with minimum extended process time at the recovered retort temperature. This would be accomplished with use of the same alternative process "look-up tables" that would normally be used with currently accepted methods of on-line correction of process deviations, but with a "proportional correction" applied to the alternative process time that would reduce it to a minimum without compromising safety. In order to fully understand this strategy, it will be helpful to first review the currently accepted method that is in common practice throughout the industry. Commercial systems currently in use for on-line correction of process deviations do so by extending process time to that which would be needed to deliver the same final lethality had the entire process been carried out with an alternative lower constant retort temperature equal to that reached at the lowest point in the deviation. These alternative retort temperature-time combinations that deliver the same final process lethality (F_0) are called equivalent lethality processes. When these equivalent time-temperature combinations are plotted on a graph of process time versus retort temperature, they fall along a smooth curve called an equivalent lethality curve. These curves are predetermined for each product from heat penetration tests and thermal process calculations carried out for different retort temperatures.

In practice, the new process times obtained from these curves at such low alternative temperatures can be as much as two or three times longer than the originally scheduled process time required to reach the same final target lethality, resulting in considerable quality deterioration and costly disruption to scheduled retort operations. Nonetheless, these systems are versatile because they are applicable to any kind of food under any size, type or container shape, as well as mode of heat transfer (Larkin 2002). These consequences are particularly painful when, as in most cases the deviation recovers quickly, and the alternative extended process time is carried out at the recovered original retort temperature. Canned food products subjected to such "corrected" processes become severely over processed, with final lethalities far in excess of that required and quality deterioration often reaching levels below consumer acceptance, a safe correction, but by no means optimal or efficient (Alonso *et al.* 1993a; Von Oetinger 1997; Simpson 2004). To avoid these painful corrections processors normally operate at retort temperatures 3 to 4°C over the registered retort temperature.

$$t_D = t_{TRT} + \sum_{i=1}^{n} (t_{D_i} - t_{TRT}) \left(\frac{\Delta t_i}{t_{TRT}} \right); t_{D_i} \geq t_{TRT}$$

The "proportional-correction" strategy developed in this research significantly avoids such excessive over processing by taking advantage of the short duration of most recovered retort temperature deviations, and the lethality delivered by carrying out the corrected process at the recovered retort temperature. The strategy will calculate the corrected process time (t_D) as a function of the temperature drop experienced during the deviation, but also the time duration of the deviation. The following expression illustrates mathematically how this "proportional-corrected" process time would be calculated for any number (n) of deviations occurring throughout the course of a single process:

$$t_D = t_{TRT} + \sum_{i=1}^{n} (t_{D_i} - t_{TRT}) \left(\frac{\Delta t_i}{t_{TRT}} \right); t_{D_i} \geq t_{TRT}$$

Where
n: Number of deviations occurring during the process.
t_D: Corrected process time.
t_{TRT}: Pre-established process time at retort temperature *TRT*.
Δt_i: Duration of deviation i.
t_{D_i}: Process time at the deviation temperature *TRT*$_i$
TRT$_i$: Lowest temperature during the deviation i.
TRT : Retort temperature.

For example, in the case of a single deviation, the corrected process time would be calculated by first finding the alternative process time that would be required to deliver the same final lethality had the entire process been carried out with an alternative lower constant retort temperature equal to that reached at the lowest point in the deviation (t_D). This would be done by use of the equivalent lethality process curve or look-up table described earlier. The difference between this longer alternative process time and the originally scheduled process time $(t_D - t_{TRT})$ is the extra time that would normally be added to the original process time to correct the process according to current industry practice. However, in the new strategy proposed here, this extra process time differential $(t_D - t_{TRT})$ is multiplied by a proportionality factor consisting of the ratio, [time duration of the deviation]:[originally scheduled process time], or expressed mathematically as $(\Delta t / t_{TRT})$. This proportionality factor is always less than or equal to one and always results in a corrected process that delivers no less than the final target lethality specified for the original process, but with near minimum extended process time.

The logic behind this "proportional-correction" strategy stems from the following rationale:

- The current industry practice is necessary only when the deviation fails to recover, and retort temperature remains at the lowest point for the duration of the process.
- This practice is unnecessary when the deviation recovers and processing resumes at the original scheduled retort temperature over the "corrected" extended time.
- If the extended process time is chosen to be in proportion to the duration of the deviation as a fraction of original scheduled process time, we are making the assumption that the amount of lethality lost during the deviation duration time is the amount that would have accumulated at retort temperature.
- In reality, this amount of lost lethality is much less, since the actual retort temperature had fallen during the deviation to some lower level where lethality would accumulate at a slower rate.
- Therefore, the "proportional-correction" should always deliver total final lethality greater than that originally specified for the scheduled process.
- With the implementation of this novel and efficient on-line strategy it will be unnecessary for processors to operate at higher retort temperatures.

Task 2 – Performance Demonstration

This task consisted of demonstrating the performance of these strategies by simulating the occurrence of process deviations happening at different times during the process (early, late and randomly) to both solid and liquid canned food products, calculating the alternative corrected process times, and predicting the outcomes of each corrected process in terms of final lethality and quality retention. For each deviation, three different alternative corrected process times were calculated:

- "Exact correction," giving corrected process time to reach precisely the final target lethality specified for the scheduled process, using computer simulation with heat transfer models;
- "Proportional-correction," using the strategy described in this research with look-up tables; and,

TABLE 9.2. Product and process conditions used for on-line correction strategy simulations.

Product Simulated	Dimensions (cm)			Properties		Normal Process	
	Major	Intermedium	Minor	alfa (m²/s)	f_h	time (min)	TRT (°C)
Pure Conduction Can, Biot > 40	11.3	–	7.3	1.70E-07	44.4	64.1	120
Forced Convection Can, Biot < 1	11.3	–	7.3	–	4.4	15.6	120

- "Commercial correction," using current industry practice with look-up tables (manually or computerized).

The heat transfer models were explicitly chosen to simulate the two extreme heat-transfer cases encountered in thermal processing of canned foods. The rationale behind this decision was that canned foods possess heating characteristics between these two extreme situations. Conclusions extracted from these simulations will be extended to all canned foods. First was the case of pure conduction heating of a solid product under a still-cook retort process. The second was the case of forced convection heating of a liquid product under mechanical agitation. In both cases, the container shape of a finite cylinder was assumed, typical of a metal can or wide-mouthed glass jar. However, suitable models appropriate for a true container shapes can be used as required for this purpose. Examples of such models can be found in the literature (Teixeira et al. 1969; Manson et al. 1970; Manson et al. 1974; Datta et al. 1986; Simpson et al. 1989; Simpson et al. 2004). The product and process conditions chosen to carry out the demonstrated simulations for each case are given in Table 9.2.

Task 3 – Demonstration of Safety Assurance by Complex Optimization Search Routine

This on-line correction strategy was validated and tested for safety assurance by executing a strict and exhaustive search routine with the use of the heat-transfer models selected in Task 2 on high-speed computer. The problem to be solved by the search routine was to determine if the minimum final lethality delivered by all the corrected processes that could be found among all the various types of deviations and process conditions considered in the problem domain met the criterion that it had to be greater than or equal to the lethality specified for the original scheduled process. This criterion can be expressed mathematically:

$$\underset{U}{\text{Min}} \lfloor F_{\text{proportional}} - F_{\text{Tol}} \rfloor \geq 0$$

Where

$F_{\text{proportional}}$: F-value calculated with the proportional correction
F_{TOL}: F-value specified for normal scheduled process
U: Universe of feasible process conditions in search routine

TABLE 9.3. Problem domain for search routine.

Process Variable	Description of Process Variable	Minimum Value	Maximum Value
TRT	Scheduled Retort Temperature	110°C	135°C
TRT_i	Lowest retort temperature reached during deviation i	100°C	$TRT - 0.5°C$
t_{CUT}	Initial come up time of retort to reach TRT	5 [min]	15 [min]
t_{dev-i}	Time during the process at which the deviation i begins	t_{cut}	t_{TRT}
t_i	Time duration of the deviation i	0.5 [min]	$t_{TRT} - t_{dev-i}$
T_{ini}	Initial product temperature	20°C	70°C

Table 9.3 identifies the various types of deviations and process conditions that were explored and evaluated in the search routine (problem domain). The search routine was designed as an attempt to find a set of conditions under which the required search constraint was not met. The table gives the symbol used to represent each variable and a description of that variable, along with the minimum and maximum values limiting the range over which the search was conducted.

Task 4 – Economic and Quality Impact of On-line Correction Strategy

The purpose of this task was to estimate the impact on plant production capacity of alternative corrected process times used in response to the same frequency and type of deviations occurring annually in a typical cannery. The rationale behind Task 4 stems from the realization that every time a process deviation is corrected in this way with any given retort in the cook room of a canning factory, the number of batches processed that day will be less than normal capacity, which can be translated into cost of lost productivity over the course of a canning season. This can be approached mathematically with the algorithm described below.

The following expression allows the calculation of the number of batch per autoclave at the processing plant:

$$N_{Bi} = \frac{H}{t_{ci} + t_{oj} + t_{di}} \tag{9.1}$$

Where:

N_{Bi}: Number of batches processed per retort **i** during the season.
H: Operating time of the plant during the season (h)
t_{ci}: Time to load retort **i** with product **j** (h)
t_{oj}: Time to operate retort **i** (process cycle time) with product **j** (h)
t_{di}: Time to download retort **i** with product **j** (h)

To simplify the analysis and to be able to have an estimate of the impact of operating time on plant production capacity, the following assumptions were made:

(a) The plant has N_A retorts and all of them are of equal size.
(b) The number of containers (units) processed in each retort is the same ($N_{CBi} = N_{CB}$)
(c) The plant is processing a single product ($t_{oj} = t_o$; $t_{ci} = t_c$ and $t_{di} = t_d$)

Therefore, the total number of units that can be processed in the whole season (N_t) can be expressed by the following equation:

$$N_t = \frac{N_A^* N_{cB}^* H}{t_c + t_0 + t_d} \tag{9.2}$$

According to equation (9.2) an extension in process time (t_o) will decrease plant production capacity (N_t). In addition, considering processing time as a variable and utilizing equation (9.2), it is possible to quantify the impact of processing time in terms of plant production capacity. Another way to assess the impact of the adopted strategy will be to consider that processors are operating at much higher retort temperature to avoid deviant processes. For the purposes of analysis it was considered that processors operate each batch process at a temperature that is 2–3°C (common practice in the United States, although some plants even operate at higher temperatures) higher than the registered process, and with this practice processors are completely avoiding deviant processes. As was mentioned before the product and process conditions chosen to carry out the simulations to evaluate the impact on product quality are given in Table 9.3.

This research has described a practical, simple and efficient strategy for on-line correction of thermal process deviations during retort sterilization of canned foods. The strategy is intended for easy implementation in any cannery around the world. This strategy takes into account the duration of the deviation in addition to the magnitude of the temperature drop. It calculates a "proportional" extended process time at the recovered retort temperature that will deliver the final specified target lethality with very little over processing in comparison to current industry practice. Results from an exhaustive search routine using the complex method support the logic and rationale behind the strategy by showing that the proposed strategy will always result in a corrected process that delivers no less than the final target lethality specified for the originally scheduled process. Economic impact of adopting this strategy over that currently used in industry practice can be a significant increase in production capacity for a typical cannery, as well as, canned products that are of much higher quality.

9.3.3. Control of Hydrostatic Sterilizers

The general features of a hydrostatic sterilizer were discussed in Chapter 8. The steam flow to the central sterilizing chamber is regulated by a controller, which uses the temperature measured in the sterilizing chamber, about half way up the side. Two other controllers are used to control the feed leg and the discharge leg, with temperatures measured near to the base of the feed leg. A fourth controller measures the level of water in the sterilizing chamber and maintains it at a constant level; this also sets the levels in the hydrostatic legs. The control system provides a fresh supply of water to the incoming level and operates a drain valve at the base of the discharge leg. When there is an additional spray cooling section to the sterilizer, the flow of fresh cooling water is regulated by the temperature at the base of the second cooling section.

Guidelines for the operation of hydrostatic sterilizers have been produced by Austin and Atherton (1981, 1984a, b). Recently has been published an interesting paper analyzing the on-line control of hydrostatic sterilizers (Chen *et al.* 2007).

9.3.4. Control of Continuous Reel-And-Spiral Pressure Cookers

The initial stages of venting and come-up of continuous cooker/coolers are done without any containers going through the system. The cookers are fitted with several vent valves and lines (5 cm diameter pipes). The venting cycle depends on the size of the units and steam flows, but is usually of the order of 6–8 min to a temperature of 105°C with the vent valves wide open and the drain valve partially open. An alternative labor-saving method involves only the use of the bleeders, drain and purge lines, rather than the main vent valves (Adams *et al.* 1985). The temperature is regulated by automatic PID controllers and measured by thermocouples in the shell of the cooker and the cooler, and the flow of steam and water regulated accordingly. Bratt *et al.* (1990) have produced detailed guidelines for the operation of reel-and-spiral cooker/coolers.

9.3.5. Derived-Value Control

Figure 9.1 shows a sequence of operations that could be used to control a process based on the lethality developed in the product (Holdsworth 1974, 1983, 1985). Essentially the cooling cycle would be activated when the heating cycle had produced a specified F-value. The system would require knowledge of the temperature–time relationship as measured at the point of slowest heating in the food in the container. This could only be obtained, under commercial processing conditions, by calculation from known heating characteristics of the pack. In the computer system developed by Bown (see Section 9.3.1.2), it was originally intended that this type of calculation would be carried out, and the microprocessor would implement a control strategy based on the mathematically derived F-value. At the time of installation this was not possible because the program running this part was too slow. Consequently, to use this type of derived value control it is necessary to be able to compute temperatures in real time. One of the important advantages of this type of system is to be able to deal with process deviations and to make judgements on the need for further processing or otherwise.

The technique has been developed further and extended to the control of hydrostatic cookers. (Scott 1992). An important conduction-heating model, which can be used for any j-value, has been developed (Teixeira *et al.* 1992; Teixeira & Tucker 1997). A method of dealing with process deviations using on-line retort control has been discussed by Akterian (1997). Optimal control theory has been used to develop a model to determine the optimum retort temperature for conduction-heating retort pouches, which involves a 1-D heat conduction model. (Terajima & Nonaka 1996). An account of the application of control strategies to thermal processing operations has been presented by

Ryniecki and Jayas (1997). A sophisticated closed-loop receding horizon optimal control system has been developed for controlling the temperature profile, which will minimize costs and loss of quality (Chalabi *et al.* 1999). The problems of retro fitting retorts for on-line control has been discussed by Kumar *et al.* (2001). The system was used for the retorting of canned vegetables, and the results showed that using the on-line system the process time could be reduced from 17 to 33%.

9.3.6. Guidelines for Computer Control

The NFPA (2002) in the United States have published revised guidelines for the validation of computer control systems. Whereas the original *Bulletin 43L* dealt with aseptic processing systems only, this edition includes retort processing. The object of the current procedures is to establish documented evidence that provides a high degree of reassurance that a specific process or system will produce a product meeting its predetermined specifications and quality attributes. It includes life-cycle definition and planning, system design and development, system qualification, system operation, maintenance and on-going verification, system security requirements, and documentation (Deniston 2004).

9.4. Industrial Automation of Batch Retorts

Many of the most recent advances made in the design of industrial batch retorts has come about in response to the increasing popularity of flexible retort pouches and retortable semi-rigid microwavable plastic dinner trays and lunch bowls. These flexible and semi-rigid containers lack the strength of traditional metal cans and glass jars to withstand the large pressure differences experienced across the container during normal retort operations. To safely process these types of flexible packages, careful control of overriding air pressure is needed during retort processing, and pure saturated steam, alone, cannot be used as the heat exchange medium. Instead, new retorts designed to be used with pressure-controlled steam-air mixtures, water spray, or water cascade have been recently developed for this purpose (Blattner 2004). Examples of some of these new retort designs are given in Figure 9.2. A close-up view of some of the specially designed racking configurations used to hold flexible retortable packages in place during retorting is shown in Figure 9.3.

Perhaps the most significant advances made in the food canning industry to date have been in the area of automated materials handling systems for loading and unloading batch retorts. Traditionally, the loading and unloading of batch retorts has been the most labor-intensive component in food canning factories. Unprocessed sealed containers would be manually stacked into baskets, crates or carts. Then, the baskets or crates would be loaded into empty vertical retorts with the aid of chain hoist, or wheeled carts would be loaded into horizontal retorts with the aid of track rails for this purpose. In recent years leading manufacturers of retort systems have been hard at work designing and offering a host of new

FIGURE 9.2. New retort systems (rotating and still-cook) with specially designed racking configurations for processing flexible and semi-rigid packages (Courtesy of ALLPAX, Covington, LA).

automated materials handling systems to automate this retort loading and unloading operation.

Most of the new automated systems available to date are based on the use of either automated guided vehicles (Heyliger 2004), or orthogonal direction shuttle systems (Blattner 2004; Heyliger 2004). Both types of systems are designed for use with horizontal retorts. The automated guided vehicles (AGV) work like robots. They carry the loaded crates of unprocessed product from the loading

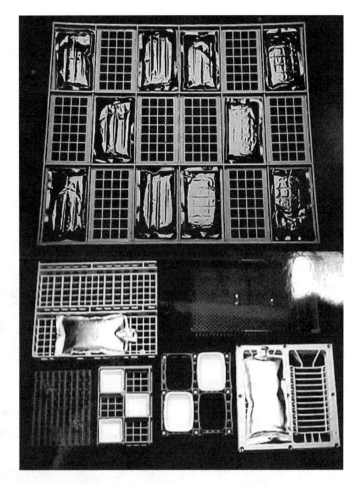

FIGURE 9.3. Rack designs for flexible and semi-rigid retortable packaging systems, (Courtesy, ALLPAX, Covington, LA).

station to any designated retort on the cook room floor that is ready to be loaded. They also carry the loaded crates of finished processed product from the unloaded retort to the unloading station for discharge as out-going product exiting the cook room to the case-packing operations. These robotic AGV's are designed to integrate with the loading station in such a way that sealed product containers arriving on a conveyor automatically stack into the crate carried by the AGV, which later inserts the entire crate into the designated retort.

Unloading at the unloading station for finished product discharge is likewise accomplished in a similar automated way, but in reverse. The AGVs are guided by an underground wire tracking system buried beneath the cook room floor. This leaves the cook room floor space open and free of any rail tracks or guide rails that would otherwise impede the safe movement of factory workers in their normal

FIGURE 9.4. Automated batch retort system with use of automated guided vehicles in large Cook room operation (Courtesy, FMC Food Tech., Madera, CA).

work flow operations. A panoramic view of a large cook room operation using an automated batch retort system with automated guided vehicles is shown in Figure 9.4 (Heyliger 2004), and a close-up view of an automated guided vehicle in the process of loading or unloading a horizontal retort is shown in Figure 9.5.

FIGURE 9.5. Automated guided vehicle for batch retort loading/unloading (Courtesy, FMC Food Tech., Madera, CA).

An alternative to the automated guided vehicle (AGV) system is the shuttle system offered by several retort manufacturers. Unlike the AGV system, the shuttle system relies upon a set of tracks or rails that are fixed in place on the cook room floor. These rails span the length of the cook room along the row of horizontal retorts, allowing a shuttle carrying loaded crates to slide along these rails until it has aligned itself in front of the designated retort waiting to be loaded. In a similar fashion, when a retort is ready for unloading, an empty shuttle slides along these rails until it has aligned itself with that retort to receive the loaded crates of processed product. Then the shuttle slides along the rails to the far end of the cook room where unloading of processed product takes place for discharge out of the cook room. Normally, the unprocessed product loading station and the processed product unloading stations are located at opposite ends of the cook room (Figure 9.6). Figures 9.7 and 9.8 illustrate the shuttle systems offered by ALLPAX and FMC, respectively.

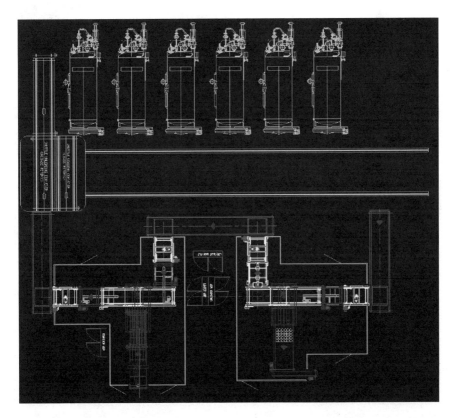

FIGURE 9.6. Automated shuttle-based batch retort control system (Courtesy, ALLPAX, Covington, LA).

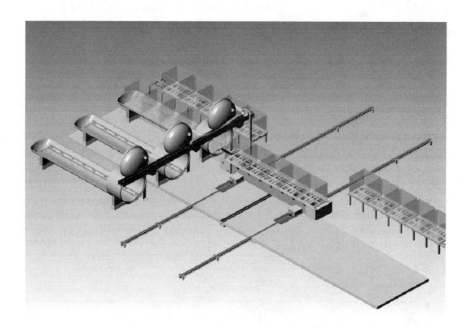

FIGURE 9.7. Automated shuttle batch retort system (Courtesy of ALLPAX, Covington, LA).

FIGURE 9.8. FMC shuttle system for automated batch retort loading/unloading (Courtesy, FMC Food Tech., Madera, CA).

References

Adams, H. W., Heyliger, T. L., Penly, D. H., & Wert, R. E. (1985). Venting of continuous rotary sterilizers. Paper presented at the 45th Annual Meeting, IFT, Atlanta, GA.

Akterian, S. G. (1999). On-line control strategy for compensating for arbitrary deviations in heating-medium temperature during batch thermal sterilization processes. *J. Food Eng., 39*(1), 1–7.

Alonso, A. A., Perez-Martin, R. I., Shukla, N. V., & Deshpande, P. B. (1993a). On-line quality control of non-linear batch systems: application to the thermal processing of canned foods. *J. Food Eng., 19*, 275–289.

Alonso, A. A., Banga, J. R., & Perez-Martin, R. I. (1993b). A new strategy for the control of pressure during the cooling stage of the sterilization process in steam retorts: Part I. A preliminary study. *Food and Bioproducts Processing. Trans. IChemE, 71*(C), 197–205.

Alonso, A. A., Banga, J. R., & Perez-Martin, R. I. (1994). Different strategies for controlling pressure during the cooling stage of batch retorts. In T. Yano, R. Matsuno, & K. Nakamura (Eds.), *Developments in food engineering ICEF6*, Part 2, 724–726.

Austin, G., & Atherton, D. A. (1981). Process control in hydrostatic cookers. Part 1. Validification of cooker operating conditions. *Technical Manual* No. 5. Chipping Campden, Glos., UK: Campden & Chorleywood Food Research Association.

Austin, G., & Atherton, D. A. (1984a). Process control in hydrostatic cookers. Part 2. Factors affecting heat penetration rates. *Technical Manual No. 5*. Chipping Campden, Glos., UK: Campden & Chorleywood Food Research Association.

Austin, G., & Atherton, D. A. (1984b). Process control in hydrostatic cookers. Part 3. Guidelines on emergency procedures. *Tecnical Manual No. 5*. Chipping Campden, Glos., UK: Campden & Chorleywood Food Research Association.

Baines, T. S. (1985). Technical review of valve types presently marketed for process control of steam. *Technical Bulletin No. 55*. Chipping Campden, Glos., UK: Campden & Chorleywood Food Research Association.

Bermudez de Castro, A., & Martinez Varela, A. M. (1989). Optimal control of industrial sterilization of canned foods. *Proc. 5th IFAC symp. Control of distributed parameter control systems*. Perpignan, France.

Bhowmik, S. R., Vichnevetsky, R., & Hayakawa, K. (1985). Mathematical model to estimate steam consumption in vertical still retort for thermal processing of canned foods. *Lebensm.-Wiss. u. -Technol., 18*, 15–23.

Biro, G. (1978). A study of autoclave control. *Prumysl Potravin, 29*(II), 176–178 [in Czech].

Blattner, M. F. (2004). Advances in automated retort control, and today's new packaging. Presentation at IFT Symposium, 2004 IFT Meeting, Las Vegas, NV.

Bown, G. (1982). Application of microcomputer technologies in the food and allied industries. *Measurement & Control, 15*(11), 409–410.

Bown, G. (1983). Computer helps improve product and process. *Food Eng., 55*(1), 130–131.

Bown G. (1984). Recommendations on the specification and calibration of platinum resistance thermometry for use in thermal processing. *Technical Bulletin No. 54*. Chipping Campden: Glos., UK: Campden & Chorleywood Food Research Association.

Bown, G. (1985a). Microcomputer controlled batch sterilization in the food industry. *Chemistry & Industry*, 359–362.

Bown, G. (1985b). Retort control – the application of a microcomputer based control system. *Technical Memorandum No. 391*. Chipping Campden, Glos., UK: Campden & Chorleywood Food Research Association.

Bown, G. (1986). Process control microcomputers in the food industry. In *Developments in food preservation Vol. 4* (pp. 35–58). Barking: Elsevier Applied Science.

Bown, G., & Duxbury, D. D. (1986). Microcomputer control system automates 24 batch retorts. *Food Processing, 55*, 82–84.

Bown, G., & Slight, H. (1980). Investigations into retort instrumentation – phase 1. Long term reliability and stability. *Miscellaneous Reports No. 1*. Chipping Campden, Glos., UK: Campden & Chorleywood Food Research Association.

Bown, G., & Slight, H. (1982). Investigations into retort instrumentation – phase 2. Trials using platinum resistance digital thermometers. *Miscellaneous Reports No.2*. Chipping Campden, Glos., UK: Campden & Chorleywood Food Research Association.

Bown, G. (2004). Modelling and optimising retort temperature control. In P. Richardson (Ed.), *Improving the thermal processing of foods* (pp. 103–123). Cambridge: Woodhead Publishing.

Bratt, L., Williams, A., Seager, A., & May, N. (1990). Process control in reel and spiral cooker/coolers. Part 1: The operation of continuous cooker/coolers. *Technical Manual No. 26*. Chipping Campden, Glos., UK: Campden & Chorleywood Food Research Association.

Chalabi, Z. S., Van Willigenbeurg, L. G., & Van Straten, G. (1999). Robust optimal receding control of the thermal sterilization of canned foods. *J. Food Eng., 40*(3), 207–218.

Chen, G., Campanella, O., Corvalan, C., & Haley, T. (2007). On-line corrections of process temperature deviations in continuous retorts. In Press On-line. *J. of Food Engineering.*

Cossey, R. S. (1988). Performance of steam control valves. *Process Eng.*, December, 37–39.

Cossey, R. S. (1990). Algorithm assessment: the study of control valve performance as new and after 15 months in service. *Technical Memorandum No. 579*. Chipping Campden: Campden & Chorleywood Food Research Association.

Cossey, R. S. (1991). The use of constant temperature blocks for calibration of master temperature indicators. *Technical Memorandum No. 613*. Chipping Campden, Glos., UK: Campden & Chorleywood Food Research Association.

Datta, A. K., Teixeira, A. A., & Manson, J. E. (1986). Computer-based retort control logic for on-line correction of process deviations. *J. Food Sci., 51*(2), 480–483, 507.

David, J. R. D., & Shoemaker, C. F. (1985). A transducer for the direct measurement of rates of lethality during thermal processing of foods. *J. Food. Sci., 50*(1), 223–225.

Deniston, M. (2004). Validation guidelines for automated control of thermal processing systems. In G. S. Tucker (Ed.), *Third international symposium on thermal processing – Process and package innovation for convenience foods.* (Session 3:3). Chipping Campden, Glos., UK: Campden & Chorleywood Food Research Association.

Department of Health (1994). *Guidelines for the safe production of heat preserved foods.* London: HMSO.

Earle, R. L., & Navankasattusas, S. (1974). A control method for sterilization processes. *Proc. 4th int. congress Food science & technology, 4*, 426–433.

Fastag, J., Koide, H., & Rizvi, S. S. H. (1996). Variable control of a batch retort and process simulation for optimization studies. *J. Food Process Eng., 19*(1), 1–14.

FDA (1983). *Thermally processing low-acid foods packaged in hermetically sealed containers*, Regulations 21, Part 113. Washington, DC: Food and Drug Administration.

Felmingham, J. D., & Leigh, R. J. (1966). Cannery retort operation and procedure to cope with modern can-handling methods. *J. Food Technol., 1*, 345–358.

Getchell, J. R. (1980). *Computer process control of retorts, Paper No. 80–6511*. St Joseph, MI: ASAE.

Gill, T. A., Thompson, J. W., & Leblanc, G. (1989). Computerized control strategies for a steam retort. *J. Food Eng.*, *10*, 135–143.

Heyliger, T. L. (2004). Advances in retort control for batch and continuous systems. Presentation at IFT Symposium, 2004 IFT Meeting, Las Vegas, NV.

Hildenbrand, P. (1980). An approach to solving the optimal temperature control problem for sterilization of conduction-heating foods. *J. Food Process Eng.*, *3*, 123–142.

Holdsworth, S. D. (1974). Instrumentation developments for heat processing. *Food Manufacture*, *49*(11), 35–38.

Holdsworth, S. D. (1983). Developments in the control of sterilizing retorts. *Process Biochem.*, *18*(5), 24–27.

Holdsworth, S. D. (1985). Developments in the control of sterilizing retorts. *Packaging*, *56*(662), 13–17.

Holdsworth, S. D. (1988). Controlling batch versus continuous processes. In *Managing and controlling of food processes* (pp. 8–20). London: Sterling Conferences.

Holdsworth, S. D. (1989). Controlling batch versus continuous processes. *Food Technol. Europe*, 354–356.

Hughes, P. (1971a). Semi-automatic retort control systems. *Food Manufacture*, *46*(12), 32–35, 47.

Hughes, P. (1971b). Fully automatic retort control systems. *Food Manufacture*, *47*(2), 27–32, 61.

Kumar, M. A., Ramesh, M. N., & Nagara Rao, S. (2001). Retrofitting of a vertical retort for on-line control of the sterilization process. *J. Food Eng.*, *47*(1), 89–96.

Lappo, B. P., & Povey, M. J. W. (1986). A microprocessor control system for thermal sterilization operations. *J. Food Eng.*, *5*, 31–53.

Larkin. 2002. Personal Communication. Branch Chief, National Center for Food Safety and Technology, Food and Drug Administration (FDA/NCFST), Chicago, IL., USA.

Laurenceau, J., Dousset, X., & Pesche, D. (1991). Prediction of the sterilization value of thermal processes and automatic control. *Cerevisia Biotechnol.*, *16*(2), 33–45.

Loucka, M. (1997). Simulation of dynamic properties of batch sterilizer with external heating. *J. Food Process Eng.*, *20*(2), 91–106.

Manson, J. E., Zahradnik, J. W., & Stumbo, C. R. (1970). Evaluation of lethality and nutrient retentions of conduction-heating food in rectangular containers. *Food Technol.* *24*(11), 1297–1301.

Manson, J. E., Zahradnik, J. W., & Stumbo, C. R. (1974). Evaluation of thermal processes for conduction heating foods in pear-shaped containers. *J. Food Sci.,* *39*(2), 276–281.

Martens B. & Clifton, D. (2002). Advances in retort control. In Tucker, G. S. (Ed.), *Second international symposium on thermal processing – Thermal processing: alidation challenges.* (Session 4:2). Chipping Campden UK: Campden & Chorleywood Food Research Association.

May, N., & Withers, P. M. (1993). Data acquisition units for thermal process evaluation. *Food Technol. Europe*, 97–99.

Mihori, T., Watanabe, H., & Kaneko, S. (1991). A control system for achieving correct heat sterilization processes: a one dimensional approach to packaged conduction heating foods. *J. Food Process Eng.*, *15*(2), 135–156.

Mulvaney, S. J. (1987). Dynamic process modeling and evaluation of computer control of a retort for thermal processing. PhD Thesis. Ithaca NY: Cornell University.

Mulvaney, S. J., & Rizvi, S. S. H. (1984). A microcomputer controller for retorts. *Trans. ASAE, 27*, 1964–1969.

Mulvaney, S. J., Rizvi, S. S. H., & Johnson, C. R., Jr. (1990). Dynamic modeling and computer control of a retort for thermal processing. *J. Food Eng.*, *11*, 273–289.

Navankasattusas, S., & Lund, D. B. (1978). Monitoring and controlling thermal process by on-line measurement of accomplished lethality. *Food Technol.*, *32*(3), 79–82.

NFPA (2002). *Validation guidelines for automated control of food processing systems, used for the processing and packaging of preserved foods.* 2nd Edition, Dublin, CA, USA: National Food Processors Association.

Overington, W. J. G. (1975). Analogue simulation of heat transfer in the thermal processing of food. Master's thesis. Birmingham, University of Aston.

Pederson, L. D. (1984). Control of steam: optimization by oxygen tester. In *Annual Report*, Berkeley, CA.: NFPA.

Ramesh, M. N., Kartik, V., Navneth, L. V., & Bhanuprakash, K. N. (1994). Microprocessor based retort control system for optimum sterilization of foods. *Proc. int. convention – TEPEM 94*, Madras, India.

Richardson, P. S., & Bown, G. (1987). A standard for platinum resistance thermometers, for use on food industry sterilisers and pasteurisers. *Technical Bulletin No. 61*. Chipping Campden, Glos., UK: Campden & Chorleywood Food Research Association.

Richardson, P. S., & Holdsworth, S. D. (1989). Mathematical modelling and control of sterilization processes. In R. W. Field and J. A. Howell (Eds.), *Process Engineering in the Food Industries* (pp. 169–187). London: Elsevier Applied Science.

Richardson, P. S., Kelley, P. T., & Holdsworth, S. D. (1988). Optimization of in-container sterilization processes. *Proc. progress in food preservation processes Vol. 2* (pp. 1–11). Brussels: CERIA.

Ryniecki, A., & Jayas, D. S. (1993). Automatic determination of model parameters for computer control of canned food sterilization. *J. Food Eng.*, *19*, 75–94.

Ryniecki, A., & J ayas, D. S. (1997). Process control of thermal processing. In G. S. Mittal (Ed.), *Computerized control systems in the food industry*. New York: Marcel Dekker

Saguy, I., & Kiploks, E. M. (1989). Revolution counter for food containers in an agitating retort. *Food Technol.*, *43*, 68–70.

Scott, G. M. (1992). Simulation and modelling of container sterilization processes and the application of derived-value control technique. In Hamm, W. & Holdsworth, S. D. (Eds.), *Food engineering in a computer climate*. Symposium Series No. 126. (pp. 209–214). Rugby, U.K: Institution of Chemical Engineers.

Simpson, R., Aris, I., & Torres, J. A. (1989). Sterilization of conduction-heated foods in oval-shaped containers. *J. Food Sci.*, *54*(5), 1327–1331, 1363.

Simpson, R., Almonacid-Merino, S. F., & Torres, J. A. (1993). Mathematical models and logic for the computer control of batch retorts: conduction-heated foods. *J. Food Eng.*, *20*, 283–295.

Simpson, R., Figueroa, I., & Teixeira, A. (2007). Simple, practical, and efficient on-line correction of process deviations in batch retort through simulation. *Food Control, 18*(5), 458–465.

Simpson, R. (2004). Control logic for on-line correction of batch sterilization processes applicable to any kind of canned food. Symposium of Thermal processing in the 21st century: Engineering modeling and automation, at IFT Meeting, 2004, Las Vegas, NV, USA.

Singh, R. P. (1977). Energy consumption and conservation in food sterilization. *Food Technol.*, *31*(3), 57–60.

Smith, C. A., & Corripio, A. B. (1985). *Principles and practice of automatic process control.* New York: Wiley.

Steele, D. J. (1980a). Microprocessor applied to retort control. *Food Eng. Int.*, 5(12), 28–32.

Steele, D. J. (1980b). Microprocessors and their application to the control of a horizontal batch retort. *IFST Proc.* (UK), *13*(3), 183–193.

Stephanopoulos, G. (1984). *Chemical process control: An introduction to the theory and practice.* Englewood Cliffs, NJ: Prentice Hall,.

Sumer, A., & Esin, A. (1992). Effect of can arrangement on temperature distribution in still retort sterilization. *J. Food Eng., 15*, 245–259.

Teixeira, A., Dixon, J., Zahradnik, J., & Zinsmeiter, G. (1969). Computer optimization of nutrient retention in the thermal processing of conduction-heated foods. *Food Technol. 23* (6), 845–850.

Teixeira, A. A. & Manson, J. E. (1982). Computer control of batch retort operations with on-line correction of process deviations. *Food Technol., 36*(4), 85–90.

Teixeira, A. E., & Tucker, G. S. (1996). Computer simulation control of canned food sterilization processes. In Fito, P., Ortega-Rodríguez, E., & Barbosa-Cánovas, G. V. (Eds.), *Food engineering 2000* (pp. 291–307). New York: Chapman & Hall.

Teixeira, A. A., & Tucker, G. S. (1997). On-line retort control in thermal sterilization of canned foods. *Food Control, 8*(1), 13–20.

Teixeira, A. A., Tucker, G. S., Balaban, M. O., & Bichier, J. (1992). Innovations in conduction-heating models for in-line retort control of canned foods with any j-value. In R. P.Singh & A. Wirakarakusumah (Eds.), *Advances in food engineering* (pp. 561–74). Boca Raton, FL: CRC Press.

Terajima, Y., & Nonaka, Y. (1996). Retort temperature profile for optimum quality during conduction heating of foods in retortable pouches. *J. Food Sci., 61*(4), 673–678, 682.

Thorpe, R. H., Atherton, D. A., & Steele, D. A. (1975). Canning retorts and their operation. *Technical Manual No. 2.* Chipping Campden, Glos., UK: Campden & Chorleywood Food Research Association.

Tracey, R. J. (1982). The Westinghouse oxygen analyser as a potential rapid means of checking on the correct venting of retorts. *Technical Note.* Chipping Campden, Glos., UK: Campden & Chorleywood Food Research Association.

Valle, C. E. del, & Soule, C. L. (1987a). Modelling the temperature histories during venting of still retorts. *J. Food Process Eng., 9*, 213–220.

Valle, C. E. del, & Soule, C. L. (1987b). Modeling of venting time, temperature distribution and steam consumption during venting of retorts. *J. Food Process Eng., 9*, 287–298.

Von Oetinger, K. 1997. Lógica para el control en línea del proceso de esterilización comercial. Tesis Escuela de Alimentos. Pontificia Universidad Católica de Valparaíso, Valparaíso, Chile.

Withers, P. M. (1991). A can rotation counter for use in reel and spiral sterilizers. *Technical Memorandum No. 604.* Chipping Campden, Glos., UK: Campden & Chorleywood Food Research Association.

Withers, P. M., Tucker, G. S., & Holdsworth, M. D. (1993). A comparison of conduction errors in type T and type K thermocouples for food application. *Technical Memorandum No. 680.* Chipping Campden, Glos., UK: Campden & Chorleywood Food Research Association.

Withers, P. & Richardson, P. S. (1992). Measurement of can rotation during continuous rotary processes and its influence on process value. In *Food engineering in a computer climate*, Symposium Series No. 126 (pp. 191–197). Rugby, UK: Institution of Chemical Engineers.

10
Safety Aspects of Thermal Processing

10.1. Introduction

The aim of this text has been to consider some of the engineering aspects of the thermal processing of packaged foods. These have included the establishment of a thermal process through mathematical and experimental techniques, and the delivery of the thermal process, involving the correct operation and control of the process equipment. Many aspects of the complete process outlined in Figure 1.1 have not been dealt with, in particular the food preparation operations, filling and seaming or sealing, can-handling procedures and hygienic aspects of equipment design and operation. However, an attempt has been made to cover the main aspects of the thermal process and to review developments. In conclusion, it is necessary to consider some of the procedures used in what is today referred to as "good manufacturing practice" (GMP) in terms of legislation and processing guidelines.

10.2. Information Sources

10.2.1. Legislation and Codes of Practice

One of the most important legislative documents relating to the canning industry is the USA Food and Drug Administration document titled *Thermally Processed Low-acid Foods Packaged in Hermetically Sealed Containers* (Regulations 21, Part 113, 1983), which deals with definitions, GMP, establishing scheduled processes, operations in the thermal processing room, instrumentation and control for batch, agitating and hydrostatic retorts, processing and production records, and process deviations. Part 114 of the same regulation relates to acidified foods.

In the UK a code of practice has been established by the industry titled *Guidelines for the Safe Production of Heat Preserved Foods*, published by HMSO in 1994 for the Department of Health. This covers similar topics to the US legislation; it is not mandatory but recommended.

Another legislative document, adopted by the European Commission, is the Codex Alimentarius document Vol. G CAC/RCP23–1978, titled *Recommended International Code of Practice for Low-acid and Acidified Low Acid Canned Foods*, published in Rome by the World Health Organization.

10.2.2. GMP Guidelines and Recommendations

In the United States, The National Food Processors' Association produces a number of booklets of which the following are important: "Thermal Processes for Low-Acid Canned Foods fn Metal Containers" (Bulletin 26L); "Thermal Processes for Low-Acid Canned Foods in Glass Containers" (Bulletin 30L); "Flexible Package Integrity Bulletin" (Bulletin 41L) and "Guidelines For Validation Of Automatic Retort Control" (Bulletin 43L 2nd Ed. 2002).

The American Society for Testing Materials has produced a useful guide to the flexible packages: Designation: F 1168–88 "Standard Guide For Use In The Establishment Of Thermal Processes For Foods Packaged In Flexible Containers" (F1168–88).

A major reference work, published annually, for US legislation and standards is *The Almanac of the Canning, Freezing Preserving Industries*, published by Edward E. Judge & Sons, Inc., Westminster, MD.

The United Kingdom has developed a considerable number of GMP guidelines, many of which have already been mentioned. These include a range of documents from Campden and Chorleywood Food Research Association, in particular, the following Technical Manuals: "Canning Retorts and Their Operation" (No. 2); "Guidelines for the Establishment of Sheduled Processes for Low-Acid Foods" (No. 3); "Guidelines on GMP for Sterilisable Flexible Packaging Operations for Low-Acid Foods" (No. 4); "Process Control in Hydrostatic Cookers" (No. 5); "The Heat Processing Of Uncured Canned Meat Products" (No. 6); "Guidelines for the Establishment of Procedures for the Inspection of Canneries" (No. 12); "Operation of Batch Retort Systems–Fully Water Immersion, Raining Water and Steam/Air" (No.13); "Process Control in Reel and Spiral Cooker/Coolers" (No. 26); "Food Pasteurization Treatments" (No. 27); "The Processing of Canned Fruits and Vegetables" (No. 29); "The Shelf Stable Packaging of Thermally Processed Foods in Semi-Rigid Plastic Barrier Containers" (No 31); and "HACCP: A Practical Guide' (No. 38).

10.2.3. Technical Training

The technical complexity of the canning operation demands that all types of operatives are thoroughly trained in the aspects of the subject to the appropriate level. This is particularly true as new methods of processing and packaging develop. The training is carried out at appropriate institutions that have close connections with the industry. At the present time courses, cover factory operations, canning principles and thermal process calculations. Most institutions produce documentation for their courses of instruction, which is generally confidential to the recipients. However, a number of general training documents are available from US sources. The Food Processors' Institute publishes a joint NFPA/FPI document *Canned Foods – Principles of Thermal Process Control, Acidification and Container Closure Evaluation*, the latest edition being the 5th (1989). The Institute of Food Technologists publishes a course book titled *Introduction to the Fundamentals of Thermal Processing*. Finally, the Environmental Sterilization

Laboratory publishes I. J. Pflug's *Textbook for an Introductory Course in the Microbiology and Engineering of Sterilization Processes* (6th edition, 1987).

10.3. Some Techniques for the Implementation of GMP

10.3.1. HACCP Techniques

The technique known as hazard analysis critical control points (HACCP) involves several stages:

(a) Hazard identification and analysis;
(b) Identification of critical control points (CCPs);
(c) The establishment of CCP criteria;
(d) The monitoring of CCPs;
(e) The development of protocols for CCP deviations and verification.

The basis for HACCP is hazard analysis, and this deals with all aspects of the manufacturing and distribution operations. It includes the determination of the hazards and assessment of risks and the effects of each hazard as determined by using an engineering flow sheet of the operations. The CCPs are then selected on the basis of the degree of risk to which the consumer would be exposed, and where a breakdown in control would lead to a consumer being subjected to risk. The number of CCPs should be about six: more would indicate an increase in the degree of complexity of the process, and less would indicate that something may have been missed. The choice is critical to the success of the operation of the system. Each CCP should be capable of being effectively monitored, preferably in real time; this method replaces end-of-line sampling. This is not always possible with microbiological techniques, which take time to deliver results. A monitoring system should include careful recording to identify deviations and their correction. The protocols that must be followed when deviations occur must then be established, and plans implemented to correct the system. Verification should also be included, since this is often required by regulatory agencies or process audit investigators. The stages of HACCP analysis are as follows. First, the details of the process, factory, operation and management routines are obtained and clearly displayed. Second, the essential characteristics of the product are obtained and analysis made of the requirements to safeguard the product at various stages in the operation. Third, all the process operations are examined for process deviations and then the CCPs can be identified. To determine whether a control point is critical or not, two parameters are used to determine the "severity" of the process, i.e. the magnitude of the potential hazard and an estimate of the likely occurrence of the hazard and the risk. Risk analysis may be achieved by one of three methods: probabilistic, comparative or pragmatic.

The probabilistic method of risk analysis involves an assessment of the probability of an event occurring, e.g. a failure. It is also necessary to be able to define the chance accurately; this can be done with machinery failures if performance records are kept. The probabilistic concept is embodied in the classical 1 in 10^{12} concept for the survival of spores of *Clostridium botulinum*, used in the canning of low-acid foods.

In cases where information on risks is not available, e.g. in the case of a new product formulation or a new process, it is possible to use the comparative method, comparing the product with related products that have proved by long experience to be acceptable. The related property must show a relationship or accepted process criteria, e.g. pH or water activity.

The pragmatic method is used when information on the process is not available; it relies on expert judgement as to whether a risk is acceptable or not. Stage 3 requires a high degree of expertise and experience and should only be carried out by suitably qualified persons.

In the fourth stage the various control options are decided. The education and training element is particularly important. Monitoring procedures can also involve visual inspection without the use of measurements, and the observations should be carefully recorded using suitable checklists. These should show details of the location of the CCP, the monitoring procedures, the frequency of monitoring, and satisfactory compliance criteria.

Further details of the basic aspects of the technique are given in the following texts and documents: Bauman (1974), NRC (1985), CCFRA (1987), ICMSF (1988), CCFRA (1989), Buchanan (1990), USDA (1990), and Williams (2004).

Another technique, process analysis critical control points (PACCP), extends the former concepts to include mechanical, physical, chemical and contamination aspects (Peri 1991).

10.3.2. Process Audits

An audit is defined in BS 4778/ISO 8402 (1986) as "a systematic and independent examination to determine whether quality activities and related results comply with planned arrangements and whether these arrangements are implemented effectively and are suitable to achieve objectives."

Factory audits involve investigating the operation of the factory to ensure the effectiveness of the factory system. It is an objective method which uses comparison with recognized standards, and it can be used internally as well as with factory customers. The system requires auditors free from any pressures that may bias the report.

10.4. Aspects of GMP

10.4.1. Identification of Critical Factors

While the principles of HACCP apply to the whole process, it is important in relation to this text to look at the components of the process to which this technique may be applied. The delivery of the scheduled process is paramount in ensuring the safety of the processed packaged food. It is therefore important to consider all relevant factors in assessing the relative importance and the risks involved:

1. Food product characteristics including:
   (a) Formulation of components and weight distribution;
   (b) Filled weight and volume;
   (c) Consistency/viscosity of sauce or covering liquid;
   (d) Drained weight before process;
   (e) Size of particulates;
   (f) Size of conglomerate;
   (g) pH, a_w and other physical properties;
   (h) Location of slowest point of heating in the product.
2. Microbiological load on the product: This is what the process sets out to inactivate and eliminate; consequently it must be a major factor. The microbiological status of all the ingredients must be assessed, especially when new product formulations are being developed. Delays in handling the products and the filled cans must be evaluated for increased microbiological load. While the conditions of handling material in factories have greatly improved over the last four decades, and on-site analytical facilities have become available and more sophisticated, it is still necessary to consider this as a factor of high priority. The scheduled process must be capable of achieving a shelf-stable product.
3. Headspace and filling control: The headspace in cans being retorted should be kept constant and at a predetermined value. This is especially critical in cans that will undergo an agitated process, the severity of which is determined by the motion of the headspace bubble. The effect of deviations in the filling process should be noted.
4. For steam retorting in batch retorts, the following are critical to the delivery of the scheduled process:
   (a) Correct venting schedule as predetermined for can stacking pattern;
   (b) Correct process schedule for appropriate container/product;
   (c) Correct steam temperature;
   (d) Correct time;
   (e) Cooling water temperature;
   (f) Appropriate cooling time;
   (g) Correct functioning of controller/valves;
   (h) Steam distributor unblocked and adequate condensate removal facilities. Other factors such as compressed air supply, chlorinated cooling water and can handling operations are important factors for the integrity of the container and product.
5. Other heating media and batch retorts: steam–air mixtures and hot water. In addition to the points raised above, it is essential that the correct steam–air mixture at the scheduled temperature is used, and that in the case of both heating media, the flow rates, directions of flow and flow distribution are correctly maintained.

6. Agitating batch retorts: In addition to points 3–5 above it is essential to maintain the rate of rotation at the prescribed rate.
7. For continuous reel-and-spiral pressure cooker/coolers, all the above factors are appropriate. An important point is the prevention of excessive condensate build-up due to incorrect operation of steam condensate valves, so that the cans passing through the lower part of the cooker are not immersed in water.
8. For hydrostatic cooker/coolers, again all the above points are relevant, with the additional requirement of having the legs at the correct levels and maintaining the correct speed of the carrier bars.

10.4.2. Process Deviations

In the event of a process being incorrectly performed due to, for example, steam supply failure, thermometer inaccuracies, time curtailment or power supply interruption, it is necessary to determine what course of action is necessary. This generally requires expert judgement, following reassessment of the F-value achieved by the deviant process. Studies range from deviant processes (Houtzer & Hill 1977); to conduction-heating packs (Giannoni-Succar & Hayakawa 1982); to convection-heating packs, (McConnell 1952; Huang & Hayakawa 1983); to convection-heating packs with agitation (Denys *et al.* 1996): and to broken-heating packs (Stroup 1982).

Tucker *et al.* (1996) made a study of the effect of deviant processes on the sterilization of a number of canned foods (conduction heating – beans in tomato sauce, broken heating – cream of chicken soup and convection heating – peas in brine) using the *CTemp* model see page 24 (first edition). This work showed that both the conduction model and the *CTemp* model were suitable for predicting heating profiles for the canned foods.

Chen and Ramaswamy (2003) studied a range of critical processes variables, e.g., retort temperature (RT), initial temperature T_i, cooling water temperature T_w, f_h, j_h and j_c, using article neural networks. The results showed that for the process lethality F, $RT > f_h \cdot j_h \cdot T_i T_w > T_i \cdot RT \; T_i > T_i f_h > RT \; Ti > j_c RT j_h$; for process time PT, $RT > f_h \gg j_h > T_i > j_c > T_i T_w$; for cooling time CT, $j_c > T_w > f_h$ and for Total process time TT, $RT > f_h. > j_h > j_c > T_w > T_i > T_i j_c > T_i T_w$.

With the development of computer control this will ultimately be done automatically (see Chapter 9).

10.5. Thermal Process Validation

10.5.1. Process Establishment

The first stage in thermal process validation is to consider process evaluation. Some of the major factors that need to be considered in the establishment of scheduled processes have already been listed in Section 10.4. The identification of retorting and product preparation conditions is necessary in order to determine the likely extremes that will occur during production. These are known as "worst case" conditions and must be evaluated to establish the ultimate safety of the process. For heat penetration determinations, these include product formulation and properties; initial temperature; container size and shape; thermocouple location and point of slowest heating in the product. For retorting operations they include container stacking; temperature distribution; slowest heating location in retort; retort control conditions; thermal load; and process service supplies (steam, water and air pressure) (Tucker 2001).

10.5.2. Lethality Assurance

This is the second stage in process validation. It requires comparison of the scheduled process as established by experimental trials and the delivered process in the production environment. The various factors to be considered include raw material specification and monitoring to the standard required, monitoring the food preparation stages, and establishing the correct thermal process conditions.

The time–temperature conditions should be recorded on a chart recorder for future reference, as well as for process operation, and a manual record of the time and temperature of the master temperature indicator should be kept. The importance of calibration of the temperature indicator against a known traceable standard is paramount.

It is important to ensure in batch operations that each retort crate of cans has been processed. This is done by using either striped retort tape, the stripes going black/brown when the tape has been heated, or Cook-Chex indicators, cardboard tags impregnated with a purple indicator ink that turns green when exposed to wet steam heating (no reaction to dry heating). These indicators can also be altered to have an F_0 range from 2 to 90 min, and by selecting a particular value the process can be controlled to a particular value.

10.5.3. Records

It is necessary to keep full and accurate records of the sterilization history of any production batch or run as part of the quality assurance routine. These records should contain details of all the critical factors concerned with the scheduled process time and temperature, product and production details and container.

The Food and Drug Administration of the United States has an elaborate system of process filing which requires complete information about the equipment, installation, and product, packaging and process details, including critical factors and scheduled process. Electronic records and signatures are the legal equivalents of paper records and handwritten signatures and are required to comply with *FDA Guidance Document 21* CFR Part 11. (www.fda.gov/ora/compliance_ref/part11) (Spinak 2002).

Cook-room audits should also be performed at regular intervals, so that there is an independent confirmation of safety assurance. A periodic survey or compliance audit should be made of equipment, procedures and personnel training. For further details see NFPA/FPI (1989), Shapton (1986), Shapton and Shapton (1991), and BSI (1987).

References

Bauman, H. E. (1974). The HACCP concept and microbiological hazard categories. *Food Technol., 35*, 78–87.

BSI (1987). *Quality assurance – Handbook No. 22*, London: British Standards Institution. [This contains BS 5760 *Quality systems*: ISO 9000–9004, and other related standards].

Buchanan, R. L. (1990). HACCP: A re-emerging approach to food safety. *Trends in Food Science and Technology, 1*(5), 104–106.

CCFRA (1987). Guidelines to the establishment of hazard analysis critical control point. *Technical Manual No. 19*. Chipping Campden, Glos., UK: Campden & Chorleywood Food Research Association.

CCFRA (1989). the Microbiological Aspects of Commissioning and Operating Production Processes. *Technical Manual No. 24*. Chipping Campden, Glos., UK: Campden & Chorleywood Food Research Association.

Chen, C. R., & Rhamaswamy, H. S. (2003). Analysis of critical control points in deviant thermal processes using artificial neural networks. *J. Food Eng., 57*(3), 225–236.

Denys, S., Noronha, J., Stoforos, N. G., Hendrickx, M., & Tobback, P. (1996). Evaluation of process deviations, consisting of drops in rotational speed, during thermal processing of foods in rotary water cascading retorts. *J. Food Eng., 30*(3/4), 327–338.

Giannoni-Succar, E. B., & Hayakawa, K. (1982). Correction factors of deviant thermal processes applied to packaged heat conduction food. *J. Food Sci., 47*, 642–646.

Houtzer, R. L., & Hill, R. C. (1977). Effect of temperature deviations on process sterilization value with continuous agitating retorts. *J. Food Sci., 42*(3), 755–777.

Huang, F., & Hayakawa, K. (1983). A deviant heat process applied to canned or packaged liquid food. *J. Food Sci., 48*, 1229–1234, 1241.

ICMSF (1988). Application of the hazard analysis critical control point (HACCP). System to ensure microbiological safety and quality. In *Microorganisms in Food 4*, Oxford: Blackwell Scientific Publications.

McConnell, J. E. W. (1952). Effect of a drop in retort temperature upon the lethality of processes for convection heating products. *Food Technol., 6*(2), 76–78.

NFPA/FPI (1989). *Canned foods – Principles of thermal process control, acidification and container closure evaluation* (5th Edition). Washington, DC: The Food Processors' Institute.

NRC (1985). *An evaluation of the role of microbiological criteria for foods and food ingredients*. Washington, DC: Food Protection Committee, National Research Council.

Peri, C. (1991). PACCP: process analysis critical control point in food technology. *Ital. J. Food Sci., 3*(1), 5–10.

Shapton D. A. (1986). Canned and bottled food products (soups, mayonnaise and sauces). In S. M. Herschdoerfer (Ed.), *Quality control in the food industry Vol. 3* (pp. 261–322). London: Academic Press.

Shapton, D. A., & Shapton, N. F. (Eds.) (1991). *Principles and practices for the safe processing of foods*. Oxford: Butterworth-Heinemann.

Spinak, S. (2002). FDA approaches to process filing. In G. S. Tucker (Ed.). *Third International symposium on thermal processing – Process and package innovation for convenience foods* (Session 3:1). Chipping Campden, Glos., UK: Campden & Chorleywood Food Research Association.

Stroup, W. H. (1982). Compensating for temperature drops in still retorts. *J. Food Protection*, *45*(9), 1022–1027.

Tucker, G. S. (2001). Validation of heat processes. In P. Richardson (Ed.), *Thermal technologies in food processing* (pp. 75–90). Cambridge: Woodhead Publishing.

Tucker, G. S., Noronha, J. F., & Heydon, C. J. (1996). Experimental validation of mathematical procedures for the evaluation of thermal process deviations during the sterilization of canned foods. *Food and Bioproducts Processing, Trans. IChemE., 74*C3, 140–148.

USDA (1990). *Hazard analysis and critical control points system.* Washington, DC: National Advisory Committee on Microbiological Criteria, US Dept of Agriculture – Food Safety and Inspection Service.

Williams, A. (2004). Role of HACCP in managing thermal processing risks. In G. S. Tucker (Ed.), *Third international symposium on thermal processing – Process and package innovation for convenience foods.* (Session 3:2). Chipping Campden, Glos., UK: Campden & Chorleywood Food Research Association.

Appendix A: Kinetic Factors for Microbial Inactivation

TABLE A.1. Kinetic factors: microbial inactivation by moist heat.

Organism	Temperature range, T (°C)	pH	T_{ref} (°C)	Reaction rate k_T (s^{-1})	Activation energy, E_A (kJ mol^{-1})	Frequency factor, log A (s^{-1})	Thermal destruction rate, D_T (s)	z-value (°C)	Reference (see end of Appendix A)
Bacillus									
stearothermophilus									
TH 24 aqueous	120–160	—	120	2.303×10^{-3}	423	53.7	1000	7.3	Davies *et al.* (1977)
TH 24 aqueous	120–160	—	160	10.49	513	63.0	0.22	7.3	Perkin *et al.* (1977)
TH 24 milk	120–160	—	120	4.92×10^{-3}	276	345	468	11.2	
TH 24 milk	120–160	—	160	7.19	335	41.3	0.32	11.2	
FS 7954 phosphate buf.	127–144	7	121	9.38	374	49.2	6.0	8.3	Wang *et al.* (1964)
FS 7954 phosphate buf.	121	6.9	121	0.17	383	50.4	13.2	8.1	Scharer (1965)
NCIB 8919 water	121	—	121	0.01	443	56.8	186	7.0	Briggs (1966)
ATCC 7953 water	121	—	121	0.068	384	49.8	33.6	8.0	Miller and Kandler (1967)
ATCC 7953 phos. buf.	111–125	7	121	0.018	365	46.7	126	8.5	Jonsson *et al.* (1977)
NCIB 8710 phos. buf.	100–140	7	121	0.01	256	32	210	12.1	Navani *et al.* (1970)
NCIB 8919 phos. buf.	100–140	7	121	0.015	217	27	149	14.25	Navani *et al.* (1970)
NCIB 8923 phos. buf.	100–140	7	121	0.13	252	31.5	170	12.3	Navani *et al.* (1970)
NCIB 8924 phos. buf.	100–140	7	121	0.01	265	33.2	226	11.7	Navani *et al.* (1970)
– glucose sol.	120–127		121	0.012	478	61.5	189	6.5	Reichert (1979)
IAM 1053 phos. buf.	120		120	6.1×10^{-3}	374	47.6	378	8.3	Matsuda *et al.* (1981)
– heat-activated	100–145		120	0.23	485	63.9	10	6.4	Srimani *et al.* (1990)
C 953 aqueous	143–152		143	1.95	342	43.3	1.18	10.1	Dodeja *et al.* (1990)
C 953 aqueous	143–152		145	3.07	345	43.6	0.75		Dodeja *et al.* (1990)
C 953 aqueous	143–152		150	9.20	354	44.7	0.25		Dodeja *et al.* (1990)

(continue)

TABLE A.1. (Continued)

Organism	Temperature range, T (°C)	pH	T_{ref} (°C)	Reaction rate k_T (s^{-1})	Activation energy, E_A (kJ mol^{-1})	Frequency factor, $\log A$ (s^{-1})	Thermal destruction rate, D_T (s)	z-value (°C)	Reference (see end of Appendix A)
C 953 aqueous	143–152	—	152	11.5	358	45.1	0.20	—	Dodeja et al. (1990)
TH 24 (NCDO 1096)	115–155	—	121	0.26	333	43.6	8.7	9.3	David and Merson (1990)
FS 122A (NCIB 8919)									
80% pea purée/water	121	—	121	5.3×10^{-3}	—	—	435	—	Atherton and Brown (1973)
20% pea purée/water	121	—	121	5.0×10^{-3}	—	—	462	—	Brown (1974)
100% pea purée/water	121	—	121	8.0×10^{-3}	—	—	282	—	Brown (1974)
FS 128 buffer	110–135	7	110	2.3×10^{-4}	386	49	10^4	7.7	Brown (1974)
FS 128 buffer	110–135	7	135	0.2	—	—	11.4	—	Brown (1974)
FS 128 buffer	104.5–138	6	104.5	8.3×10^{-5}	343	43.5	2.5×10^4	8.7	Brown (1975)
FS 128 buffer	104.5–138	6	138	0.265	—	—	8.7	—	Brown (1975)
FS 128 buffer	110–138	7	110	9.0×10^{-4}	294	37	2.5×10^3	10.25	Brown (1975)
FS 128 buffer	110–138	7	138	0.29	—	—	7.92	—	Brown (1975)
NCA 1518 skim milk	128–140	—	128	0.17	436	56	134	7.8	Busta (1967)
NCA 1518 skim milk	128–140	—	140	4.6	—	—	0.5	—	Busta (1967)
NCA 1518 milk conc.	120–128	—	—	—	255	—	—	12.16	Segner et al. (1963)
FS 210 (TH 24) phosph.	110–130	7	120	8.0×10^{-3}	302	38	288	10.26	Gaze et al. (1990)
FS 210 (TH 24) phosph.	110–130	7	130	0.064	—	—	36	—	Gaze et al. (1990)
FS 218 (NCIB 8919) phosph.	110–130	7	120	6.2×10^{-3}	370	47	385	8.4	Gaze et al. (1990)
FS 218 (NCIB 8919) phosph.	110–130	7	130	0.076	—	—	30	—	Gaze et al. (1990)
FS 219 (NCA 1518) phosph.	110–130	7	120	4.3×10^{-3}	387.5	49.2	533	8.0	Gaze et al. (1990)
FS 219 (NCA 1518) phosph.	110–130	7	130	0.082	—	—	26.4	—	Gaze et al. (1990)
NCA 1518 water	120–130	—	121	8.2×10^{-3}	361	45.8	281	8.6	Neaves and Jarvis (1978a)

										Reference
FD 7954 water	105–120	—	287.5	0.017	121	36.4	138.2	10.43		Aiba et al. (1965)
ATCC 7553 water	105–130	—	—	0.038	121	—	60	—		Abraham et al. (1990)
Bacillus subtilis										
5230 aqueous	105–132	6.6	374	0.038	121	48.2	6.0	8.3		Jacobs et al. (1973)
5230 aqueous	100–110	—	367	3.3×10^{-3}	100	49	698	7.6		Briggs (1966)
5230 aqueous	124–140	—	220	0.076	121	28	30.3	14.1		Fox and Pflug (1968)
– Phosphate buffer	127–144	7	350	0.105	121	45.5	21.9	8.8		Wang et al. (1964)
– Aqueous	124–130	—	330	0.08	121	42.7	28.8	9.4		Jacobs et al. (1965)
9372 aqueous	100–148	—	839	2.3×10^{-3}	120	109	1×10^{-3}	3.7		Srimani et al. (1990)
A skim milk	112.5–135	—	352	0.025	112.5	46.7	20	8.8		Edwards et al. (1965)
A skim milk	112.5–135	—	—	3.29	135	—	0.7	—		Edwards et al. (1965)
Clostridium beijerinckii										
NCIB 9362 phosphate	80–97	<4.5	330	0.063	90	46.3	36	8.0		Gaze et al. (1987)
Clostridium botulinum										
Type C aqueous	93–104	—	507	0.038	104	69	60	5.6		Segner and Schmidt (1971)
Type A aqueous	115.6–121	—	310	1.9	121	41.5	1.2	10		Schmidt (1964)
NCTC 7272	115–135	—	86	0.72	121	11.3	3.2	36.1		Neaves and Jarvis (1978b)
NCTC 7273	115–135	—	120	0.72	121	15.8	3.2	26.0		Neaves and Jarvis (1978b)
BOT 44 phosphate buffer	100–120	—	378	0.68	120	50.2	3.36	8.2		Gaze et al. (1990)
BOT 44 phosphate buffer	101–150	—	310	1.53	120	40	1.5	9.8		Simpson and Williams (1974)
213B phosphate buffer	104–116	—	376	0.027	110	49.8	85.3	7.8		Kaplan et al. (1974)
213B beets	104–116	—	264	0.035	110	34.6	65.8	11.1		Kaplan et al. (1974)
213B carrots	104–116	—	259	0.016	107	33.86	143.9	11.3		Kaplan et al. (1974)
213B corn	104–116	—	264	0.025	110	34.5	92.1	11.1		Kaplan et al. (1974)
213B snap beans	104–113	—	333	0.038	110	44.0	60.6	8.8		Kaplan et al. (1974)
213B spinach	107–110	—	302	0.026	110	39.6	88.5	9.7		Kaplan et al. (1974)
213B sea food	100–113	5.9	367	0.76	121	48.6	3.0	7.4		Perkin et al. (1975)

(continue)

TABLE A.1. (Continued)

Organism	Temperature range, T (°C)	pH	T_{ref} (°C)	Reaction rate k_T (s^{-1})	Activation energy, E_A (kJ mol^{-1})	Frequency factor, log A (s^{-1})	Thermal destruction rate, D_T (s)	z-value (°C)	Reference (see end of Appendix A)
213B poultry	100–113	—	121	0.76	367	48.6	3.0	7.4	Perkin et al. (1975)
213B meat and vegetable	100–113	—	121	0.34	384	50.5	6.6	9.8	Perkin et al. (1975)
213B phosphate buffer	120–140	7	120	0.26	310	39.7	8.75	10	Gaze and Brown (1988)
213B phosphate buffer	120–140	7	140	14.4	310	40.0	0.16	10	Gaze and Brown (1988)
213B corn purée	110–115.5	7	115.5	4.7×10^{-3}	300	38.0	489	10.1	Sognefast et al. (1948)
213B pea purée	107–112.7	7.2	115.5	3.4×10^{-3}	364	46.54	669	8.3	Sognefast et al. (1948)
213B spinach purée	110–118.3	8.6	115.5	4.2×10^{-3}	293	37.0	549	10.3	Sognefast et al. (1948)
213B 1% fat milk	104–113	7	113	0.16	351	46.8	14.3	8.5	Denny et al. (1980)
213B neutral phosphate	104–113	7	113	0.28	260	34.7	8.2	11.4	Denny et al. (1980)
213B phosphate buffer	110	7	110	0.024	284	37.2	96	10.3	Matsuda et al. (1981)
62A macaroni creole	110–118	4	110	0.078	306	40.7	29.5	9.6	Xezones and Hutchings (1965)
62A macaroni creole	110–118	7	110	0.015	313	—	153.5	—	Xezones and Hutchings (1965)
62A Spanish rice	110–118	4	110	0.088	306	40.75	26.2	9.6	Xezones and Hutchings (1965)
62A Spanish rice	110–118	7	110	0.016	304	—	143.9	—	Xezones and Hutchings (1965)
62A phosphate	104–127	7	121	0.288	357	46.8	7.98	9.0	Stumbo et al. (1950)
62A water	104–127	7	121	0.75	334	44.2	3.06	8.5	Stumbo et al. (1950)
62A puréed peas	104–127	7	121	0.44	374	49.3	5.34	8.3	Stumbo et al. (1950)
62A strained squash	104–127	7	121	0.64	373	49.3	3.6	8.2	Stumbo et al. (1950)
A35B phosphate buffer	105–115.5	6.8	121	0.12	287.5	37.3	19.2	10.8	Knock and Lambrechts (1956)
A35B rock lobster	105–115.5	6.6	121	0.13	287.5	9.18	18.0	10.8	Knock and Lambrechts (1956)

Clostridium butyricum									
NCIB 7423 phosphate	80–97	<4.5	0.038	90	329	46	60	8.1	Gaze et al. (1987)
Strain 1 buffer and milk	85	<4.5	2.74×10^{-3}	85	—	—	840	—	Cerf et al. (1967)
Strain 2 buffer and milk	85	<4.5	3.2×10^{-3}	85	—	—	720	—	Cerf et al. (1967)
Clostridium thermosaccharolyticum									
S9 McIlvaine spore form	99–127	—	0.045	121	211	26.7	51	14.7	Gillespy (1948)
S9 acid spore form	99–127	—	0.012	121	318	40.4	192	9.76	Gillespy (1948)
S9 water	124–132	—	0.523	132	476	61.2	4.4	6.89	Xezones et al. (1965)
S9 molasses	124–132	—	0.698	132	322	41.4	3.3	10.2	Xezones et al. (1965)
Tree bark compost	115.5–127	—	5.6×10^{-4}	121	270	32.7	4080	11.5	Alcock and Brown (1982), Brown and Alcock (1983)
Desulfotomaculum nigrificans									
ATCC 7946 infant food	121–131	—	1.48×10^{-3}	121	463	58.8	1550	6.7	Donnelly and Busta (1980)
RG1 66	121–131	—	7.06×10^{-4}	121	327	40.4	3260	9.5	Donnelly and Busta (1980)
Escherichia coli									
Nutrient broth	56	—	8.6×10^{-3}	56	442	68.2	270	4.9	Chambers et al. (1957)
Aqueous	55	—	9.6×10^{-3}	55	—	—	240	—	Lemcke and White (1959)
Aqueous	57.2	—	0.32	57.2	—	—	72	—	Goepfert et al. (1970)
Aqueous	55–60	—	5.7×10^{-3}	55	604	94.0	402	3.6	Reichert (1979)
Aqueous	55–60	—	0.14	60	—	—	16.2	—	Reichert (1979)
Milk	62–82	—	66.9	82	378	57.5	—	—	Evans et al. (1970)
Milk solution 10% solids	52–62	—	0.03	58	462	71.5	76.8	—	Dega et al. (1972)
Milk solution 51% solids	52–62	—	2.85×10^{-3}	58	245	—	800	—	Dega et al. (1972)
ATCC 9637 raw milk	52–80	—	0.03	57	375	58	76.8	—	Read et al. (1961)
ATCC 9637 chocolate milk	52–80	—	0.015	57	375	57.6	153.5	—	Read et al. (1961)

(continue)

TABLE A.1. (Continued)

Organism	Temperature range, T (°C)	pH	T_{ref} (°C)	Reaction rate k_T (s^{-1})	Activation energy, E_A (kJ mol^{-1})	Frequency factor, log A (s^{-1})	Thermal destruction rate, D_T (s)	z-value (°C)	Reference (see end of Appendix A)
ATCC 9637 40% raw cream	52–80	—	57	0.015	375	57.6	153.5	—	Read et al. (1961)
ATCC 9637 ice cream mix	52–80	—	57	7.5×10^{-3}	371	56.7	307	—	Read et al. (1961)
Putrefactive anaerobe									
PA 3679 white corn purée	110–127	4.5	k-119:D-121	0.01	390	50	230	8.8	Sognefast et al. (1948)
PA 3679 white corn purée	110–127	7.1	k-119:D-121	—	322	—	—	11.1	Sognefast et al. (1948)
PA 3679 yellow corn purée	110–127	5.0	k-119:D-121	7.45×10^{-3}	390	49.8	309	7.8	Sognefast et al. (1948)
PA 3679 yellow corn purée	110–127	6.8	k-119:D-121	—	262	—	—	11.6	Sognefast et al. (1948)
PA 3679 pea purée	110–140	4.5	k-119:D-121	9.25×10^{-3}	305	38.6	249	6.6	Sognefast et al. (1948)
PA 3679 pea purée	110–140	8.0	k-119:D-121	—	—	—	—	10.0	Sognefast et al. (1948)
PA 3679 sweet potato purée	110–127	5.0	k-119:D-121	9.25×10^{-3}	494	63.7	249	7.8	Sognefast et al. (1948)
PA 3679 sweet potato purée	110–127	8.0	k-119:D-121	—	275	—	—	11.7	Sognefast et al. (1948)
PA 3679 pumpkin purée	110–127	5.0	k-119:D-121	9.25×10^{-3}	494	63.7	249	6.1	Sognefast et al. (1948)
PA 3679 pumpkin purée	110–127	9.0	k-119:D-121	—	390	—	—	7.8	Sognefast et al. (1948)
PA 3679 skim milk	115–135	—	115	0.28	180	24	8.0	18.8	Busta (1967)
PA 3679 skim milk	115–135	—	135	2.30	—	—	1.0	—	Busta (1967)
PA 3679 3:1 milk concentrate	120–128	—	—	—	277	—	—	11.2	Segner et al. (1963)
PA 3679 pea purée	115.5–143.3	—	121	0.38	317	40.8	60	9.8	Secrist and Stumbo (1958)

Product									Reference
PA 3679 pea purée	115.5–143.3	—	143.3	3.8	—	—	0.6	—	Secrist and Stumbo (1958)
PA 3679 distilled water	115.5–143.3	—	115.5	0.06	310	40	39	10.4	Secrist and Stumbo (1958)
PA 3679 distilled water	115.5–143.3	—	143.3	—	—	—	—	0.3	Secrist and Stumbo (1958)
PA 3679 phosphate buffer	100–120	—	110	0.44	309	40.8	5.2	9.5	Townsend et al. (1938)
PA 3679 asparagus buffer	100–120	—	110	0.225	309	40.5	10.2	9.4	Townsend et al. (1938)
PA 3679 phosphate buffer	77–157	7.0	121	0.038	113.5	13.6	60	26.4	Alcock et al. (1981)
PA 3679 aqueous	110–132.2	—	121	0.048	312	40.3	48	9.8	Stumbo et al. (1950)
PA 3679 pea purée	110–132.2	—	121	0.025	317	40.6	91.2	9.6	Stumbo et al. (1950)
PA 3679 strained spinach	110–132.2	—	121	0.045	302	38.8	51.0	10.1	Stumbo et al. (1950)
PA 3679 whole kernel corn	110–132.2	—	121	0.032	330	42.5	70.8	9.2	Stumbo et al. (1950)
PA 3679 uncooked pork	110–132.2	—	121	0.027	305	39.0	83.4	10.0	Stumbo et al. (1950)
PA 3679 cooked pork	110–132.2	—	121	0.045	302	38.8	92.6	9.7	Stumbo et al. (1950)
PA 3679 white sauce	110–132.2	—	121	0.031	330	42.5	73.8	9.2	Stumbo et al. (1950)
PA 3679 evaporated milk	110–132.2	—	121	0.029	281	35.9	77.4	11.1	Stumbo et al. (1950)
PA 3679 20% cream	110–132.2	—	121	0.017	292	37.3	91.2	10.4	Stumbo et al. (1950)
PA 3679 beef	110–121	—	121	0.017	263	33.3	133.8	11.8	Kaplan et al. (1974)
PA 3679 corn	110–121	—	121	0.019	282	35.8	120.0	11.0	Kaplan et al. (1974)
PA 3679 lima beans	110–121	—	121	0.015	290	36.8	153.6	10.72	Kaplan et al. (1974)
PA 3679 mushrooms	110–121	—	121	0.019	335	43.0	118.8	9.27	Kaplan et al. (1974)
PA 3679 peas	110–121	—	121	0.017	314	40.0	133.8	9.88	Kaplan et al. (1974)
PA 3679 pork	110–121	—	121	0.017	310	39.5	128.4	10.0	Kaplan et al. (1974)
PA 3679 asparagus	110–127	5	121	9.4×10^{-3}	543	70.3	245.0	5.5	Sognefast et al. (1948)
PA 3679 asparagus	110–127	9	121	—	342	—	—	8.8	Sognefast et al. (1948)
PA 3679 green bean purée	110–135	5	121	9.4×10^{-3}	419	53.8	245.0	7.3	Sognefast et al. (1948)
PA 3679 green bean purée	110–135	9	121	—	332	—	—	10.0	Sognefast et al. (1948)
PA 3679 beet purée	110–138	5.5	121	9.4×10^{-3}	336	42.7	245.0	10.0	Sognefast et al. (1948)
PA 3679 carrot purée	110–127	9	121	9.4×10^{-3}	342	43.5	245.0	8.8	Sognefast et al. (1948)

(continue)

TABLE A.1. (Continued)

Organism	Temperature range, T (°C)	pH	T_{ref} (°C)	Reaction rate k_T (s^{-1})	Activation energy, E_A (kJ mol^{-1})	Frequency factor, log A (s^{-1})	Thermal destruction rate, D_T (s)	z-value (°C)	Reference (see end of Appendix A)
PA 449 phosphate buffer (NCIB 8053 Cl. sporogenes)	100–120	7.0	120	0.15	340	44.5	15	9.11	Gaze et al. (1990)
PA 3679 pea purée	115 & 121	6.13	121				380	10.41	Körmendy and Mohácsi-Farkas (2000).
PA 3679	121.0	6.7–4.65	121.0				40.2–43.2	10.16–10.26	Ocio et al. (2000)
PA ATTC 7533	121.1		121.1				240		Le Jean et al. (1994).

References

Abraham, G., Debray, E., Candau, Y., & Piar, G. (1990). Mathematical model of thermal destruction of *Bacillus stearothermophilus* spores. *Applied and Environ. Microbiol.*, *56*(10), 3073–3080.

Aiba, S., Humphrey, A. E., & Millis, N. (1965). *Biochemical Engineering.* New York: Academic Press.

Alcock, S. T., & Brown, K. L. (1982). Spore studies in relation to heat processed foods – growth and heat resistance of *Clostridium sporogenes* and extremely heat resistant strain of *Clostridium thermosaccharolyticum. Technical Memorandum No. 310*, Chipping Campden, Glos., UK: Campden & Chorleywood Food Research Association.

Alcock, S. T., Brown, K. L., & Thorpe, R. H. (1981). Spore studies. Sporulation of PA 3679 and heat resistance of PA 3679 and mesophilic bacilli. *Technical Memorandum No. 280.* Chipping Campden, Glos., UK: Campden and Chorleywood Food Research Association.

Atherton, D., & Brown, K. L. (1973). Aseptic packaging of vegetable products. In *1st AMDEC Progress report* (pp. 9–13). Chipping Campden, Glos., UK: Campden & Chorleywood Food Research Association.

Briggs, A. (1966). The resistance of spores of the genus *Bacillus* to phenol, heat and radiation. *J. Appl. Bact.*, *29*(3), 490–504.

Brown, K. L. (1974). Aseptic packaging of vegetable products. In *2nd AMDEC Progress report* (pp. 2–23). Chipping Campden, Glos., UK: Campden & Chorleywood Food Research Association.

Brown, K. L. (1975). Aseptic packaging of vegetable products. In *3rd AMDEC Progress report* (pp. 5–12). Chipping Campden, Glos., UK: Campden & Chorleywood Food Research Association.

Brown, K. L., & Alcock, S. J. (1983). Spore studies in relation to heat processed foods. *Technical Memorandum No. 344.* Chipping Campden, Glos., UK: Campden & Chorleywood Food Research Association.

Busta, F. F. (1967). Thermal inactivation characteristics of bacterial spores at high temperatures. *Appl. Microbiol.*, *15*(3), 640–645.

Cerf, O., Bergere, J. L., & Hermier, J. (1967). Thermal resistance of spores of *Clostridium tyrobutyricum* and *Clostridium butyricum. J. Dairy Res.*, *34*, 221–229.

Chambers, C. W., Tabak, H. H., & Kabler, P. W. (1957). Effect of Kreb's cycle metabolites on the viability of *Escherichia coli* treated with heat and chlorine. *J. Bact.*, *73*, 77–84.

David, J. R. D., & Merson, R. L. (1990). Kinetic parameters for inactivation of *Bacillus stearothermophilus* at high temperatures. *J. Food Sci.*, *55*(2), 488–495.

Davies, F. L., Underwood, H. M., Perkin, A. G., & Burton, H. (1977). Thermal death kinetics of *Bacillus stearothermophilus* spores at ultra high temperatures. I. Laboratory determination of temperature coefficients. *J. Food Technol.*, *12*(2)., 115–129.

Dega, C. A., Goepfert, J. M., & Admundson, C. H. (1972). Heat resistance of salmonellae in concentrated milk. *Appl. Microbiol.*, *23*, 415–420.

Denny, C. B., Shafer, B., & Ito, K. (1980). Inactivation of bacterial spores in products and on container surfaces. In *Proc. int. conf. on UHT processing* (pp. 82–88). Raleigh, NC.

Dodeja, A. K., Sarma, S. C., & Abichandani, H. (1990). Thermal death kinetics of *B. stearothermophilus* in thin film heat exchanger. *J. Food Process. Preserv.*, *14*, 221–230.

Donnelly, L. S., & Busta, F. F. (1980). Heat resistance of *Desulfotomachulum nigrificans* spores in soy protein infant food preparations. *Appl. Environ. Microbiol.*, *40*(4), 721–725.

Edwards, J. L. Jr, Busta, F. F., & Speck, M. L. (1965). Thermal inactivation characteristics of *Bacillus subtilis* spores at high temperatures. *Appl. Microbiol.*, *13*, 851–857.

Evans, D. A., Hankinson, D. G., & Litsky, W. (1970). Heat resistance of certain pathogenic bacteria in milk using a commercial plate heat exchanger. *J. Dairy Sci.*, *53*(2), 1659–1665.

Fox, K., & Pflug, I. J. (1968). Effect of temperature and gas velocity on dry heat destruction rate of bacterial spores. *Appl. Microbiol.*, *16*(2), 343–348.

Gaze, J. E., & Brown, K. L. (1988). The heat resistance of spores of *Clostridium botulinum* 213B over the temperature range 120–140 deg C. *Int. J. Food Sci. Technol.*, *23*, 373–378.

Gaze, J. E., Brown, K .L., Brown, G., & Stringer, M. F. (1987). Construction of a thermoresistometer and studies on the heat resistance of bacterial spores. *Technical Memorandum No. 568*. Chipping Campden, Glos., UK: Campden & Chorleywood Food Research Association.

Gaze, J. E., Brown, G. D., & Brown, K. L. (1990). Comparative heat resistance studies on spores of *Clostridium botulinum, Clostridium sporogenes* and *Bacillus stearothermophilus*, thermoresistometer studies and Bacillus data file. *Technical Memorandum No. 588*. Chipping Campden, Glos., UK: Campden & Chorleywood Food Research Association.

Gillespy, T. G. (1948). The heat resistance of spores of theromophilic bacteria. In *Annual Report 1948* (pp. 34–43). Chipping Campden, Glos., UK: Campden & Chorleywood Food Research Association.

Goepfert, J. M., Iskander, I. K., & Admundson, C. H. (1970). Relation of the heat resistance of salmonallae to the water activity of the environment. *Appl. Microbiol.*, 19, 429–438.

Jacobs, R. A., Nicholas, R. C., & Pflug, I. J. (1965). Heat resistance of *Bacillus subtilis* in atmospheres of different water content. *Quart. Bull. Michigan Agric. Expt Station*, *48*(2), 238–246. East Lancing, MI: Michigan State University.

Jacobs, R. A., Kemp, L. L., & Milone, N. A. (1973). High-temperature short-time (HTST). processing of suspensions containing bacterial spores. *J. Food Sci.*, *38*, 168–172.

Jonsson, U., Snygg, B. G., Harnulv, B. G., & Zachrisson, T. (1977). Testing two models for the temperature dependence of the heat of inactivation rate of *Bacillus stearothermophilus* spores. *J. Food Sci.*, *42*(5)., 1251–1252, 1263.

Kaplan, A. M., Reynolds, H., & Lictenstein, H. (1974). Significance of variations in observed shapes of thermal death time curves of putrefactive anaerobes. *Food Res.*, 19, 173–181.

Knock, G. G., & Lambrechts, M. S. J. (1956). A note on the heat resistance of a South African strain of *Clostridium botulinum* type B. *J. Sci. Food Agr.*, *7*(4), 244–248.

Körmendy, I., & Mohácsi-Farkas, C. (2000). Heat treatment and defective units' ratio: surviving active spores in pea puree. *J. Food Eng.*, *45*(4), 225–230.

Le Jean, G., Abraham, G., Debray, E., Candau, Y., & Piar, G. (1994). Kinetics of thermal destruction of Bacillus stearothermophilus spores using a two reaction model. *Food Microbiology, 11*, 229–241.

Lemcke, R. M., & White, H. R. (1959). The heat resistance of *Escherichia coli* from cultures of different ages. *J. Appl. Bact.*, *22*, 193–201.

Matsuda, N., Komaki, M., & Matsunawa, K. (1981). Thermal death characteristics of spores of *Clostridium botulinum* 62A and *Bacillus stearothermophilus* subjected to non-isothermal heat treatment. *J. Food Hygiene Soc. Jap.* 22(2), 125–134.

Miller, J., & Kandler, O. (1967). Temperatur- und zeitabhangigkeit der Sporenabtotung in Bereich der Ultrahocherhitzung. *Milchwissenschaft*, 22, 686–691.

Navani, S. K., Scholfield, J., & Kirby, M. R. (1970). A digital computer program for the statistical analysis of heat resistance data applied to *Bacillus stearothermophilus* spores. *J. Appl. Bact.*, 33(4), 609–620.

Neaves, P., & Jarvis, B. (1978a). Thermal inactivation kinetics of bacterial spores at ultra high temperatures with particular reference to *Clostridium botulinum. Research report No. 280.* Leatherhead, UK: Leatherhead Food International.

Neaves, P., & Jarvis, B. (1978b). Thermal inactivation kinetics of bacterial spores at ultra high temperatures with particular reference to *Clostridium botulinum.* 2nd Report. *Research report No. 286.* Leatherhead, UK: Leatherhead Food International.

Ocio, M. J., Sánchez, T., Fernandez, P. S., Rodrigo, M., & Martinez, A. (2000). Thermal resistance characteristics of PA 3679 in temperature range of 110–121°C as affected by pH type and acidulant and substrate. *Int. J. Food Microbiology,* 22, 239–247.

Perkin, A. G., Burton, H., Underwood, H. M., & Davies, F. L. (1977). Thermal death kinetics of *Bacillus stearothermophilus* spores at high temperatures. II Effect of heating period on experimental results. *J. Food Technol.*, 12(2), 131–148.

Perkin, W. E., Ashton, D. H., & Evancho, G. M. (1975). Influence of the z value of *Clostridium botulinum* on the accuracy of process calculations. *J. Food Sci.*, 40(6), 1189–1192.

Read, R. B. Jr, Schwartz, C., & Litsky, W. (1961). Studies on the thermal destruction of *Escherichi coli* in milk and milk products. *Appl. Microbiol.*, 9(5), 415–418.

Reichert, O. (1979). A new experimental method for the determination of the heat destruction parameters of micro-organisms. *Acta Alimentaria*, 8(2), 131–155.

Scharer, J. M. (1965). *Thermal death behaviour of bacterial spores.* Doctoral thesis, University of Pennsylvania.

Schmidt, C. F. (1964). Spores of *Cl. botulinum* – formation, resistance and germination. In K. H. Lewis and K. Casel Jr (Eds.), *Proc. symposium on botulism* (pp. 69–73). Publication No. 999–FP–1. Washington, DC: Public Health Service.

Secrist, J. L., & Stumbo, C. R. (1958). Some factors influencing thermal resistance values obtained by the thermoresistometer method. *Food Res.*, 23, 51–60.

Segner, W. P., Frazier, W. C., & Calbert, H. E. (1963). Thermal inactivation of heat resistant bacterial spores in milk concentrate at ultra high temperatures. *J. Dairy Sci.*, 46, 891–896.

Segner, W. P., & Schmidt, C. F. (1971). Heat resistance of spores of marine and terrestrial strains of *Clostridium botulinum* type C. *Appl. Microbiol.*, 22, 1030–1036.

Simpson, S. G., & Williams, M. C. (1974). An analysis of high temperature short time sterilization during laminar flow. *J. Food Sci.*, 39, 1047–1054.

Sognefast, P., Hays, G. L., Wheaton, E., & Benjamin, H. A. (1948). Effect of pH on thermal process requirements of canned foods. *Food Res.*, 13(5), 400–416.

Srimani, B., Stahl, R., & Loncin, M. (1990). Death rate of bacterial spores at high temperatures. *Lebensm.-Wiss. u. -Technol.*, 13, 186–189.

Stumbo, C. R., Murphy, J. R., & Cochran, J. (1950). Nature of thermal death time curves for PA 3679 and *Clostridium botulinum. Food Technol.*, 4, 321–326.

Townsend, C. T., Esty, J. R., & Baselt, F. C. (1938). Heat resistance studies on spores of putrefactive anaerobes in relation to determination of safe processes for canned foods. *Food Res.*, 3, 323–345.

Wang, D. I.-C., Scharer, I., & Humphrey, A. E. (1964). Kinetics of thermal death of bacterial spores at elevated temperatures. *Appl. Microbiol.*, *12*(5), 451–455.

Xezones, H., & Hutchings, I. J. (1965). Thermal resistance of *Clostridium botulinum* (62A) as affected by fundamental food constituents. *Food Technol.*, *19*(6), 113–115.

Xezones, H., Segmiller, J. L., & Hutchings, I. J. (1965). Processing requirements for heat tolerant anaerobe. *Food Technol.*, *18*, 1001–1002.

Appendix B: Kinetic Factors for Quality Attributes

TABLE B.1. Kinetic factors: vitamin degradation.

Product	Temperature range, T (°C)	Reference temperature, T_{ref} (°C)	Reaction rate, k_T ($\times 10^3$ s^{-1})	Activation energy, E_A (kJ mol^{-1})	Frequency factor, log A (s^{-1})	Thermal destruction rate, D_T ($\times 10^{-3}$ s)	z-value (°C)	Reference
Vitamin A, beta carotene								
Beef liver purée	103–127	122	0.95	112.6	12.0	2.40	23.0	Wilkinson et al. (1981, 1982)
Carrot juice	104–132	104	0.09	106.4	10.2	23.60	25.5	Hojilla et al. (1985)
Vitamin B$_1$, thiamin								
Range	84–150	–	–	84–130	–	–	23–35	Mauri et al. (1989)
Buffer solution pH 6	109–150	109	0.24	127.0	13.3	9.50	24.0	Feliciotti and Esselen (1957)
	109–150	150	12.0	127.0	13.8	0.20	24.0	Feliciotti and Esselen (1957)
Carrots	109–150	150	2.80	155.4	16.6	0.83	22.0	Feliciotti and Esselen (1957)
Green beans	109–150	150	2.90	155.4	16.5	0.80	22.0	Feliciotti and Esselen (1957)
Spinach	109–150	150	3.80	155.4	16.8	0.61	22.0	Feliciotti and Esselen (1957)
	70–100	100	1.80	41.0	3.0	1.28	67.8	Paulus et al. (1975)
Pea purée	109–150	150	2.76	155.4	16.7	0.83	22.0	Feliciotti and Esselen (1957)
	121.1	121.1	0.23	95.8	9.1	10.0	31.3	Bendix et al. (1951)
	122	122	0.16	112.5	11.2	14.0	26.6	Mulley et al. (1975)
	–	126.6	0.166	100.5	–	–	–	Lenz and Lund (1977)
	103–116	121.1	–	–	–	18.0	30.3	Nasri et al. (1993)
Beef heart	109–150	150	3.62	135.0	14.3	0.64	25.0	Feliciotti and Esselen (1957)
Beef liver	109–150	150	3.76	131.1	13.8	0.59	26.0	Feliciotti and Esselen (1957)
Beef purée	122	122	0.16	112.5	11.2	14.60	26.6	Mulley et al. (1975)
Lamb purée	109–150	150	3.20	135.0	15.4	0.71	25.0	Feliciotti and Esselen (1957)
Pork purée	109–150	150	2.80	135.0	14.15	8.82	25.0	Feliciotti and Esselen (1957)

(cont.)

TABLE B.1. (Continued)

Product	Temperature range, T (°C)	Reference temperature, T_{ref} (°C)	Reaction rate, k_T ($\times 10^3$ s^{-1})	Activation energy, E_A (kJ mol^{-1})	Frequency factor, log A (s^{-1})	Thermal destruction rate, D_T ($\times 10^{-3}$ s)	z-value (°C)	Reference
Pork luncheon meat	100–127	127	0.36	86.0	8.0	6.30	35.0	Greenwood et al. (1944)
	–	126.6	0.64	108.8	–	–	–	Lenz and Lund (1977)
Meat loaf	70.5–98	98	4.18	113.3	11.40	0.55	23.8	Skjöldebr et al. (1983)
Meat purée	120	–	–	118.1	–	–	25.4	Weir (1948)
Salmon	118–130	121	0.39	71.9	6.16	5.76	41.7	Greene (1983)
Milk	–	–	–	100	–	–	av.30.0	Bayoumi and Reuter (1980)
	35–150	120	0.30	100.8	9.80	7.70	29.7	Kessler and Fink (1986)
	120–150	120	0.30	100.4	10.33	7.65	28.8	Horak and Kessler (1981); Horak (1980)
Aqueous solution	110–150	120	0.12	72.0	5.67	19.00	38.0	Ramaswamy et al. (1990)
M/15 phosphate buffer	112–142	140	4.40	127.3	13.80	0.52	25.6	Matsuda et al. (1981)
Vitamin B6								
Cauliflower	106–138	121	0.09	66.7	4.82	24.00	45.00	Navankattusas and Lund (1982)
Pyridoxine hydrochloride:								
Pyridoxie model sol.	105–133	118	0.04	119.7	12.64	54.8	22.0	Gregory and Hiner (1983)
Pyridoxamine	105–133	118	0.11	100.0	7.6	20.9	26.0	Gregory and Hiner (1983)
Pyridoxal	105–133	118	0.15	86.6	7.8	15.0	30.0	Gregory and Hiner (1983)
Folic acid:								
Apple juice	100–140	140	0.23	83.4	6.9	100.0	31.0	Mnkeni and Beveridge (1982)
Pteroyl glutamic acid								
Tomato juice	100–140	140	0.20	83.4	6.86	115.0	31.0	Mnkeni and Beveridge (1982)

Product								Reference
Citrate buffer	100–140	140	0.65	93.8	3.5	3.5	28.0	Mnkeni and Beveridge (1982)
	100–140	140	0.06	81.2	6.60	100.0	35.0	Mnkeni and Beveridge (1982)
	100–140	140	0.06	74.5	5.20	410.0	35.0	Mnkeni and Beveridge (1982)
Pantothenic acid								
Beef purée pH 4.5	118–143	121.1	0.017	83.7	6.34	138.0	35.8	Hamm and Lund (1978)
Pea purée pH 7	118–143	121.1	0.017	154.8	15.70	135.0	18.3	Hamm and Lund (1978)
Buffer pH 4	118–143	121.1	0.165	83.7	7.3	14.0	35.8	Hamm and Lund (1978)
pH 5	118–143	121.1	0.024	92.0	7.60	96.0	32.5	Hamm and Lund (1978)
pH 6	118–143	121.1	0.020	113.0	10.3	117.0	26.5	Hamm and Lund (1978)
Vitamin C, ascorbic acid								
Cupuaçu nectar pH 3.2	60–99	80	0.53	74		17.76	130.0	Vieira et al. (2000)
Grapefruit juice 11.2° brix	61–96	96	0.04	20.8	1.4	8.2	57.5	Saguy et al. (1978c)
62.5° brix	61–69	96	0.03	47.3	3.15			Saguy et al. (1978c)
Grape fruit juice	20–92	85				50.57	44	Alvarado and Viteri (1989)
Lemon juice (Sierra)	20–92	85				68.82	72	Alvarado and Viteri (1989)
Lime juice	20–92	85				44.46	44	Alvarado and Viteri (1989)
Mandarin juice (Sierra)	20–92	85				24.36	47	Alvarado and Viteri (1989)
Mandarin juice (Costa)	20–92	85				32.52	58	Alvarado and Viteri (1989)
Peas	110–132	121.1	0.04	164.5	17.5	50.0	18.2	Lathrop and Leung (1980)
Spinach var. Fruhjahr	70–100	100	2.92	36.4	2.57	25.9	74.4	Paulus et al. (1975)
Spinach var. Herbst	70–100	100	2.16	30.5	1.61	1.07	91.2	Paulus et al. (1975)
Beef, restructured	100–30	121.1	6.7	111.3				Street and Tong (1994)
Model solution:								
Buffer pH 4	30–100	100	0.56	87.0	8.97	4.07	31.9	Huelin (1953)
	110–127	120	0.07	78.6	–	31.2	39.4	Ghazala et al. (1989)

TABLE B.2. Kinetic factors: proteins.

Product	Temperature range, T (°C)	Reference temperature, T_{ref} (°C)	Reaction rate, k_T ($\times 10^3$ s^{-1})	Activation energy, E_A (kJ mol^{-1})	Frequency factor, $\log A$ (s^{-1})	Thermal destruction rate, D_T ($\times 10^{-3}$ s)	z-value (°C)	Reference
Whey protein denaturation								
Milk	70–90	–	–	208–212	–	–	10.6–14.4	Agrawala and Reuter (1979)
Milk	120–160	–	–	98.4	–	–	30–37	Agrawala and Reuter (1979) see also Mulvihill and Donovan (1978)
Casein								
Milk	–	–	–	146.5	–	–	20.5	White and Sweetsur (1977)
Skim milk	–	–	–	104–146.5	–	–	20.5–28.8	White and Sweetsur (1977)
Haemoglobin								
Beef	–	–	–	53.2	–	–	56.4	Haurowitz et al. (1954)
Oxyhaemoglobin								
Beef	–	–	–	90.8	–	–	33.0	Haurowitz et al. (1954)
Methionine								
Model systems	60–90	90	0.53	125.7	14.8	4.3	20.0	Martens (1980)
Milk	110–126	121.1	0.01	129.0	12.2	209	25.3	Martens (1980)
Methylmethionine sulphonium chloride								
Buffer solution pH 6	121.1–132.2	121.1	5.10	135.4	15.8	451	22.0	Berry et al. (1989)
Model system pH 5	126.7	126.7	2.51	121.2	13.2	916	24.7	Berry et al. (1989)
pH 4	82–99	99	0.18	125.6	13.9	120	20.0	Williams and Nelson (1974)
pH 8	82–99	99	750[a]	117.6	16.4	2.7×10^{-3}	21.3	Williams and Nelson (1974)
Sweet corn pH 7	82–99	99	0.78	132.2	16.2	2.9	19.0	Williams and Nelson (1974)
Tomato pH 7	82–99	99	0.20	116.3	12.65	1.0	21.6	Williams and Nelson (1974)
Lysine								
Milk	130–160	160	0.03	109.0	10.29	69.8	32.8	Kessler and Fink (1986)

TABLE B.3. Kinetic factors: enzymes*.

Product	Temperature range, T (°C)	Reference temperature, T_{ref} (°C)	Reaction rate, k_T ($\times 10^3$ s^{-1})	Activation energy, E_A (kJ mol^{-1})	Frequency factor, $\log A$ (s^{-1})	Thermal destruction rate, D_T ($\times 10^{-3}$ s)	z-value (°C)	Reference
Peroxidase								
Horseradish pH 7	85–150	85	0.269	105	11.81	8.56	22.8	Joffe and Ball (1962)
pH 5.6	60–160	120	2.8	100	10.76	0.83	27.8	Adams (1978)
In aq. methanol	76–87	82	0.9	88.3	8.7	2.56	27.1	Ling and Lund (1978)
Corn on the cob	70–100	100	1.5×10^{-3}	75	9.7	15.5	37.0	Naveh et al. (1982)
	93	93	3.48	84	9.5	0.66	30.0	Yamamoto et al. (1962)
	80–100	100	23.0	83	10.0	0.10	33.5	Luna et al. (1986)
Green bean purée	104–177	104	51.0	76	9.25	4.5×10^{-2}	36.6	Resendre et al. (1969)
	104–132	104	13.2	82	9.5	3.3×10^{-2}	33.9	Zoueil and Esslen (1958)
	70–110	110	2.6	82	8.6	0.885	34.15	Adams (1978)
Pea purée	70–110	110	4.4	84.5	9.2	0.523	33.16	Adams (1978)
Potato purée	100–140	120	33.0	80	9.16	7.0×10^{-2}	35.0	Svensson (1977)
Spinach purée	77–92	77	327	138	20.14	7.0×10^{-3}	18.0	Resendre et al. (1969)
Turnip purée	82–132	104	95.0	205	25.4	2.4	13.56	Zoueil and Esselen (1958)
Catalase								
Spinach extract	60	60	38	253	38.5	6.0×10^{-2}	8.3	Sapers and Nickerson (1962)
Lipoxygenase								
Pea/soya extract	50–80	77	3.2	676	98.6	0.72	3.4	Farkas and Goldblith (1962)
Pea/buffer pH 6	60–72	60	370	584	89.3	62	3.6	Svensson and Eriksson (1974)

(cont.)

TABLE B.3. (Continued)

Product	Temperature range, T (°C)	Reference temperature, T_{ref} (°C)	Reaction rate, k_T ($\times 10^3$ s^{-1})	Activation energy, E_A (kJ mol^{-1})	Frequency factor, log A (s^{-1})	Thermal destruction rate, D_T ($\times 10^{-3}$ s)	z-value (°C)	Reference
O-diphenyl oxidase								
Fruits pH 3.4–6.2	30–100	75	1.0	210	29.5	0.21	10.0	Dimick et al. (1951)
Pectinesterase								
Guava pH 4	74–94	96	6.0	157.6	21.0	3.5×10^{-2}	16.5	Nath and Ranganna (1983)
Mandarin Orange juice pH 3.6	82–94	85				0.13	11.4	Nath and Ranganna (1977)
Papaya (acidified) pH 3.8	77	77				0.12	7.8	Dos et al. (1999)
Papaya nectar pH 3.8	75–85	85				0.30	15.1	Argáiz (1994)
Papaya pulp	75–80	80	–	72.7	–	16.7–3.7	7.8	Massaguer et al. (1994)
Papaya pulp pH 4.0	82–102	85				0.23	15.0	Nath and Ranganna (1981)
Papaya purée pH 3.5	75–85	85				0.28	14.8	Argáiz (1994)
pH 3.8	75–85	85				0.23	14.7	Argáiz (1994)
pH 4.0	75–85	85				0.43	14.2	Argáiz (1994)
Polyphenol oxidase								
Potato	80–100	89	23	322	44.9	0.10	7.8	Dagerskog (1977)

* See Aylward and Haisman (1969) for an excellent review of earlier work.

TABLE B.4. Kinetic factors: overall sensory quality.

Product	Temperature range, T (°C)	Reference temperature, T_{ref} (°C)	Reaction rate, k_T ($\times 10^3$ s^{-1})	Activation energy, E_A (kJ mol^{-1})	Frequency factor, log A (s^{-1})	Thermal destruction rate, D_T ($\times 10^{-3}$ s)	z-value (°C)	Reference
Beef purée	110–134	121.1	–	136–158	–	–	19–22	Ohlsson (1980)
Fish cake	110–134	121.1	–	103–130	–	–	23–29	Ohlsson (1980)
Liver paté	110–134	121.1	–	88.2–120	–	–	25–34	Ohlsson (1980)
Tomato sauce	110–134	121.1	–	111–187	–	–	16–27	Ohlsson (1980)
Vanilla sauce	110–134	121.1	–	136–230	–	–	13–22	Ohlsson (1980)
Vegetable purée	110–134	121.1	–	125–167	–	–	18–24	Ohlsson (1980)
Beetroot	80–110	121.1	19.0	142	17.2	0.12	19	Mansfield (1974)
Broccoli	100–121	121.1	8.7	54.4	5.16	0.26	44.4	Lund (1975)
Carrots	80–116	121.1	27.0	160	19.7	0.084	16.7	Lund (1975)
Corn, whole kernel	100–121	121.1	16.0	67	7.1	0.15	36.6	Lund (1975)
Corn, whole kernel	80–148	121.1	70.0	94.6	11.4	0.26	31.7	Hayakawa et al. (1977)
Green beans	80–148	121.1	10.0	104	11.8	0.20	28.8	Hayakawa et al. (1977)
Green beans	84–116	121.1	38.0	171.6	21.3	0.06	15.6	Mansfield (1974)
Peas	100–121	121.1	16.0	81.6	9.01	0.15	32.2	Mansfield (1974)
Peas	80–148	121.1	23.0	106	12.45	0.096	28.3	Hayakawa et al. (1977)
Potatoes	72–121	121.1	35.0	115	13.76	0.072	23.0	Mansfield (1974)
Squash	84–116	121.1	38.0	171.6	21.36	0.06	15.6	Mansfield (1974)

TABLE B.5. Kinetic factors: texture and softening

Product	Temperature range, T (°C)	Reference temperature, T_{ref} (°C)	Reaction rate, k_T ($\times 10^3$ s^{-1})	Activation energy, E_A (kJ mol^{-1})	Frequency factor, log A (s^{-1})	Thermal destruction rate, D_T ($\times 10^{-3}$ s)	z-value (°C)	Reference
Fish cake	110–134	121.1	47.0	130	15.9	0.049	23	Ohlsson (1980)
Asparagus (green)	70–98	84	26–45	100.8				Lau et al. (2000)
Asparagus (green)	115	115	3.65	56.4				Rodrigo et al. (1998)
Beetroot	104.4–121.1	121.1	7.13	65.3av.	6.52	0.32	46	Huang and Bourne (1983)
Beets	60–100	–	61.6 & 112.8	–			–	Buckenhuskes et al. (1990)
Beans, black	98–127	120	6.4	148.6	17.6	0.36	19	Quast and da Silva (1977)
Beans, brown	98–127	120	10.0	156.9	18.9	0.22	18	Quast and da Silva (1977)
Beans, white	90–122	100	0.028	130.8	–	84.9	21.3	Van Loey et al. (1955)
Brussels sprouts	100–150	120	1.8	125.7	13.9	1.28	23.8	Tijskens and Schijvens (1987)
Carrot, Nantes	104.4–121.1	121.1	3.9	63.6	6.0	0.59	47	Huang and Bourne (1983)
Carrot, –	79–99	121.1	83	153.5	19.3	0.027	19.5	Tijskens and Schijvens (1987)
Carrot, –	–	121.1		112.5		–	2.66	Paulus and Saguy (1980)
Carrot, –	100	100	k_1: 5.34 k_2: 1.5					Ramesh et al. (1998)
Corn, whole kernel	80–148	121.1	9.6	81.6	8.8	0.24	34.3	Hayakawa et al. (1977)
Dry white beans	104.4–121.1	120	66	104.2	10.7	0.22	28.7	Huang and Bourne (1983)
Green beans	79–99	121.1	19	177	21.7	0.12	16.9	Tijskens and Schijvens (1987)
Green beans, cut	80–148	121.1	9.6	92	10.2	0.24	32.5	Hayakawa et al. (1977)
Haricot beans	115–130	120	7.4	118.5	13.6	0.31	25.0	Michiels (1973)
Khol-khol	100	100	k_1: 8.34 k_2: 1.98					Ramesh et al. (1998)
Peas	80–148	121.1	16	94.2	10.7	0.14	31.8	Hayakawa et al. (1977)
	90–122	100	0.07	94.9	–	33.24	28.5	Van Loey et al. (1995)
	100	100	k_1: 8.46 k_2: 1.5					Ramesh et al. (1998)

Product							Reference
Radish	100	100	3.00	182	22.0	0.17	16.0	Ramesh et al. (1998)
Soya beans	98–127	125	13	163.6	21.63	0.048	17.0	Quast and da Silva (1977)
Potato (3 var.)	80–110	100	48	148	18.5	0.100	18.8	Dagerskog (1977)
	110	110	23	176.6	–	–	17.0	Harada et al. (1985)
	–	–	–	176.6	–	–	17.0	Pravisani et al. (1985)
	–	–	–	176.4	21.35	0.26	17.0	Ohlsson (1986)
	88–100	99	8.6	114	15.2	0.36	24.2	Loh and Breene (1981)
	80–100	100	6.4	126	15.1	0.72	21.2	Kubota et al. (1978)
(4 var.)	74–104	100	3.16	151–152	6–8.3	–	–	Kozempel (1988)
	60–110	100	8.9	58.2–74.1	12–19	–	–	Rahardjo and Sastry (1993)
	110–128	100	9.1	–	–	0.24	27.7	Rahardjo and Sastry (1993)
Potato cylinders	80–100	100	–	79.5	8.65	4.6	35.2	Linders et al. (1990)
Rice	75–110	75	0.5	36.8	2.54	0.69	75.0	Suzuki et al. (1976)
	110–150	110	3.3	98–106	–	–	–	Suzuki et al. (1976)
Starch gel.	60–100	100	0.0125–0.0144	131	18.1	0.17	21.2	Verlinden et al. (1995)
Sweet potato	80–100	100	15	–	–	0.97	109.7	Kobota et al. (1978)
Vegetable particles	80–100	–	–	66.4	5.5	9.6	42.2	Linders et al. (1990)
Apple, Cortl and	70–120	108	0.24	113	11.9	9.6	42.2	Anantheswaran et al. (1985)
Spigold	70–120	108	0.24	11	13.3	0.41	25.0	Anantheswaran et al. (1985)
Gould Reinette	80–100	100	5.6	87	9.8	0.56	32.0	Tijskens and Schijvens (1987)
Golden Delicious	80–100	100	4.1	14.5	–	0.48	22.0	Tijskens and Schijvens (1987)
5 varieties	80–98	98	4.5	–	–	–	–	Lacroix and Bourne (1990)
Shrimp	115–140	121.1	0.66	102	10.4	3.45	29.4	Ma et al. (1983)
Meat	55–60	–	–	586	–	–	3.75	Lund (1982)

TABLE B.6. Kinetic factors: colour degradation and browning.

Product	Temperature range, T (°C)	Reference temperature, T_{ref} (°C)	Reaction rate, k_T ($\times 10^3$ s^{-1})	Activation energy, E_A (kJ mol^{-1})	Frequency factor, log A (s^{-1})	Thermal destruction rate, D_T ($\times 10^{-3}$ s)	z-value (°C)	Reference
Green pigments, chlorophylls								
Asparagus	80–140	121.1	2.25	71	6.79	1.02	41.6	Hayakawa and Timbers (1971)
	70–89	84	0.11	54.6				Lau et al. (2000)
Cupaça	100			31–36				Silva and Silva (1999)
Green beans	80–148	121.1	1.8	72	6.81	1.26	38.8	Hayakawa and Timbers (1971)
Peas	80–148	121.1	1.5	76	7.26	1.50	39.4	Hayakawa and Timbers (1971)
	121–148	–	–	52.3	–	–	57.3	Epstein (1959)
	–	100	1.23	67.9	6.11	31.1	42.9	Shin and Bhowmik (1995)
	90–122	100	0.09	102.4	–	25.8	26.4	Van Loey et al. (1995)
Pea purée	94–132.2	121.1	0.34	92.0	8.74	6.90	32.5	Lenz and Lund (1980)
Spinach, chlorophyll A	127–148	148.8	0.10	66.9	6.3	0.21	51.1	Gupta et al. (1964)
	116–126	121	2.8	114.2	12.6	0.82	26.2	Schwartz and von Elbe (1983)
chlorophyll B	127–148	148.8	5.0	34.8	2.0	0.46	98.3	Gupta et al. (1964)
	116–148	121	1.36	103.4	10.9	1.70	29.0	Schwartz and von Elbe (1983)
chlorophylls	94–132.2	121.1	0.23	79.5	6.9	9.80	17.7	Lenz and Lund (1980)
Red pigments								
Betanin pH 4.8	61.5–100	100	1.8	76.2	7.9	1.28	36.5	Saguy et al. (1978a), Saguy (1979)

								Reference
pH 6.2	61.5–100	100	1.96	83.4	8.0	1.17	33.4	Saguy et al. (1978a), Saguy (1979)
Vulgaxanthin	61.5–100	100	2.2	64.5	6.4	1.05	43.1	Saguy et al. (1978a), Saguy (1979) see also Huang and von Elbe (1985)
Blackberry juice	24–70	70	0.02	62.8	4.8	117	35.8	Debicki-Pospisil et al. (1983)
Pomegranate	70–92	92	0.3	104.6	10.47	7.5	24.8	Mishkin and Saguy (1982)
Grapes	77–121	121	0.3	54.8	3.75	7.2	54.7	Mishkin and Saguy (1982)
	85–95	95	0.08	79.1	12.1	2.76	31.8	Calvi and Francis (1978)
Raspberry juice	78–108	108	0.32	92.1	9.15	7.0	30.4	Tanchev (1983)
Other juices	–	–	–	–	–	–	–	Ioncheva and Tanchev (1974)
Browning reactions								
Chestnut paste	105–128	121.1	0.16	122	11.4	141	24.6	Nunes et al. (1988)
Grapefruit juice, 11.2° brix	60–96	95	0.4	63.2	5.6	5.6	42.7	Saguy et al. (1978b)
63.5° brix	60–96	95	2.3	99.6	11.5	0.9	27.1	Saguy et al. (1978b)
Milk	50–160	130	184	139	17.3	0.012	26.7	Kessler and Fink (1986)
Peach puree	110–135	122.5	2.9	106		0.03		Avila and Silva (1999)
Pear puree 11° brix	80–98	100		62.5–126.5	5.0–9.5		1.6	Ibarz et al. (1999)

References

Adams, J. B. (1978). The inactivation and regeneration of peroxides in relation to HTST processing of vegetables. *J. Food Technol.*, *13*, 281–297.

Agrawala, S. P., & Reuter, H. (1979). Effects of different temperatures and holding times on whey protein denaturation in a UHT plant. *Milchwissenschaft*, *34* (12), 735–737.

Alvarado, J. D., & Viteri, N. P. (1989). Effect of temperature on the degradation of ascorbic acid in citrus fruit juices. *Arch. Latinoam. Nutricion*, *39*(4), 601–612.

Anantheswaran, R. C., McLellan, M. R., & Bourne, M. C. (1985). Thermal degradation of texture in apples. *J. Food Sci.*, *50*, 1136–41.

Argáiz, A. (1994). Thermal inactivation kinetics of pectinesterase in acidified papaya nectar and purées. *Rev. Esp. Cienc. Tecnol. Alimentos*, *34* (3), 301–309.

Ávila, I. M. L. B., & Silva, C. L. M. (1999). Modelling kinetics of thermal degradation of colour in peach puree. *J. Food Eng.*, *39* (2),161–166.

Aylward, F. X. & Haisman, D. R. (1969). Oxidation systems in fruits and vegetables. *Adv. Food Res.*, *17*, 1–76.

Bayoumi, E. -S. & Reuter, H. (1980). Destruction of vitamin B_1 during UHT treatment of milk. *Milchwissenschaft*, *35*, 278.

Bendix, G. H., Herberlein, D. G., Ptak, L. R., & Clifcorn, L. E. (1951). Stabilization of thiamine during heat sterilization. *Food Res.*, *16*, 494–503.

Berry, M. F., Singh, R. K., & Nelson, P. E. (1989). Kinetics of methylmethionine sulfonium in buffer solutions for estimating the thermal treatment of liquid foods. *J. Food Process. Preserv.*, *13*, 475–488.

Buckenhuskes, H., Rausch, C., Spiess, W. E. L., & Gierschner, K.-H. (1990). Influence of heat treatment on the texture of red beets. In W. E. L. Spiess & H. Schubert (Eds.), *Engineering and food Vol. 2* (pp. 185–191). London: Elsevier Applied Science.

Calvi, J. P., & Francis, P. J. (1978). Stability of Concord grape (*V. labrusca*) anthocyanins in model systems. *J. Food Sci.*, *43*(5), 1488–1501.

Dagerskog, M. (1977). Calculation of heat transfer and quality changes during industrial processing of whole potatoes. In EFCE Mini-Symposium *Mathematical modelling in food processing* (pp. 269–288). Sweden: Lund University of Technology.

Debicki-Pospisil, J., Lovric, T., Trinajstic, N., & Sabljic, A. (1983). Anthocyanin degradation. *J. Food Sci.*, *48*, 411–415.

Dimick, K. P., Ponting, J. D., & Makower, R. (1951). Heat inactivation of polyphenoloxidase in fruit juices. *Food Technol.*, *5*, 237–241.

Dos, M. M., Magalhaes, A., Tosello, R. M., & Massaguer, P .R. (1999). Thermal design process for acidified canned papaya (*Carica papaya L.*). *J. Food Process Eng.*, *22*,103–112.

Epstein, A. I. (1959). A study of the stability of HST green peas during processing and storage. PhD thesis. New Brunswick, NJ: Rutgers University.

Farkas, D. F., & Goldblith, S. A. (1962). Studies on the kinetics of lipoxidase inactivation using thermal and ionizing energy. *J. Food Sci.*, *27*, 262–276.

Feliciotti, E., & Esselen, W. B. (1957). Thermal destruction of thiamine in puréed meat and vegetables. *Food Technol.*, *11*, 77–84.

Ghazala, S., Ramaswamy, H. S., van de Voort, F. R., Prasher, S. O., & Barrington, S. (1989). Evaluation of a conduction heating food model for ascorbic acid retention and colour formation during thermal processing. *Can. Inst. Food Sci. Technol. J.*, *22*(5), 475–480.

Greene, R. N. (1983). Heat resistance of thiamine in Sockeye salmon. *NFPA Annual Report*. Washington, DC: National Food Processors' Association.

Greenwood, D. A., Kraybill, H. R., Feaster, J. F., & Jackson, J. M. (1944). Vitamin retention in processed meat. *Ind. Eng. Chem.*, *36*, 922–927.

Gregory, J. F. III, & Hiner, M. E. (1983). Thermal stability of vitamin B6 compounds in model liquid food systems. *J. Food Sci.*, *48*, 1323–1328.

Gupta, S. M., El-Bisi, H. M., & Francis, F. J. (1964). Kinetics of thermal degradation of chlorphyll in spinach purée. *J. Food Sci.*, *29*, 379–384.

Hamm, D. J., & Lund, D. B. (1978). Kinetic parameters for the thermal inactivation of pantothenic acid. *J. Food Sci.*, *43*, 631–634.

Harada, T., Tortohusodo, H., & Paulus, K. (1985). Influence of temperature and time on cooking kinetics in potatoes. *J. Food Sci.*, *50*, 459–463.

Haurowitz, F., Hardin, R. L., & Dicks, M. (1954). Denaturation of haemoglobin by alkali. *J. Phys. Chem.*, *58*, 103–108.

Hayakawa, K., & Timbers, G. E. (1971). Influence of heat treatment on the quality of vegetables: changes in visual green colour. *J. Food Sci.*, *42*, 778–781.

Hayakawa, K., Timbers, G. E., & Stier, E. F. (1977). Influence of heat treatment on the quality of vegetables: organoleptic quality. *J. Food Sci.*, *42*, 1286–1289.

Hojilla, M. P., Garcia, V. V., & Raymundo, L. C. (1985). Thermal degradation of beta-carotene in carrot juice. *ASEAN Food J.*, *1*, 157–161.

Horak, F. P. (1980). The reaction kinetics of spore destruction and chemical degradation during the thermal treatment of milk optimization of the heat process. Doctoral thesis, Technical University Munich.

Horak, F. P., & Kessler, H. G. (1981). Thermal destruction of thiamine – a second order reaction. *Z. Lebensm. Untersuch. Forsch.*, *172*, 1–6.

Huang, A. S., & von Elbe, J. H. (1985). Kinetics of the degradation and regeneration of betanine. *J. Food Sci.*, 50, 1115–1118.

Huang, Y. T., & Bourne, M. C. (1983). Kinetics of thermal softening of vegetables. *J. Texture Studies.*, *14*, 1–9.

Huelin, F. E. (1953). Studies on the anaerobic decomposition of ascorbic acid. *Food Res.*, *18*, 633–639.

Ibarz, A., Pagán, J., & Garza, S. (1999). Kinetic models for colour changes in pear puree during heating at relatively high temperatures. *J. Food Eng.*, *39*(4), 415–422.

Ioncheva, N., & Tanchev, S. (1974). Kinetics of the degradation of some anthocyanins. *Z. Lebensm. Untersuch. Forsch.*, *155*, 257–60.

Joffe, F. M., & Ball, C. O. (1962). Kinetics and energetics of thermal inactivation and regeneration rates of a peroxidase system. *J. Food Sci.*, *27*, 587–592.

Kessler, H. G., & Fink, R. (1986). Changes in heated and stored milk with an interpretation by reaction kinetics. *J. Food Sci.*, *51*, 1105–1111, 1155.

Kozempel, M. F. (1988). Modeling the kinetics of cooking and precooking potatoes. *J. Food Sci.*, *53*, 753–756.

Kubota, K., Oshita, K., Hosokawa, Y., Suzuki, K., & Hosaka, H. (1978). Cooking rate equations of potato and sweet-potato slices. *J. Fac. Fish Anim. Husb., Hiroshima University.* 17, 97–106.

Lacroix, P., & Bourne, M. C. (1990). Rate of softening of apple flesh during cooking. In W. E. L. Spiess & H. Schubert (Eds.), *Engineering and food Vol. 2* (pp. 201–210). London: Elsevier Applied Science.

Lathrop, P. J., & Leung, H. K. (1980). Rate of ascorbic acid degradation during thermal processing of canned peas. *J. Food Sci.*, *45*, 152–155.

Lau, M. H., Tang, J., & Swanson, B. G. (2000). Kinetics of textural and color changes in green asparagus during thermal treatments. *J. Food Eng., 45*(4), 231–236.

Lenz, M. K., & Lund, D. B. (1977). The lethality-Fourier number method. Confidence intervals for calculated lethality and mass-average retention of conduction-heating canned foods. *J. Food Sci., 42*, 1002–1009.

Lenz, M. K., & Lund, D. B. (1980). Experimental procedures for determining destruction kinetics of food components. *Food Technol., 34*, 51–55.

Linders, C., Foroni, G., Taeymans, R., & Taeymans, D. (1990). Experimental study of mechanical damage of heat-processed vegetables. In W. E. L. Spiess and H. Schubert (Eds.), *Engineering and food Vol. 2* (pp. 192–200). London: Elsevier Applied Science.

Ling, A. C., & Lund, D. B. (1978). Determining the kinetic parameters for thermal inactivation of heat resistant and heat-labile isozymes from thermal destruction curves. *J. Food Sci., 43*, 1307–1310.

Loh, J., & Breene, W. M. (1981). The thermal fracturability loss of edible plant tissue: pattern and within-species variation. *J. Texture Studies, 12*, 457–471.

Luna, J. A., Garrote, R .L., & Bressan, J. A. (1986). Thermokinetic modelling of peroxidase inactivation during blanching–cooling of corn-on-the-cob. *J. Food Sci., 51*, 141–145.

Lund, D. B. (1975). Effects of blanching, pasteurization and sterilization on nutrients. In R. S. Harris & E. Karmas (Eds.), *Nutritional evaluation of food processing.* Westport, CT: AVI Publishers.

Lund, D. B. (1982). Quantifying reactions influencing quality of foods: texture, flavour, appearance. *J. Food Process. Preserv.*, 6, 133–37.

Ma, L. Y., Deng, J. C., Ahmed, E. M., & Adams J. P. (1983). Canned shrimp texture as a function of its heat history. *J. Food Sci., 48*, 360–364.

Mansfield, T. (1974). *A brief study of cooking.* San Jose, CA: Food Machinery Corporation.

Martens, T. (1980). Mathematical model of heat processing in flat containers. PhD Thesis, Catholic University of Leuven, Belgium.

Massaguer, P. R., Magalhaes, M. A., & Tosello, R. M. (1994). Thermal inactivation of pectin esterase in papaya pulp (pH 3.8). In T. Yano, R. Matsuno & Nakamura, K (Eds.), *Developments in food engineering Part 1* (pp. 495–497). London: Blackie Academic & Professional.

Matsuda, N, Lomaki, M., & Matsunawa, K. (1981). Thermal death characteristics of *Clostridium botulinum* 62A and *Bacillus stearothermophilis* subjected to non-isothermal heat treatment. *J. Food Hygenic Soc. Jap., 22*(2), 125–134.

Mauri, L. M., Alzamora, S. M., Chirife, J., & Tomio, M. J. (1989). Review: kinetic parameters for thiamine degradation in foods and model solutions of high water activity. *Int. J. Food Sci. Technol., 24*, 1–9.

Michiels, L. (1973). Establishment of a thermal process as a function of cooking of white beans used in canning. *Compt. Rend. Acad. Agric. France*, pp. 891–900.

Mishkin, M. & Saguy, I. (1982). Thermal stability of pomegranate juice. *Z. Lebensm. Untersuch. Forsch., 163*, 37.

Mnkeni, A. P., & Beveridge, T. (1982). Thermal destruction of pteroyl glutamic acid in buffer and model systems. *J. Food Sci., 47*, 2038–2041.

Mulley, E. A., Stumbo, C. R., & Hunting, W. M. (1975). Kinetics of thiamine degradation by heat. *J. Food Sci., 40*, 985–989.

Mulvihill, D. M., & Donovan, M. (1987). Whey proteins (review). *Irish J. Food Sci. Technol., 11*, 56–59.

Nasri, H., Simpson, R., Bouzas, J., & Torres, J.A. (1993). An unsteady-state method to determine kinetic parameters for heat inactivation of quality factors: conduction-heated foods. *J. Food Eng.*, *19*, 291–301.

Nath, N., & Rangana, S. (1977). Time-temperature relation ship for the thermal inactivation of pectinesterase in Manderin orange (*Citrus reticulata blanco*) *J. Food Technol.*, *12*, 411–419.

Nath, N., & Rangana, S. (1981). Determination of the thermal schedule for acidified papaya. *Food Technol.*, *46*, 201–206, 211.

Nath, N., & Ranganna, S. (1983). Determination of a thermal process schedule for guava. *J. Food Technol.*, *18*, 301–309.

Navankattusas, S., & Lund, D. B. (1982). Thermal destruction of vitamin B_6 vitamers in buffer solution and cauliflower purée. *J. Food Sci.*, *47*, 1512–1515.

Naveh, D., Mizrahi, S., & Kopelman, I. (1982). Kinetics of peroxidase deactivation in blanching corn-on-the-cob. *J. Agric. Food Chem.*, *30*, 967–970.

Nunes, J., Louro, L., Viera, J., & Melo. R. (1988). Determination of kinetic parameters of chestnut paste darkening. In *Conf. proc. Progress in food preservation processes Vol. 2* (pp. 21–27). Brussels.

Ohlsson, T. (1980). Temperature dependence on sensory quality changes during thermal processing. *J. Food Sci.*, *45*, 836–839, 847.

Ohlsson, T. (1986). Boiling in water. In *Progress in Food Preparation Processes*, Tylosand, Sweden.

Paulas, K., & Saguy, I. (1980). Effect of heat treatment on the quality of cooked carrots. *J. Food Sci.*, *45*, 239–242.

Paulus, K., Duden, R., Fricker, A., Heintze, K., & Zohm, H. (1975). Influence of heat treatments of spinach at temperatures up to 100°C on important constituents. II. Changes in drained weight and contents of dry matter, vitamin C, vitamin B_1 and oxalic acid. *Lebensm.-Wiss. u. -Technol.*, *8*, 11–16.

Pravisani, C. I., Califano, A. N., & Calvelo, A. (1985). Kinetics of starch gelatinization in potato. *J. Food Sci.*, *50*, 657–661.

Quast, D. C., & Silva, S. D. da (1977). Temperature dependence of the cooking rate of dry legumes. *J. Food Sci.*, *42*, 370–374.

Rahardjo, B., & Sastry, S. K. (1993). Kinetics of softening of potato tissue during thermal processing. *Food and Bioproducts Processing Trans. IChemE*, *71*(C4), 235–241.

Ramaswamy, H., Ghazala, S., & Van de Voort, F. (1990). Degradation kinetics of thiamine in aqueous systems at high temperature. *Can. Inst. Food Sci. Technol. J.*, *22*(2/3), 125–130.

Ramesh, M. N., Sathyanarayana, K., & Girish, A. B. (1998). Biphasic model for the kinetics of vegetable cooking at 100°C .*J. Food Eng.*, *35*(1), 127–133.

Resendre, R., Francis, F. J., & Stumbo, C. R. (1969). Thermal destruction and regeneration of enzymes in green and spinach purée. *Food Technol.*, *23*, 63–66.

Rodrigo, C., Matesu, A., Alvarruiz, A., Chinesta, F., & Rodrigo, M. (1998). Kinetic parameters for thermal degradation of green asparagus texture by unsteady-state method. *J. Food Sci.*, *63*(1), 126–129.

Saguy, I. (1979). Thermostability of red beet pigments (betanine and vulgaxanthin – 1: Influence of pH and temperature. *J. Food Sci.*, *44*, 1554–1555.

Saguy, I., Kopleman, I. J., & Mizrahi, S. (1978a). Thermal kinetic degradation of betanine and betalamic acid. *J. Agric. Food Chem.*, *26*, 360–365.

Saguy, I., Kopleman, I. J., & Mizrahi, S. (1978b). Extent of nonenzymatic browning in grapefruit juice during thermal and concentration processes. *J. Food Process. Preserv.*, *2*(3), 175–212.

Saguy, I., Kopleman, I. J., & Mizrahi, S. (1978c). Simulation of ascorbic acid stability during heat processing and concentration of grapefruit juice. *J. Food Process. Preserv.*, *2*(3), 213–219.

Sapers, G. M., & Nickerson, J. T. (1962). Stability of spinach catalase II. Inactivation by heat. *J. Food Sci.*, *27*, 282–286.

Schwartz, S. J., & Elbe, J. H. von (1983). Kinetics of chlorophyll degradation to pyropheophytin in vegetables. *J. Food Sci.*, *48*(4), 1303–1305.

Shin, S., & Bhowmik, S. R. (1995). Thermal kinetics of colour changes in pea purée. *J. Food Eng.*, *24*, 77–86.

Silva, F. M., & Silva, C. L. M. (1999). Colour changes in thermally processed cupaça (*Theobroma grandiflorum*) puree: critical times and kinetic modelling. *Int. J. Food Sci. Technol.*, *34*(1),87–94.

Skjöldebrand, C., Anas, A., Oste, R., & Sjodin, P. (1983). Prediction of the thiamine content in convective heated meat products. *J. Food Technol.*, *18*, 61–5.

Street, J. A., & Tong, C. H. (1994).Thiamin degradation kinetics in pureed restructured beef. *J. Food Process. Preserv.*, *18*(3), 253–262.

Suzuki, K., Kubota, K., Omichi, M., & Hosaka, H. (1976). Kinetic studies on cooking of rice. *J. Food Sci.*, *41*, 1180–1184.

Svensson, S. (1977). Inactivation of enzymes during thermal processing. In T. Hoyem and O. Kvale (Eds.), *Physical, chemical and biological changes in food caused by thermal processing* (pp. 202–206). London: Applied Science Publishers.

Svensson, S. G., & Eriksson, C. E. (1974). Thermal inactivation of lipoxygenase from peas. III. Inactivation energy obtained from single heat treatment experiments. *Lebensm.-Wiss. u. -Technol.*, *7*, 142–144.

Tanchev, S. (1983). Kinetics of thermal degradation of anthocyanins. *Proc. 6th. int. cong. food sci. technol.*, *2*, 96.

Tijskens, L., & Schijvens, E. (1987). Preservation criteria based on texture kinetics. In K. O. Paulus (Ed.), *Influence of HTST treatments on product quality and nutritive value of food and feed* (pp. 84–102). COST 91 bis Meeting, Wageningen, Netherlands.

Van Loey, A., Fransis, A., Hendrickx, M., Maesmans, G., & Tobback, P. (1995). Kinetics of quality changes in green peas and white beans during thermal processing. *J. Food Eng.*, *24*, 361–377.

Verlinden, B. E., Nicolai, B. M., & De Baerdamaeker, J. (1995). The starch gelatinization in potatoes during cooking in relation to the modelling of texture kinetics. *J. Food Eng.*, *24*, 165–179.

Vieira, M. C., Teixeira, A. A., & Silva, C. L. M. (2000). Mathematical modeling of the thermal degradation of vitamin C in cupuaçu *(Theobroma grandiflorum)* nectar. *J. Food Eng.*, *43*(1), 1–7.

Weir, E. (1948). The thermal stability of thiamine in strained foods. Doctoral thesis, Amherst, MA: University of Massachusetts.

White, J. C. D., & Sweetsur, A. W .M. (1977). Kinetics of heat induced aggregation of milk protein. *J. Dairy Res.*, *44*, 237.

Wilkinson, S. A., Earle, M. D., & Cleland, A. C. (1981). Kinetics of vitamin A degradation in beef liver purée on heat processing. *J. Food Sci.*, *46*, 32–37.

Wilkinson, S. A., Earle, M. D., & Cleland, A. C. (1982). Effects of food composition, pH and copper on the degradation of vitamin A in beef liver purée. *J. Food Sci.*, *46*, 844–848.

Williams, M. P., & Nelson, P. E. (1974). Kinetics of the thermal degradation of methylmethionine sulfonium salts. *J. Food Sci.*, *39*, 457–460.

Yamamoto, H. Y., Steinberg, M. P., & Nelson, A. I. (1962). Kinetic studies of heat inactivation of peroxidase in sweet corn. *J. Food Sci.*, *27*, 113–119.

Zoueil, M. E., & Esselen, W. B. (1958). Thermal destruction rates and regeneration of peroxidase in green beans and turnips. *Food Res.*, *24*, 119–133.

Appendix C: Heat Penetration Protocols

IFTPS

The Institute for Thermal Processing Specialists (*IFTPS*) is a not-for-profit corporation of the State of Virginia (USA) organized to serve the needs of those working in thermal processing. The Institute was conceived in 1981 by a group of individuals working in the field of thermal processing to advance professionalism through educational presentations and the development of uniform standards of procedure. More information about *IFTPS* can be found at www.iftps.org

The referred WEB site has important information on thermal processing, mainly in the area of protocols for thermal processing studies.

IFTPS Protocols

- Nomenclature for Studies in thermal Processing.
- Protocol for Carrying out Neat Penetration Studies.
- Temperature distribution Protocol for Processing in Steam Still Retorts, Excluding Crateless Retorts.
- Temperature Distribution Protocol for Processing in still, Water Immersion Retorts, including Agitating Systems Operated in a Still Mode.
- Protocol for Conducting Temperature Distributions Studies in Water-Cascade and Water-Spray Retorts Operated in a Still Mode, Including Agitating Systems Operated in the Still Mode.
- Nomenclature Pour Les Études De Traitement Thermique.
- Protocole De Mesure De La Pénétration De La Chaleur.
- Protocole De Mesure De La Distribution De Température Lors De Traitement Thermique En Autoclaves Statiques Traditionnels, A L' exclusión Des Autoclaves Sans Panier.

IFTPS protocols can be accessed at: http://www.iftps.org/protocols.html

Appendix D: FDA Food Process Filing

Important information can be obtained through the Web's page of Food and Drug Administration (FDA). In particular the "Importer's Guide Low-Acid Canned & Acidified Foods" can be accessed through: http://www.cfsan.fda.gov/~lrd/lacf.html

Index

M

Mass and energy balance
  during holding time, 296, 297
  during venting, 293, 294
Mass and energy consumption, venting
      and holding time, 295, 296
Mass-average cook value, 242
Mass-average sterilizing values, 213,
      248
  methods of calculation of, 214
Mass-average temperatures, 22
Mass balance equation, quality of
      attribute j through system, 258
Master temperature indicator (MTI),
      326
Mathematical formulation,
      simultaneous sterilization for,
      313, 314
Mean or volume average temperatures,
      22, 23
Mechanical agitation and rotation of
      cans
  axial rotation and spin cooking, 300,
      304
  end-over-end agitation, 300
  Shaka™ retort process, 304
  steritort and orbitort processes, 304
Metal containers, 4, 5
Microbial behavior, nature of, 87
Microbial death, statistical nature of,
      94–96
Microbial inactivation, approaches
  absolute reaction rate theory, 107,
      108
  Arrhenius kinetic approach: $k–E$
      model, 105, 106
  biphasic model, 104
  $D–z$ model, 104, 105
  equivalence of models, 109, 110
  Log-logistic model, 103
  Prentice model, 103–104
  quotient indicator method, 108, 109
  sigmoidal model, 104
  Weibull distribution, 103

Microbial inactivation, combined
      effect of pH and temperature
      methods
  Davey–linear Arrhenius model, 107,
      108
  Mafart model, 108
  nth order polynomials, 108
  square-root model, 108
Microbial inactivation curve. *See*
      Survivor curve
Microbiological methods
  biological and chemical indicators,
      219–222
  encapsulated spore method, 219
  inoculated pack method, 218
Microscribe-3D™, 3-D digitizer, 168
Microwavable products, packaging, 8
Microwave heating, 136
  heat transfer in packaged foods,
      43–45
Minimally processed foods
  acidified products, 135, 136
  electrical methods of heating, 136
  pasteurized/chilled products, 136
Model systems of heat transfer, 150,
      151
Moulded thermocouples, 145

N

National Canners' Association, 142
National Food Processors'
      Association, 1
NEURAL Networks, conduction heat
      transfer analysis programs, 69
No-capacitance surface node (NCSN)
      procedure, heat transfer
      simulation for, 29
Nodal temperatures, 29
Numerical techniques of solution, 27.
      See also Heat transfer
  finite-difference approximation
      method, 28, 29
  finite-element method, 33, 34
  other methods, 34, 35